Atmospheric Science at NASA

NEW SERIES IN NASA HISTORY
Steven J. Dick, *Series Editor*

Atmospheric Science
at NASA

A History

ERIK M. CONWAY

The Johns Hopkins University Press
Baltimore

© 2008 The Johns Hopkins University Press
All rights reserved. Published 2008
Printed in the United States of America on acid-free paper
2 4 6 8 9 7 5 3 1

The Johns Hopkins University Press
2715 North Charles Street
Baltimore, Maryland 21218-4363
www.press.jhu.edu

Library of Congress Cataloging-in-Publication Data

Conway, Erik M., 1965–
Atmospheric science at NASA : a history / Erik M. Conway.
p. cm. — (New series in NASA history)
ISBN-13: 978-0-8018-8984-4 (hardcover : alk. paper)
ISBN-10: 0-8018-8984-7 (hardcover : alk. paper)
1. Atmospheric sciences—History. 2. Satellite meteorology—History.
3. Astronautics in meteorology—History. 4. United States. National
Aeronautics and Space Administration. I. Title.
QC855.C66 2008
551.50973—dc22

2008007633

A catalog record for this book is available from the British Library.

Special discounts are available for bulk purchases of this book.
For more information, please contact Special Sales at 410-516-6936 or
specialsales@press.jhu.edu.

The Johns Hopkins University Press uses environmentally friendly book
materials, including recycled text paper that is composed of at least
30 percent post-consumer waste, whenever possible. All of our book papers
are acid-free, and our jackets and covers are printed on paper with
recycled content.

To my grandparents,

for the new world

A satellite vehicle with appropriate instrumentation can be expected to be one of the most potent scientific tools of the Twentieth Century.
—RAND, 1946

We did these industrial activities for perfectly understandable reasons, and we discover that we have outsmarted ourselves, that we haven't understood the fragility of the Earth's atmosphere and the power of our technology.
—Carl Sagan, 1992

Contents

Color illustrations follow page 140

Acknowledgments

This book is the result of a contract initiated by former NASA chief historian Roger D. Launius in 2002. It was jointly funded by the NASA History Office and by the Atmospheric Science Project Office at Langley Research Center, and administered by the Office of Public Services at Langley Research Center. Thanks to Donna Lawson, Michael P. Finneran, Stephen Sanford, of Langley, Carla Coombs and George Wood of Science and Technology Corporation, and Andrea Magruder, of Tessada Associates, for arranging and supporting this work.

For the first two years of this effort, I was the resident historian at Langley Research Center. Declining funding at Langley led me to seek other opportunities, and I landed at the Jet Propulsion Laboratory (JPL) in Pasadena, California, in September 2004, where I completed the work (with some delays) in my free time and with occasional use of unpaid leave periods. Blaine Baggett, Stephen Kulczycki, Laura Cinco, Jenny Coons, and Yvonne Samuels made this transition as painless as a cross-country move can be.

In writing this book, I've drawn heavily on several libraries. At Langley Research Center, Carolyn Helmetsie and facilities manager Garland Gouger of the Floyd Thompson Technical Library provided an excellent working environment. The interlibrary loan staff, Kenneth Carroll and Cecelia Grzeskowiak, rounded up veritable stacks of history books and articles that engineering libraries don't carry. At JPL, Margo Young, Barbara Amago, and Mickey Honchel provided access to the collections in Pasadena. At the Schwerdtfeger Library of the University of Wisconsin, Jean Phillips granted access to Verner E. Suomi's remaining papers and to the library's extensive collections on early meteorological satellite research. Diane Rabson, the archivist for the National Center for Atmospheric Research (NCAR) in Boulder, Colorado, hosted a research visit and provided copies of a number of oral histories from the NCAR collection.

One of the pleasures of writing about modern science has been the opportunity to interview working scientists and science managers. In researching this work, I interviewed many people. A few people deserve special mention, however. William L. Smith, Sr., and Bruce Wielicki of Langley Research Center, Robert C. Harriss from NCAR, Adrian Tuck of the National Oceanic and Atmospheric Administration's Aeronomy Lab, and Milton Halem of Goddard Space Flight Center gave lengthy interviews and also read portions of the manuscript, aiding my understanding of their areas of specialty and saving me from embarrassing errors. Edward A. Frieman, Charles F. Kennel, Shelby Tilford, Michael D. King, James E. Hansen, and Dixon Butler helped me understand the politics swirling around Earth observations in the 1990s. Finally, informal conversations with Moustafa Chahine, Eric Fetzer, and Joe Waters at JPL clarified my understanding of some of the technical challenges in space-based remote sensing.

This book has also benefited from my associations with other historians of science, many of whom I met, or re-met, at the International Commission for the History of Meteorology meeting at Wilheim, Germany, in 2004. Mott Greene helped me think about modeling in modern science. James R. Fleming and Katharine Anderson have been valuable for their knowledge of the history of meteorology. Maiken Lykke Lolck guided me to sources in the history of ice core research. Naomi Oreskes' inadvertent initiation of a conflict with climate change disinformation specialists in 2004 unearthed some unique resources, which she graciously shared. Oreskes and Fleming also kindly read and extensively critiqued this manuscript, for which I'm very grateful.

Barton C. Hacker and Margaret Vining of the National Museum of American History hosted several of my DC area research visits in their wonderful House of Clio on Capitol Hill. Thanks also to Robert Ferguson and Louise Liu for hosting research visits. Deborah G. Douglas, curator of the MIT Museum, made space for me for a research trip during a cold, wet Boston winter. Finally, Karin Matchett did admirable research in the George H. W. Bush Presidential Library's global climate collection on my behalf in 2005.

This manuscript was also read by four anonymous reviewers chosen by the NASA History Office, and by an anonymous reviewer chosen by the Johns Hopkins University Press. They offered numerous criticisms that helped me improve the work. Nonetheless, all errors herein are mine and mine alone.

Abbreviations

AAOE	Airborne Antarctic Ozone Experiment
AASE	Airborne Arctic Stratospheric Expedition
ABLE	Atlantic (Arctic) Boundary Layer Experiment
ABMA	Army Ballistic Missile Agency
ACRIM	Active Cavity Radiometer Irradiance Monitor
AEC	Atomic Energy Commission
AIRS	Atmospheric Infrared Sounder
AMSU	Advanced Microwave Sounding Unit
APARE	East Asia/North Pacific Regional Study
APT	Automatic Picture Transmission
ARM	Atmospheric Radiation Measurement
ARPA	Advanced Research Projects Agency
ATMOS	Atmospheric Trace Molecule Spectroscopy
ATS	Advanced Technology Satellite
AVCS	Advanced Vidicon Camera System
AVHRR	Advanced Very High Resolution Radiometer
BIC	Balloon Intercomparison Campaigns
BOMEX	Barbados Oceanographic and Meteorological Experiment
BUV	Backscatter Ultraviolet Spectrometer
CALIPSO	Cloud Aerosol Lidar and Infrared Pathfinder Satellite Observation
CERES	Clouds and the Earth's Radiant Energy System
CFCs	chlorofluorocarbons
CIAP	Climate Impact Assessment Program
CITE	Chemical Instrumentation Test and Evaluation
CLIMAP	Climate: Long range Investigation, Mapping, and Prediction
CLIMSAT	Climate Satellite

CNES	Centre National d'Études Spatiales
COSPAR	Committee on Space Research
CrIs	Crosstrack Infrared Sounder
CRYSTAL-FACE	Cirrus Regional Study of Tropical Anvils and Cirrus Layers–Florida Area Cirrus Experiment
DAAC	Distributed Active Archive Center
DIAL	Differential Absorption Lidar
DMSP	Defense Meteorological Satellite Program
DST	Data Systems Test
ENSO	El Niño/Southern Oscillation
EOS	Earth Observing System
EOSDIS	Earth Observing System Data and Information System
EOSP	Earth Observing Scanning Polarimeter
EPA	Environmental Protection Agency
ERBE	Earth Radiation Budget Experiment
ESMR	Electrically Scanned Microwave Radiometer
ESSA	Environmental Science Services Administration
ESSP	Earth System Science Pathfinder
FGGE	First GARP Global Experiment
FIRE	First ISCCP Regional Experiment
GARP	Global Atmospheric Research Program
GATE	GARP Atlantic Tropic Experiment
GFDL	Geophysical Fluid Dynamics Laboratory
GHOST	Global Horizontal Sounding Technique
GISS	Goddard Institute for Space Studies
GLEEM	Global Environment and Ecology Mission
GOES	Geosynchronous Operational Environmental Satellite
GPM	Global Precipitation Mission
GRACE	Gravity Recovery and Climate Experiment
GTE	Global Tropospheric Experiment
GWE	Global Weather Experiment
HIRIS	High-Resolution Imaging Spectrometer
HIS	High resolution Infrared Sounder
HMMR	High-Resolution Multifrequency Microwave Radiometer
HRIR	High Resolution Imaging Radiometer
IAS	Institute for Advanced Study
ICSU	International Council of Scientific Unions
IGAC	International Global Atmospheric Chemistry Program

IGBP	International Geosphere/Biosphere Program
IGY	International Geophysical Year
IPCC	Intergovernmental Panel on Climate Change
IPO	Integrated Project Office
IRIS	InfraRed Intferometer Spectrometer
IRLS	Interrogation, Recording, and Location System
ISCCP	International Satellite Cloud Climatology Project
ITOS	Improved TIROS Operational System
IUGG	International Union of Geodesy and Geophysics
JPL	Jet Propulsion Laboratory
LASA	Lidar Atmospheric Sounder and Altimeter
LIMS	Limb Irradiance Monitor of the Stratosphere
LITE	Laser In-space Technology Experiment
MAS	MODIS Aircraft Simulator
McIDAS	Man-Computer Interactive Data Access System
MLS	Microwave Limb Sounder
MODIS	Moderate Resolution Imaging Spectroradiometer
MOPITT	Measurements of Pollution In The Troposphere
MRIR	Medium Resolution Imaging Radiometer
MUSE	Monitor of Solar Ultraviolet Energy
NACA	National Advisory Committee on Aeronautics
NASDA	National Space Development Agency of Japan
NCAR	National Center for Atmospheric Research
NEMS	Nimbus E Microwave Spectrometer
NESDIS	National Environmental Satellite Data and Information Service
NOAA	National Oceanic and Atmospheric Administration
NOMS	Nimbus Operational System
NOSS	National Ocean Surveillance Satellite
NOZE	National Ozone Expedition
NPOESS	National Polar Orbiting Environmental Satellite System
NPP	NPOESS Preparatory Project
NRL	Naval Research Laboratory
NSF	National Science Foundation
OMB	Office of Management and Budget
PEM	Pacific Exploratory Mission
PLATO	Permanent Large Array of Terrestrial Observatories
PSC	Polar Stratospheric Cloud

RAVE	Research on Atmospheric Volcanic Emissions
SAFARI	Southern African Fire-Atmosphere Research Initiative
SAGE	Stratospheric Aerosol and Gas Experiment
SAM	Stratospheric Aerosol Measurement
SAR	Synthetic Aperture Radar
SBUV	Solar Backscatter Ultraviolet
SCAMS	Scanning Microwave Spectrometer
SCAR	Smoke, Clouds, and Radiation
SCEP	Study of Critical Environmental Problems
SDI	Strategic Defense Initiative
SHEBA	Surface HEat Budget of the Arctic ocean
SIRS	Satellite InfraRed Sounder
SMS	Synchronous Meteorological Satellite
SOLVE	SAGE III Ozone Loss and Validation Experiment
SSB	Space Science Board
SSEC	Space Science Engineering Center
SST	Supersonic Transport
STEP	Stratosphere-Troposphere Exchange Project
SUCCESS	SUbsonic aircraft Contrail and Cloud Effect Special Study
TARFOX	Tropospheric Aerosol Radiative Forcing Observational eXperiment
THESEO	Third European Stratospheric Experiment on Ozone
TIROS	Television-Infrared Observations Satellite
TOGA	Tropical Oceans/Global Atmosphere Program
TOMS	Total Ozone Mapping Spectrometer
TOPEX	Ocean Topography Experiment
TOS	TIROS Operational System
TOVS	TIROS Operational Vertical Sounder
TRACE	TRopospheric Aerosols and Chemistry Expedition
TREM	Tropical Rainfall Explorer Mission
TRMM	Tropical Rainfall Measuring Mission
TWERLE	Tropical Wind, Energetics, and Reference Level Experiment
UARP	Upper Atmosphere Research Program
UARS	Upper Atmosphere Research Satellite
UCAR	University Corporation for Atmospheric Research
USNC	U.S. National Committee
VAS	Visual/infrared spin-scan radiometer Atmospheric Sounder

VISSR	Visual-Infrared Spin-Scan Radiometer
VTPR	Vertical Temperature Profile Radiometer
WCRP	World Climate Research Program
WMO	World Meteorological Organization
WOCE	World Ocean Circulation Experiment

Atmospheric Science at NASA

Introduction

On 24 April 2004, the *New York Times* reported that NASA leaders had issued a gag order to agency scientists barring them from discussing an upcoming film, *Day After Tomorrow*.[1] The film's plot revolved around the sudden onset of an ice age provoked by global warming. The idea was loosely based on a 1985 hypothesis by Wallace Broecker of the Lamont-Doherty Earth Observatory in New Jersey.[2] He had postulated that gradual melting of the Arctic ice cap could cause deep ocean currents, which carry heat from the equator toward the poles, to slow or stop. This might result in cooling of western Europe and the northeast coast of North America. In popular culture, this general cooling had been twisted into an ice age. Indeed, the ice age thesis had become so entrenched in the public mind that Donald Kennedy, editor of *Science*, the premier scientific journal in the United States, had decided to greet the movie by publishing an article setting out why an ice age was a highly unlikely outcome of global warming. The existence of the NASA order was leaked by an unspecified scientist unhappy about the overt censorship.[3]

Global warming wasn't the first time NASA had become involved in politically controversial science. NASA had also been involved in controversy around whether supersonic transports (SSTs) or NASA's Space Shuttle or chlorofluorocarbons (CFCs) would cause ozone depletion.[4] But it had entered into such controversies indirectly, by developing technological capabilities relevant to the ozone and climate problems in its meteorological satellite and planetary science programs of the 1960s. While these issues hadn't been politically controversial in the 1960s, both had been areas of active scientific interest. In the 1970s, the agency substantially expanded its laboratory and instrument development efforts to foster new research into these areas.

During the 1980s, it advocated for, and finally won approval of, an Earth Observing System (EOS) to study global change, defined very broadly. Among many other things, EOS was to examine land use change, ocean circulation and heat storage, atmospheric temperature, chemistry, circulation, and radiative properties. At a projected $17 billion cost in its first ten years, it was to be the most expensive science program in American history, exceeding even the $11 billion Superconducting Super Collider, the Biggest of Big Science.[5] As political opposition to climate science mounted in the 1990s, however, NASA's grand ambitions were cut severely, reducing EOS to a "mere" $7 billion. By 2004, EOS had become an orphan, with NASA's own leaders no longer supporting it. Instead, they sought to transfer the politically contentious system, and its scientific responsibilities, to some other agency.

NASA AND LATE TWENTIETH-CENTURY SCIENCE

How did NASA become so immersed in environmental controversy? NASA is, according to its own traditions and views, the *space* agency. Most of its history series reflects this institutional focus on space-oriented enterprises. The history of human spaceflight dominates the series. There is extensive coverage of the Mercury, Gemini, and Apollo programs, and a growing series on its Space Shuttle. There is even a history of the decision to establish a "permanent human presence" in space, written and published before a single bit of space station hardware flew, the cost overruns were totted up, and the whole thing was descoped, built anyway, descoped again while under construction, and then marked for abandonment in pursuit of a new Moon base. Fortunately, there is a growing body of literature on NASA robotic science.[6]

There is, however, virtually no literature in the NASA history series on Earth science. There is a only a single book on the related subject of Earth remote sensing, Pamela E. Mack's history of the Landsat program, published in 1990.[7] Landsat, which was based on the Nimbus series of meteorological satellites, was designed to provide a civilian land-imaging capability for agricultural and urban planning, resource detection, and other uses. Like the larger applications program it was a part of, it was not really a science program. NASA leaders expected other user agencies to pay for research and analysis using Landsat data. Hence, NASA did not build a scientific infrastructure of its own around Landsat.

Instead, NASA gradually built up a scientific infrastructure around atmo-

spheric remote sensing, beginning with the meteorology program (also an applications program), but expanding early in the 1970s to address atmospheric pollution. This infrastructure relatively quickly became comprehensive. New leadership
at NASA headquarters in the middle of the decade sought to improve the quality
of NASA research, and the concomitant credibility of remote sensing, by invest-
ing in new in-house scientific talent, new instrument development for both in situ
and remote sensing, and airborne science programs, and by supporting laboratory
studies and model development and verification. They also began to demand
peer review of new instrument and science proposals. By the early 1980s, the
agency had developed a powerful capacity to carry out intensive, regional-scale
atmospheric research even without space assets.

At the same time, the research NASA was carrying out was becoming more
controversial. The agency took on responsibilities for understanding atmospheric
pollution, ozone depletion, and climate change—particularly anthropogenic
change. These politically contentious realms NASA moved into also forced
agency leaders to come to grips with the demands of policymakers for reliable
knowledge. It proved to be extraordinarily difficult to find ozone depletion in the
great mass of data from ground stations and satellite instruments, for example,
because of the unreliability of some of the instruments and, more important, due
to poor calibration records. This directly impacted the design of NASA's climate
observing system, as the actual radiative changes expected from carbon dioxide
increases were quite subtle—a "one percent game," in the words of scientist Bruce
Wielicki.[8] And the one percent change would occur slowly, too. Definitive detection of small, slow changes required instruments of extraordinary stability as well
as the ability to transfer calibration, or intercalibrate in the jargon, from one
instrument to the next. It also meant the deployment of ground infrastructure to
monitor the space instruments. Thus, a climate observing system needed to be an
integrated system, comprising ground, airborne, and space-borne components.
This proved very difficult for NASA's Earth science community to justify during
the late 1990s. It was supposed to be the *space* agency, after all.

NASA, ENVIRONMENTAL POLITICS, AND THE POLITICS OF MODERN SCIENCE

After World War II, the United States began to develop a large-scale, publicly
funded research infrastructure.[9] In 1958, the Eisenhower administration transformed the National Advisory Committee for Aeronautics into the National Aeronautics and Space Administration, adding a new element to the nation's research

portfolio. In the short run, NASA's task was to develop American rocketry to overcome perceived Soviet leadership in this technology area. It was also intended to
develop space *science*. President Dwight D. Eisenhower had rejected demands
for a crash program to put men in space.[10] Instead, he favored a more measured
scientific program. His successor, John F. Kennedy, however, altered NASA's
direction into the now familiar Big Technology approach of Apollo. While scientific research was within NASA's mandate, during the 1960s the agency's focus was
on the large-scale engineering required to win the race to the Moon, bettering the
Soviet Union in human spaceflight technology.[11] Its purpose was the production
of space spectaculars. Conducting high-quality science was a secondary priority
for NASA during this period.

This changed during the 1970s. The public rapidly turned against the space
spectaculars of the Mercury/Gemini/Apollo sort. In fact, there was a larger movement against Big Technology. There was widespread opposition to large-scale,
publicly funded technologizing in general, with activist organizations opposing
SSTs, nuclear power, large public dam projects, and even NASA's Space Shuttle.[12] In its place, an alternative technology movement organized to promote
human-scale technologies.[13] Combined with the growing cost and economic
impact of the Vietnam War, these trends caused President Richard M. Nixon to
cancel the Apollo program. This left the Space Shuttle, and NASA's short-lived
Skylab, as the agency's principal human spaceflight activities during the decade.
While these maintained the agency's human spaceflight capabilities to a degree,
they were not the all-consuming mission that agency leaders had become familiar
with. They also left the agency with underutilized technical talent.

The political consensus that emerged during the early 1970s favored applying
the nation's technical and scientific resources to national needs. These were primarily defined as civilian and domestic in nature, in opposition to the heavy military bias of American science during the previous two decades.[14] There seemed
to be many domestic ills in need of curing. Suburbanization had led directly to
decaying urban cores and subsequent demands for urban renewal. The tightly
linked problem of urban unrest and consequent demands for more, or at least
more effective, policing drew public attention. Most important for atmospheric
science, however, were the growing demands for a healthier environment.

The late 1960s had seen the emergence of environmental politics as a new
feature of American life. For a number of years, environmental concern was
bipartisan, at least in the sense that both major political parties were willing to
accept that environmental degradation was real, that scientific research into this
area was legitimate, and that reasonable mitigation actions should be taken once

identified. Conflict, of course, could revolve around the definition of reasonable mitigation. But the dogmatic opposition to all things environmental that emerged during the 1980s in conservative circles was not a major force in American politics during the 1970s. President Nixon, considered a political conservative in his own time, thus signed most of the nation's primary environmental legislation: the National Environmental Policy Act, the Clear Water Act, the Endangered Species Act, and the Clean Air Act. He also created by executive order the Environmental Protection Agency (EPA).[15]

NASA leaders of the 1970s chose to see the political attention being given to environmental problems as an area to which the agency's existing talent could contribute. During the preceding decade, the agency's Office of Applications had pursued the development of applications satellites, a category that included communications and meteorological satellites. The agency's first Television-Infrared Observations Satellite (TIROS) successfully transmitted pictures of vast regions of clouds in 1960. The third TIROS spotted a hurricane forming far out over the ocean, promising much earlier storm warnings. Weather satellites had seemed to provide a clear value to people on the ground, and they garnered strong political support. This sort of technology-in-the-public-interest was a way the agency could demonstrate its continued usefulness. The applications program had not been a science program, however. Instead, its leaders had expected other organizations to pay for scientific activities and the development of utilitarian products.

Building on the capabilities it had developed in its weather satellite applications efforts and its planetary science program, NASA gradually moved itself into what has been called policy-relevant science, but with a specifically environmental bent.[16] Perceived problems included air and water pollution, depletion of the stratospheric ozone layer, and climate change. Scientific questions regarding these phenomena often appeared to require the agency's unique technological capabilities to investigate. Global changes required the global perspective satellites provided, and because NASA technologists had developed the ability to examine the chemistry and climates of the other terrestrial planets, they wanted to study Earth as well. By the mid-1970s, advocates inside the agency wanted NASA to take a leading role in these research fields, and NASA administrator James Fletcher agreed.[17] In 1975, NASA won from Congress the new mission of figuring out whether anthropogenic chemicals were reaching the stratosphere and damaging the protective ozone layer; beginning with this area, the agency gradually built its atmospheric research program around global environmental change.[18]

During the 1980s, NASA and the National Oceanic and Atmospheric Administration (NOAA) teamed up to conduct a series of expeditions to Antarctica to determine the cause of the stratospheric ozone hole that the British Antarctic Survey had discovered in 1985. NASA also began a similar series of field expeditions to characterize the chemistry of the troposphere. These research programs immersed the two agencies in political controversy, as their findings directly impacted significant industrial activities. Eventually, the joint NASA-NOAA research led directly to the elimination of CFCs by international treaty. Other authors have studied the effort to develop regulatory policy regarding CFCs and construct the international treaty regime, but they have not examined it from NASA's perspective, or from the standpoint of scientific practice, which is crucial to understanding the evolution of science in the late twentieth century.[19]

Simultaneously, NASA began advocating the construction of a small armada of EOS satellites to carry out nothing less than a reconstruction of the Earth sciences around a space-based, highly interdisciplinary concept of Earth System Science.[20] A major goal of this effort was the development of a fully integrated understanding of how the Earth's atmosphere regulates climate through its role in transmitting and distributing energy. This Mission to Planet Earth, as it was known until the late 1990s, was the largest component of the U.S. Global Change Research Program. The Global Change Research Program was launched in 1989 by the first Bush administration in an attempt to understand the consequences of the gradual, and inexorable, increase in greenhouse gas concentrations from human activities.[21] By this time, the leaders of the American atmospheric science community believed that carbon dioxide–induced warming was inevitable but could not agree on consequences; formulation of sound mitigation measures, at least to policymakers, seemed to require a better understanding of the potential consequences.

Historians James Rodger Fleming and Spencer Weart have both examined the evolution of American scientific interest in the climate change problem.[22] Fleming's work largely ends with the International Geophysical Year (IGY), when systematic carbon dioxide measurements began at Mauna Loa and NASA was created. Weart approaches the present, but largely ignores NASA's unique interest in studying planetary climates. While NASA researchers were very interested in Earth's climate, many key researchers came to that interest from studies of the climates of Venus and Mars; James Pollack at Ames Research Center, for example, wrote an influential 1979 treatise on the climate evolution of the Earthlike planets—Earth, Venus, and Mars.[23] Further, their interest in climate regulation forced them to begin examining the interactions between various elements of the climate system. If atmospheric composition controlled climate, then the chemistry of the atmosphere had to be studied in multiple contexts—planetary out-gassing, geochemical processing, and (at least on Earth) biospheric processing. This was a major influence on the space agency's scientific interest in studying climate.

The political consensus that had surrounded environmental science, if not mitigation, began to disintegrate during the 1980s, and this directly impacted NASA's research programs by the mid-1990s. Ronald Reagan had based his 1980 presidential campaign in part on anti-environmental themes, and on taking office in 1981 had embarked on efforts to weaken environmental laws. Because of this, he faced continuing attacks by environmental groups.[24] After a broad-based attack on civilian science spending in his first two years in office, however, his adminis-tration started reversing its cuts to environmentally relevant science. For the rest of the decade, while environmental policy continued to be controversial, policy-related science could largely be carried out without retaliation by government officials—although not without attacks by conservative political actors outside government.

Beginning with NASA scientist James E. Hansen's testimony in 1988 that "global warming has reached a level such that we can ascribe with a high degree of confidence a cause and effect relationship between the greenhouse effect and the observed warming," that began to change.[25] NASA had been instrumental in demonstrating that stratospheric ozone depletion was caused by manmade CFCs, leading to a worldwide ban on their production. From the perspective of some on the political right, NASA had been directly involved in the dismantlement of an entire industry by government fiat, violating a basic tenet of late twentieth-century conservatism: the superiority of "free market" economies over state-regulated ones. NASA's complicity in destroying the CFC industry, and its support for cli-mate science, which threatened the fossil-fueled American way of life at the end of the century, brought its Earth scientists increasingly under attack during the 1990s. Devoutly anti-environmental leaders took over Congress in the 1994 Re-publican Revolution. By 2003, despite overwhelming consensus among the na-tion's scientific organizations that global warming was real and likely to have sig-nificant negative consequences, a leading senator still referred to global warming as "the greatest hoax ever perpetrated on the American people."[26]

During the 1960s, physics and physicists had been under attack by the Ameri-can political left over the science's intimate involvement in America's vast mili-tary-industrial complex.[27] This had been one driver behind the downfall of phys-ics as the nation's prestige science during the 1970s, as environmentalism reached its peak. By the turn of the century, the political dynamics affecting American science had shifted dramatically, with the atmospheric sciences, tied closely to environmental concerns, suddenly out of favor.

NASA AND AMERICAN EARTH SCIENCE

Much of the history of twentieth-century American science to date has focused on physics and, more recently, biology. The field has also evolved as a history of laboratory science. There is only a small body of literature on twentieth-century field science.[28] But in the geophysical sciences, field studies and expeditions are vital: these sciences are about understanding the Earth. No laboratory can capture the complexity of the Earth's processes, hence in the Earth sciences one must go outside to study one's subject. While the satellite program of the IGY has drawn essentially all of the historical interest in that event, its purpose had been field science, and particularly polar exploration. Following this geophysical tradition of field science, NASA built field capabilities of its own.

In her history of the origins of plate tectonics, Naomi Oreskes discusses the effort to make geology a quantitative science in the early twentieth century, through laboratory studies and the gradual abandonment of mapping, the primary nineteenth-century geologic activity. Field studies became secondary, and indeed, derided as an avoidance of "serious scholarship."[29] But NASA did the opposite in the second half of the twentieth century. It not only embarked in field science, it sought to bring lab-quality measurements into the field. In other words, the agency's science leaders tried to gain for Earth sciences the best of both worlds: quantitative measurements taken in the context of the real world. Actually, their endeavors occurred in the context of several worlds, once one considers that planetary sciences are merely the Earth sciences carried out elsewhere.

NASA's leaders did this for two reasons. NASA *is* the space agency, and parts of it are dedicated to the study of other planets. But scientists can't go there to measure things themselves. With the sole exception of Moon rocks returned from the lunar surface by the Apollo program between 1969 and 1973, they couldn't bring the rocks of other planets into the laboratory either. So they sought to send their laboratories off to the other planets instead. But understanding the data sent back by their robot laboratories often meant turning the instruments loose on Earth first. That way the planetary instruments' performance could be checked via comparison by other, better understood, instruments.

Second, NASA scientific leaders sought to achieve credibility with the American science advisory community (dominated throughout this period by physicists and chemists) and with policymakers as more and more of NASA's science establishment became involved in politically controversial areas. So NASA managers focused a great deal of attention on issues of calibration and intercomparison. They mounted large-scale field experiments, as their expeditions were called, perhaps attempting to borrow some of the cultural glamour of physics or the credibility of experimental science more generally, in order to collect real-world data. The best known of these are the agency's expeditions to Antarctica to study the ozone hole problem.[30]

Methodologically, the agency's focus on large-scale data collection, analysis, and adequate calibration is recognizably what Susan Faye Cannon labeled Humboldtian science, after the efforts of nineteenth-century explorer-scientist Alexander von Humboldt.[31] Humboldt carried a huge variety of instruments on data-collecting expeditions, along with mules and porters and other assistants to help him carry them. Calibration, intercomparison, and a quantitative understanding of errors were important parts of his method. He sought patterns in data that could be masked by such errors; he was also one of the pioneers of data visualization—isobaric maps, for example. Perhaps the most successful application of his methods in the nineteenth century was to the study of tides, leading eventually to useful tide charts. Whether they knew it or not, NASA leaders replicated Humboldt's methods, although one suspects Humboldt would miss the mules and porters that NASA's airplanes have replaced.

Humboldt was also very interested in Earth's (and humanity's) place in the universe, and the spacecraft the Jet Propulsion Laboratory (JPL) sent to Mars and Venus radically altered scientists' understanding of that subject.[32] These two worlds, Earth's nearest neighbors, were largely expected by the scientific community of the 1950s to be roughly Earthlike, with Mars assumed to have at least plant life.[33] But both turned out to be enormously inhospitable. This fact forced planetary scientists to think about the relationship between *chemistry* and *climate*. Mars and Venus had not started out much differently from Earth, sharing the same basic elemental composition and receiving only slightly different amounts of energy from the Sun. How did they become so alien?

This question led NASA into one of the great scientific controversies of the late twentieth century, global warming. It was already well known by the late 1960s that human activities were changing the chemistry of Earth's atmosphere; as planetary exploration largely ended in the late 1970s and early 1980s, planetary scientists turned their interests to Earth. What would anthropogenic changes to atmospheric chemistry do to the Earth? There turned out to be quite a number of economic and political interests who didn't want to know the answer to that question.

But asking this question, and many others related to it, caused NASA to embark on a deliberate program of scientific reconstruction. Its leaders of the mid-1980s wanted to reshape the Earth sciences into a new, integrated discipline. They perceived that during the first half of the century, barriers between various components of what had once been the discipline of natural history had grown high enough to impede understanding of how Earth really worked. There were complex interactions between the oceans, land surface, atmosphere, and (perhaps most important) between and among the life forms inhabiting Earth. Earth was a *living* system, not a dead one with a thin green scum on its surface. This planetary view of Earth was not immediately welcome in the Earth sciences. Hence, NASA set out on a radical agenda in 1986. It set out to create Earth System Science, built up around a holistic, and Humboldtian, view of Earth.

This book argues that the political controversies in which NASA found itself embroiled beginning in the 1980s were a direct legacy of its scientific program of the 1960s. Its scientists developed a view of the world that put them at odds with American politics, and even at times with various communities of the Earth sciences. Their response was to try to reconstruct their sciences in the hope of resolving both the scientific questions and political controversies with the methodological tools of modern science.

All this started with the IGY.

Establishing the
Meteorology Program

Powered by solar energy equivalent to nearly seven million atomic bombs, persistent winds weave vast three-
dimensional patterns of which our daily weather charts
show mere eddies.

— *Harry Wexler, 1955*

The International Geophysical Year (IGY), which ran for the eighteen-month
period spanning 1957 and 1958, is well known as the formative event of the space
race that took place during the 1960s as well as that of the National Aeronautics
and Space Administration.[1] The new American space agency took its initial research agenda from the IGY's Earth satellite program. It also became the institutional home for a long-term effort to construct both a global dataset and a real-
time global observation system that could feed future global operational forecast
models data.

Harry Wexler, the Weather Bureau's chief of research, had been an early advocate of satellite meteorology. In fact, he had been a literal visionary. Well prior to
the first satellite, he painted a picture of what Earth might look like from space —
cloudscape and all. It now belongs to the National Air and Space Museum in
Washington. Wexler's artistry represents one of the major goals of meteorologists
of mid-century: a detailed understanding of the general circulation of Earth's
atmosphere.

NASA was tasked in its 1958 charter with the development of space applications. Communications and weather satellites were its two earliest effects, and it was in the applications directorate that the meteorology program was located. NASA, of course, was not itself responsible for weather forecasts. That responsibility belonged to Wexler's U.S. Weather Bureau. Hence the aim of the NASA program was to develop the application of satellite meteorology for the Weather Bureau's ultimate use. This made the two agencies partners, in essence, if not always comfortable ones.

The NASA meteorology program was part of a larger effort within the meteorology profession to reconstruct meteorology along the lines of the so-called hard, or physical, sciences. A handful of meteorologists had sought the ability to use the fluid dynamics equations to predict the generation and movement of weather since the early years of the twentieth century. Because the fluid dynamics equations are nonlinear, and because the atmosphere is complex, computation of the weather proved extraordinarily difficult. It could not be done successfully until the development of the stored-program digital computer after World War II. This device, even in its early, primitive form, could out-calculate rooms-full of trained human computers, as they were then known. Human computers were rapidly replaced by this new machine, which then opened new realms of computational possibility. One of the first fields of science transformed by the computer was meteorology. Global data from satellites seemed to promise a second revolution in forecasting, permitting accurate forecasts months in advance.

NUMERICAL FORECASTING

Prior to the twentieth century, meteorology was practiced two ways. Climatologists collected and analyzed regional datasets, generating mean and average values and finding regularities in these averages. They could also make reasonable statements about the frequency of major phenomena, such as hurricanes. The calculations necessary to do this were time consuming and did not permit prediction of short-term phenomena such as weather. At best, climatologists could identify long-term trends once they began. The second component of meteorology, forecasting, was more art than science. Beginning in the 1840s, the development of telegraph networks allowed rapid collection of surface data, which became the basis for synoptic meteorology.[2] Forecasters placed data on maps, and based on their training and experience they could make reasonable predictions of the next day's weather. Over many years of experience, regional forecasters developed a detailed understanding of how weather typically happened in their area of responsibility, making weather prediction a matter of both data and experience. It was not, however, a modern physical science, based upon the known laws of physics and quantitative analysis.

At the end of the nineteenth century, a few individuals began promoting reconstruction of meteorology along physical lines. Cleveland Abbe in the United States and Napier Shaw in Great Britain had each argued that the time had come to make meteorology a modern physical science, and Abbe set out a system of seven equations that could permit calculation of the weather. The most effective promoter of dynamical meteorology, as the nascent discipline was called, was Vilhelm Bjerknes, founder of the Bergen School of meteorology and pioneer of the highly successful (but still not physics-based) system of air mass analysis. Bjerknes had contended that the physical laws by which the atmosphere functioned were already known, in the principles of fluid dynamics and thermodynamics.[3] He gathered many converts, including the first person to actually try to calculate the weather, Lewis Fry Richardson.

Richardson, an Englishman, Quaker, and pacifist, carried out his effort to calculate the weather during his service as an ambulance driver during World War I. He developed a set of partial differential equations and a numerical method of solving them via approximation in order to carry out his trial. He had collected large amounts of data from throughout Europe to feed into the calculation; lastly, he had had to develop methods to fill in parts of the map where data had been unavailable. Once all this work was completed, he had then spent six weeks carrying out the calculations for a six-hour forecast for only two locations on his map. And the resulting forecast was wildly off the mark, with one of the two positions having an error significantly greater than the location's natural variability. His results were lost for several years, then found under a coal heap in Belgium and returned to him. Richardson expanded and published the results of this first attempt to calculate the weather in 1922, in a magnum opus titled *Weather Prediction by Numerical Process*.[4]

Richardson's effort, despite having produced an inaccurate forecast, was not ignored. It was widely read and commented upon, and it was highly regarded within the community of research meteorologists. It was a first try at a new method of prediction, made with inadequate data. Yet Richardson's methodology was rigorous and complete.[5] Meteorologists chose to believe that Richardson's methodology was the correct way to go about calculating the weather, but that the sheer enormity of the calculations had prevented success. Indeed, in his book Richardson had imagined the scope of the "weather factory" necessary to numerically predict the weather in real time. He envisioned a vast hall filled with 64,000 "human automata" using desk calculators and communicating via telegraph, engaged in calculating the weather.[6] Hence no one followed Richardson's lead for many years.

The development of the digital computer in the final years of World War II permitted the meteorological community to revisit numerical weather prediction. John von Neumann, an already-famous mathematician at the Institute of Advanced Studies in Princeton, had collaborated on the ENIAC computer project and in the process had developed the logical structure that eventually formed the basis of all stored-program digital computers. He sought funding for such a machine for use in scientific research. He had been introduced to Richardson's work by Carl Gustav Rossby at the University of Chicago during a meeting in 1942, and became interested in applying the digital computer to weather prediction in early 1946. After a meeting with RCA's Vladimir Zworykin and the head of the U.S. Weather Bureau, Francis W. Reichelderfer, and more than a little enthusiastic advocacy by Rossby and the Weather Bureau's chief of research, Harry Wexler, von Neumann decided to establish a meteorology project associated with the computer he was trying to build, the EDVAC.[7]

The numerical meteorology project crystallized under Jule Charney in 1948, after several other project leaders had left for other tasks. Charney, who had completed his PhD work at UCLA but had been strongly influenced by Rossby at Chicago during a nine-month stay in 1946, had carried out an examination of why Richardson's prediction had been so far off prior to joining the project; once in Princeton, Charney developed a new methodology that replaced Richardson's seven primitive equations with a single equation. This placed the computing needs within the expected performance of the EDVAC, and in fact could be solved (if slowly) by hand. Known as the barotropic model, Charney's model made several unrealistic assumptions, but it produced reasonable twenty-four-hour forecasts. It degraded quickly after that, though, primarily due to its assumption that all processes were adiabatic, in other words, occurred without energy exchange. In 1951, Norman Phillips joined the team after completing his dissertation research at Chicago. Phillips developed the first baroclinic model, a two-layer model that permitted energy exchange. This produced very successful twenty-four-hour forecasts in 1952.[8]

Charney's group believed that the baroclinic model's forecasts still had value out to forty-eight hours, a significant improvement over those achieved by state-of-the-art human forecasters most of the time. Several government organizations perceived value in the group's models as well, and the Joint Meteorology Committee, administratively under the Joint Chiefs of Staff but composed of representatives of the air force's Air Weather Service, the civilian Weather Bureau, and the Naval Weather Service, advocated formation of an operational numerical prediction center that would serve the three organizations. Known as the Joint Numerical Weather Prediction Unit, this was established in July 1954 and became operational in 1955, using models developed by Philip Thomas and George Cressman, both of whom had experience at the Institute for Advanced Study (IAS) meteorology project.[9] Cressman became the unit's head, and eventually became the director of the U.S. National Weather Service.

The models developed at the IAS project prior to 1953 were all regional in extent. Because the weather was not regional, but traveled globally, this limited their predictive range to two days or so. Longer-range prediction required models that were at least hemispheric in extent. In 1955, Phillips demonstrated the first of a series of general circulation models that accurately replicated the seemingly permanent, large-scale structures of the global atmosphere, such as the jet stream and the prevailing winds.[10] Researchers in several places began to conduct general circulation experiments after Phillips's demonstration.[11] While these were not forecast models, their success, combined with the success of the regional forecast models developed in the IAS project, indicated that global weather prediction could be accomplished.

The principal challenge facing researchers interested in developing global circulation models, and particularly those interested in extending the useful length of weather forecasts after the mid-1950s, was data. Charney's group at IAS made use of data collected in previous years to initialize their forecast models, and could compare the model output to the actual weather as recorded by the Weather Bureau. One of the most convincing experiments had been a successful retrospective prediction of an unusual winter storm on Thanksgiving Day, 1950, by Phillips's baroclinic model. This experiment had demonstrated the superiority of the baroclinic model over Charney's older barotropic model, which had not generated the storm from the same initial dataset.[12] Model researchers believed that they could only improve model performance by comparing their model's output to real data. But there was no such dataset against which to compare the detailed performance of global models. There had never been a reporting network in the Southern Hemisphere, and from the standpoint of meteorological researchers of the early 1950s the Southern Hemisphere remained a vast terra incognita. The situation in the equatorial belt was no better. Data for the Northern Hemisphere existed and was collected routinely after World War II in support of the ongoing military operations of the United States. This data, once centrally archived and evaluated for quality, permitted the construction of hemispheric prediction models during the late 1950s. But global models had to wait for a global dataset.

The small Weather Bureau modeling effort, finally, was not the only one in the United States during the 1950s. At UCLA, Yale Mintz and Akio Arakawa were also developing general circulation models. The armed services more generally were interested in atmospheric models, which could aid in everything from predicting weather for flight planning to understanding the effects of nuclear weapons. Will Kellogg, a UCLA doctoral graduate who had moved to RAND, worked on local-scale models to allow prediction of the dispersion of radioactive fallout from nuclear weapons tests during the 1950s, for example.[13] He also co-wrote one of the earliest proposals for satellite meteorology.

ROCKET RESEARCH, THE IGY, AND THE PROMISE OF GLOBAL DATA

NASA's science program of the 1960s had its roots in an informal, self-appointed group of physicists that had formed in 1946 around Germany's V-2 rocket. Calling themselves the V-2 Panel, and later the Rocket and Satellite Research Panel, this rather casual entity used many of the hundred or so V-2s assembled from parts collected in Germany at the end of the war for upper atmosphere research. They were primarily interested in investigating the ionosphere and the regions above it using new instruments and techniques that they devised for themselves. The members of the group were employed by a variety of universities, including Princeton, Johns Hopkins, Iowa State, and Harvard, and by military agencies, particularly the Naval Research Laboratory (NRL). Their funding came through various mechanisms from all three armed services, which sought better understanding of the upper atmosphere's radio characteristics to improve radar and radio performance.[14]

When the V-2s ran out, the panel had utilized U.S.-built sounding rockets to continue their research. Simultaneously, the U.S. government spent vast sums developing longer-range liquid-fueled rockets to serve as delivery systems for the atomic bomb. These new rockets offered a tantalizing future opportunity to obtain information about the Earth from outside it—a truly global dataset. The potential was not lost on the few individuals who had clearance to know about the rocket research. The RAND Corporation, founded in 1945 to serve as an advisor to the Army Air Forces, had started looking at the possibilities inherent in orbital observation posts in 1947. Writing about his experiences in the RAND project during the late 1940s, meteorologist William Kellogg explained that what RAND most needed was evidence that a higher-altitude perspective would be useful to weather prediction.[15] The V-2 Panel provided it in 1949, producing a series of cloud photographs from a V-2 launch.

Many scientists not involved with rocket research were dismissive of it. Historian David DeVorkin has argued that the rocket researchers were almost entirely outsiders, scientists who were conducting research in well-established scientific fields without being members of the relevant research communities.[16] They were members of a new technical culture interested in producing new research capabilities, new ways of doing research. They were less interested in the results, and given the unreliability of rockets, they often achieved little but frustration for their efforts. But they were by and large not deterred either by frustration at rocket and instrumentation failures or by the skepticism (and frequently outright hostility) of more traditional scientists. Because their patrons were the armed services, and the services recognized the importance of improved understanding of the high altitudes even if the rest of the scientific community did not, their funding, while hardly infinite, was assured.

Activism by the members of the rocket research panel, and particularly by Lloyd V. Berkner, aimed at increasing the stature of rocket research helped lead to a 1950 proposal that 1957 be declared an "International Geophysical Year."[17] The idea had gotten its start in James Van Allen's living room, where a number of the rocket researchers had gathered to meet geophysicist Sydney Chapman.[18] It grew out of a discussion of the need for a "Third Polar Year" to expand scientific understanding of the complex polar atmosphere.[19] The "year" was to run for eighteen months, and it was chosen to coincide with the solar sunspot maximum. During the IGY, a wide range of scientific studies would be carried out planet-wide by international teams of scientists. The organizers hoped IGY would include Antarctic studies, investigation of the airglow phenomenon that occurred at various altitudes, and deep-sea experiments, in addition to a great deal of rocket research. It would also involve extensive field expeditions, including multiyear ocean studies and Antarctic exploration. In 1952, the International Council of Scientific Unions (ICSU) accepted the American proposal, making the project international in scope.

Late in 1952, the United States started organizing its IGY effort by forming the National Committee for the International Geophysical Year (USNC). The USNC's task was coordination of what would be a very large effort encompassing a number of government agencies and many universities. It was formed under the auspices of the National Academy of Sciences, which had been chartered by

Congress in 1863.[20] The Academy represented the United States within the ICSU, the parent organization of the IGY, and it also served as a source of scientific advice to the government. The Academy was technically a nongovernmental organization, however, and placing the USNC outside the government permitted the civilian scientific mission of the IGY to remain paramount. The USNC's task was coordination and selection of experiments, which it did through a series of committees that recommended which experiments should be funded. The funding agency for the IGY's scientific effort was the National Science Foundation (NSF), which let grants and contracts to experimenters based on the USNC's recommendations.

The chairman of the committee was Joseph Kaplan and his executive secretary was Hugh Odishaw. Kaplan was a physicist at the University of California, Los Angeles, who had specialized in the spectra of diatomic molecules commonly found in the upper atmosphere. He lent strong support to the idea of orbiting a satellite during the IGY, and when President Dwight D. Eisenhower approved the satellite idea in July 1955, formed a Technical Panel on the Earth Satellite Program. The Technical Panel's function was to solicit scientific proposals and select the experiments that would fly aboard the nation's first satellites. Its members were Kaplan and Odishaw, Homer E. Newell, Jr. of NRL, William H. Pickering of the Jet Propulsion Laboratory (JPL), Athelstan Spilhaus of the University of Minnesota, Lyman Spitzer, Jr. of Princeton University, James A. Van Allen of the State University of Iowa, and Fred Whipple of the Smithsonian Astrophysical Laboratory.[21] Newell, Pickering, Spitzer, and Van Allen were also members of the Rocket and Satellite Research Panel. This network of close affiliations and overlapping committee memberships would eventually help smooth space science's transition from the Technical Panel to NASA three years later. In 1955, this group held both the scientific knowledge to recognize valuable experiments and the rocket engineering experience to select experiments that might be possible within the IGY's time horizon.

The Technical Panel hoped to orbit twelve satellites during the IGY—actually, they were more realistic in hoping for six successes out of twelve tries. By the end of 1955, the group already had five proposals in hand. The first had been Van Allen's proposal for a cosmic ray experiment, received shortly before the panel had even been appointed. Others included S. Fred Singer's proposal to measure erosion of the satellite's skin by micrometeoroids and Herbert Friedman's proposal to measure variation in the intensity of solar Lyman alpha radiation. These were proposals by insiders, people already well-connected to the IGY program, but the panel also took steps to recruit new experimenters. They began the search for more proposals by holding a symposium in Ann Arbor, Michigan, in January 1956 on "The Scientific Aspects of Earth Satellites." The symposium was organized by Van Allen and Odishaw, and was attended by about fifty geophysical scientists invited by the panel. The prospective researchers heard briefings on both the potential of satellite research and the technical constraints that satellites would impose on experiments, in order to help them plan valid experiments (and reduce the number of infeasible proposals received by the Technical Panel).

The thirty-three papers presented at the symposium reflected the breadth of possible atmospheric research programs, with a strong bias toward the upper atmosphere and magnetosphere. Three experiments were aimed at determining atmospheric pressure and density by various means, while five addressed measurements related to the Earth's magnetic field. Radiation at high altitudes drew six papers, and in two papers, experimenters proposed meteorology experiments. There were also three papers on micrometeoroids and erosion of the vehicle's surface and three more oriented at the ionosphere. The balance of the papers covered technical aspects of satellite research, including tracking and telemetry requirements.

The actual selection of experiments was made by a subset of the Technical Panel, the Working Group on Internal Instrumentation. This proved to be a dynamic process, with no single decision date.[22] Instead, the Working Group met when it had a batch of new proposals to evaluate. They also met to assess the progress of already chosen instruments. They prioritized experiments, evaluated which experiments could be packaged with others in the very small Vanguard satellite (50.8 centimeters in diameter) without causing interference, and provided an ongoing source of encouragement and advice. They did not limit their support to experiments that could be carried out during the IGY; they also recommended funding of experiments that almost certainly would not fly during IGY but would be scientifically worthwhile in some future program. One experiment that was a relative latecomer to the program is worth following to detail how experiments (and perhaps more important, experimenters) moved from the IGY effort into the NASA era.

On 31 May 1956, Harry Wexler wrote to Joseph Kaplan to propose an experiment designed to determine the Earth's heat balance—the difference between the incoming solar radiation and the outgoing infrared energy radiated by the Earth itself.[23] Wexler, who had helped push John von Neumann into setting up Charney's meteorology group at the Institute for Advanced Study (IAS) in 1946, was involved in research on weather control and modification, and he was also heavily involved in the IGY's Antarctic research efforts. He proposed this experiment because the Earth's global, short-term heat balance was of great value to meteorologists. While incoming and outgoing radiation had to match over the long term (or the Earth would heat or cool very rapidly), it might vary substantially over the short run or in localized areas, which would affect weather patterns. For the same reason, the energy balance was useful to numerical modelers, who used an average value developed by Julius London in the early 1950s from ground-based measurements as an input. London's value was widely believed to be flawed simply because he had possessed no way of determining the amount of energy that reached the top of the atmosphere from the Sun. Better predictions required a better understanding of the Earth's energy flows, and these could only be measured adequately from space.

Wexler did not provide any technical detail in this letter, but a few days later he presented more detail to the Technical Panel. At the University of Wisconsin, Madison, a group of researchers in the meteorology department headed by the department chairman, Verner Suomi, had sketched out a lightweight device that could measure the three parameters necessary to calculate the heat balance. Suomi had done his doctoral dissertation at the University of Chicago on the odd subject of measuring the heat balance of a cornfield. To measure the Earth's heat balance, he proposed using a set of three sensors. One sensor would measure the incoming solar radiation. A second sensor would detect the energy reflected off cloud tops, which had to be subtracted from the incoming radiation. Earlier ground and balloon experiments had estimated this reflected energy could be up to 35 percent of the total. A third sensor would detect the Earth's outgoing infrared energy. Wexler still did not present a proper technical proposal, which he told the group would be submitted by Suomi.

Suomi's detailed proposal was finally forwarded to the Technical Panel for evaluation in October 1958, and was promptly recommended for approval by the Working Group and assigned the number ESP-30.[24] Because it was a meteorology experiment, the National Academy of Science's Committee on Meteorology (chaired by the ubiquitous Lloyd Berkner) weighed in with a recommendation of its own, that "at the earliest feasible moment" a heat balance experiment be flown.[25] At its 5 December meeting, the USNC approved both Berkner's recommendation and Suomi's experiment, and on 16 December Suomi submitted his proposal to NSF for funding. His proposed budget was $75,000, a number that would become $131,000 as the experiment played out.

The Wexler-Suomi experiment was assigned to experiment group IV, which included another meteorological experiment designed by William G. Stroud at the U.S. Army's Signal Engineering Laboratories at Fort Monmouth, New Jersey. This sensor was designed to produce images of the Earth in the near infrared using a pair of lead sulfide detectors. But the Technical Panel did not believe that both experiments could be supported all the way through flight primarily due to funding limitations. The Technical Panel could not decide which experiment was of greater value to meteorology, however, and to help it make what it termed an "agonizing" decision, it sought the opinions of research meteorologists in several universities and air force laboratories. At a meeting to be held 5 November 1957, the panel would use their recommendations to choose between the two.[26]

Suomi's experiment was chosen, but that proved not to matter. In the very short run, Stroud's experiment continued under army funding. It would not be part of the Vanguard satellite program, but Vanguard very quickly ceased to be the only satellite program. The Soviet Union's orbiting of Sputnik 1 on 4 October 1957 disarrayed the entire IGY satellite program. As the next two years played out, both experiments were able to fly as more and more resources were devoted to finishing the IGY program and demonstrating that U.S. science was not inferior to Soviet science.

Major changes occurred around the Earth satellite program in Sputnik's vast wake and the subsequent explosion of the first Vanguard shot in December. The Army Ballistic Missile Agency (ABMA) in Huntsville, Alabama, where Wernher von Braun's rocket engineers had carried out a series of successful suborbital re-entry tests during 1957, received permission to enter the satellite race using its relatively (compared to Vanguard, anyway) mature Jupiter rocket. ABMA's Project Orbiter, proposed in conjunction with JPL in Pasadena, had been Vanguard's chief competitor in 1955 but had been shelved after President Eisenhower had chosen Vanguard. Orbiter had never been far from von Braun's mind, however, and after Sputnik he was able to convince his superior, General John B. Medaris, to let him prepare for a January 1958 launch. Medaris assigned the actual satellite to JPL, giving that organization its entry into the satellite business.[27]

The January 1958 launch gave JPL less than ninety days to produce a satellite and deliver it to Huntsville, which was possible only because one of the IGY's instruments had already been prepared. In his memoir, James Van Allen reported that Ernst Stuhlinger of ABMA had told him about the Jupiter's progress in early 1957 and in April 1957 an ABMA group had visited him in Iowa and given his instrument team specifications for a Jupiter C payload. He had used the information to produce a version of his cosmic ray instrument compatible with the Jupiter just in case Vanguard failed. Because Medaris had decided on his own authority to prepare for a Jupiter shot even before Vanguard's explosion, JPL's Pickering (who was also a member of the Vanguard Technical Panel) had Van Allen's instrument transferred to Pasadena in November 1957.[28]

JPL's satellite, Explorer 1, went into orbit successfully on 31 January 1958. In addition to being the first American satellite, it also produced radiation data that was entirely unexpected and caused JPL to put several more copies of Van Allen's instrument into orbit to verify and expand upon its results. Explorer 1's data showed much higher levels of charged particles than current theory predicted, but the dataset was very incomplete. Explorer 1 did not have the ability to store data taken during its orbit for retransmission when it passed over the ground station, limiting the utility of its data. Subsequent Explorers launched in February (which did not reach orbit) and March had tape recorders to permit capture of a full orbit's data. Explorer 3's data allowed Van Allen to determine that the high radiation levels were real and theorize that they were a product of the Earth's magnetic field, which was trapping charged particles at certain energy levels and confining them to belts around the planet.

Van Allen's radiation belts, as they quickly became known, were the first major scientific return from the IGY's Earth satellite program, and the Explorer series of satellites made JPL a major center for space research. JPL's sudden ascent out of obscurity challenged NRL's previous dominance of the budding field of space science while not yet really changing the scientific goals of the overall space program. ABMA's challenge to Vanguard also had no effect on the science agenda in the short term because Medaris had chosen to give JPL the satellite task. Because JPL director Pickering was one of the V-2 Panel veterans on the Technical Panel for the Earth satellite program, this group continued to set the scientific agenda for the nation.

That began to change in June, however. On 4 June 1958, the National Academy of Science's president, Detlev Bronk, met with Alan Waterman of NSF; Hugh Dryden, chairman of the National Advisory Committee on Aeronautics (NACA); Lloyd Berkner; and Herbert York, chief scientist of the Defense Department's Advanced Research Projects Agency (ARPA). They agreed to establish a new panel within the Academy to permanently supervise the nation's space science effort. They chose Berkner to be the chairman of the new Space Science Board (SSB), carved the amorphous term *space science* into seven scientific disciplines, and assigned SSB a set of tasks, including completion of the IGY satellite program and coordination of the nation's space program.[29] Unlike the Technical Panel, however, SSB was not to be an operating agency. Once its IGY role was over, it would not continue to actually run the science program—it would advise.

Hugh Odishaw became SSB's executive director when it began operating in June. The new organization very rapidly absorbed the satellite effort from the IGY's Technical Panel, and it began planning for an expanded post-IGY effort. On 11 June, George Derbyshire sent Odishaw a list of the instruments that needed to be completed and a handful of technical problems that needed to be solved to carry out an expanded space science program, including on-orbit stabilization, de-spinning of the satellite after launch, and recovery of instrumentation or samples dropped by satellites. Carrying over the rest of the instruments into whatever future program emerged was going to cost about $6 million, and the next day Odishaw wrote to Alan Waterman at NSF to request $2 million of the sum. He also put in a plug for money for the Weather Bureau, which had been denied a supplemental request for $75,000 to fund processing of the data it expected from Suomi's instrument.[30]

Lloyd Berkner called together the first meeting of SSB's members on 27 June. The members were Leo Goldberg, H. Keffer Hartline, Donald F. Hornig, Richard Porter, Bruno B. Rossi, Alan H. Shapley, John A. Simpson, Harold C. Urey, James A. Van Allen, O. G. Villard, Jr., Harry Wexler, Harrison S. Brown, W. A. Noyes, Jr., and S. S. Stevens. NACA's Hugh Dryden attended as an invited guest, because his organization would become the nation's new space agency within a few months. Of this group, only four were actively involved in space research already, and two were members of the self-appointed Rocket and Satellite Research Panel (Van Allen was its chair) and of the Technical Panel on the Earth Satellite Project.[31] This reflected a deliberate attempt to broaden the constituency for space research and reach out to a scientific community that was quite skeptical about the endeavor. From this meeting came a decision to immediately solicit more experiments from a wider variety of researchers and to establish subcommittees representing each of the seven newly defined space science subdisciplines plus five more devoted to various non-disciplinary technical issues. The disciplinary subcommittees would evaluate the research proposals received and recommend funding over the next several months, and they would also craft longer-range research plans that would be handed to NASA when it opened its doors.

In soliciting experiments, Berkner continued to seek expansion of the constituency devoted to space research. Whereas the Technical Panel on the Earth Satellite Project had deliberately restricted its call for experiments to those with preexisting experience, Berkner decided to cast a very wide net. He dispatched a telegram on 4 July to 150 university science departments and private research institutions asking for proposals "within a week." SSB received more than two hundred proposals and requests for more information in response. As retired

NASA scientist John Naugle pointed out, Berkner's telegram was inspirational, particularly among young scientists looking for new fields in which to make names for themselves.[32] While most of these proposals could not be funded immediately—at its second meeting on 19 July, SSB approved six for recommendation to the three extant funding agencies, NSF, ARPA, and NACA—they reflected a groundswell of interest in space science that was a product of Berkner's activism in the context of Cold War competition with the Soviet Union.

On 1 October 1958, NASA replaced NACA, initiating a period of uncertainty about who would establish the goals of the nation's space science program and choose the experiments conducted in it. The new agency had no space scientists yet. But President Eisenhower intended NASA to be a space *science* agency. He and his scientific advisors did not think that the engineering effort involved in putting men in space would be particularly useful. Robotic satellites could provide scientific and intelligence information much less expensively. And he clearly didn't yet perceive the propaganda value of Men in Space. While authorizing the Mercury program, he kept it small, and he eschewed the Moon.[33] So NASA, in his intended incarnation, was to develop space science and space applications. It needed scientists.

The new administrator, T. Keith Glennan, had kept ex-NACA chairman Hugh Dryden as his deputy and appointed the Lewis Flight Propulsion Laboratory's assistant director Abraham Silverstein head of the agency's Office of Space Flight Development. Silverstein knew he needed a chief scientist but hadn't yet offered the job to anyone. He hired NRL's Homer Newell after Newell and his NRL colleagues John W. Townsend, Jr. and John F. Clark came to see him to discuss the future of space science in the agency. Newell became assistant director for space science on 20 October, reporting to Silverstein. Newell put Clark in charge of ionospheric research and Townsend in charge of space research. Newell told SSB at its third meeting on 24–26 October that he intended to bring his entire NRL staff with him to NASA; as his efforts played out, NRL allowed him "only" fifty people.[34] Newell's fifty NRL veterans, along with others recruited from the Air Force Cambridge Research Laboratory and the Army Signals Engineering Laboratory and all of Project Vanguard, formed the core of a new research center on the site of the Beltsville Agricultural Research Center near Greenbelt, Maryland. Formally named the Goddard Space Flight Center on 1 May 1959 in honor of the American rocket pioneer Robert Goddard, this center became the focus of NASA's meteorology program.

The third meeting of SSB marked the beginning of what Newell called in his memoir his "love-hate relationship" with SSB.[35] NASA's charter made it, not SSB, responsible for the nation's space science program, but SSB expected to continue its prior role of soliciting and selecting proposals. NASA, in SSB's view, would provide engineering services and launch vehicles to outside experimenters, but it would not have scientists or a science program of its own. This led to a brief skirmish between NASA and SSB that NASA easily won. In December, the new NASA administrator, T. Keith Glennan, approved a policy document that reserved to NASA the right to establish the research priorities of its program and choose the specific experiments and schedules for them.[36] But it would use the recommendations of SSB and independent proposals in formulating its overall research program. It did not end SSB's attempt to retain for itself a larger role, which continued for almost another year. The new policy, though, did permit SSB to remain a strong influence on NASA's plans.

Writing to NASA Administrator Glennan, Alan Waterman at NSF, and the director of ARPA on 1 February 1959, Hugh Odishaw forwarded SSB's recommendations for the future program. This document established long- and short-range goals, chose experiments, recommended experimenters, and suggested satellite packages and schedules.[37] It reflected Berkner and Odishaw's ongoing attempt to secure control of the space science program, but it also was an invaluable aid to NASA in establishing its own program. The program Homer Newell described in his "National Space Sciences Program," circulated internally and to SSB on 16 April and presented to Congressman Albert Thomas's Appropriations Subcommittee on Independent Offices on 29 April, drew heavily on SSB's language.

Newell cast the NASA space sciences program into the subdisciplines SSB had created: atmospheres, ionospheres, energetic particles, magnetic and electric fields, gravity, astronomy, and biosciences. He enumerated a very similar set of existing problems scientists faced in understanding the nature and functioning of the Earth's atmosphere, including the structure, circulation, and dynamics of the high atmosphere; sources of energy within it; the relationship between it and the Van Allen radiation belts and between it and the lower atmosphere; and the high atmosphere's detailed chemical composition. He did not exclude the lower atmosphere from his presentation, noting the major problems as its radiation budget and how it affected circulation and weather, the same scientific concerns SSB had stated.[38]

In defending his request for $45 million in fiscal year 1959 to carry out his space science agenda, Newell had to repeatedly emphasize potential practical benefits from space science, an ongoing need that he knew from several years of personally defending the IGY program before Thomas's subcommittee — when Newell was out of the room, Thomas congratulated Glennan and Dryden for hiring such a "topper" away from NRL.[39] Newell also linked the NASA research program directly to the IGY effort, telling Thomas that

> our immediate program is to carry on with the momentum developed during the IGY, and to study the atmosphere at even higher altitudes. Before proceeding to show you the program, I would like to point out that in the process we have had this practical application: we have developed an engineering standard atmosphere from the rocket data obtained to date. This standard atmosphere is used in the design of aircraft and vehicles that would fly at these levels. . . . In the future we can hope to find out the relations between the upper atmosphere in this region and the lower atmospheric weather phenomena. This is the sort of thing that will be fundamental to a truly universal application of, say, meteorological satellites.[40]

In linking the NASA program to the IGY, Newell was arguing that the agency was expanding upon a successful scientific research effort, not engaging in the "wasteful duplication" that congressional funders did not like. And by pointing to a pair of economically valuable applications of the research, he drew on the notion that scientific advance produced economic gains. Finally, he cast the NASA science program as an orderly, essentially linear transition from the IGY's, ignoring the still somewhat messy relationship between SSB and NASA.

In the short run, nearly all of the IGY experiments were carried out between 1958 and 1960 using either the Vanguard rocket or Jupiters. The SSB had gotten funds from NSF after Sputnik to repackage most of its instruments for the JPL Explorer satellite; Vern Suomi's heat balance instrument was built in flight-worthy form for both Vanguards and Explorers. It ultimately flew aboard Explorer 7 on 13 October 1959, and was celebrated by Wexler as the first completely successful meteorological experiment in space. Stroud's cloud-cover experiment, which had been launched aboard Vanguard 2 on 17 February 1959, had worked as designed but the satellite's motion in space was not what the experiment required. Its data could not be resolved into images as intended.[41]

The IGY itself concluded in 1960. The data generated by the event's scientists was deposited in three World Data Centers for safekeeping and public access. Beyond its data and the space race, the IGY left a legacy of more formalized international scientific cooperation through a series of new committees attached to ICSU. One of these concerned space cooperation, the Committee on Space Research (COSPAR).[42] The IGY also left a permanent imprint on the Earth's most inaccessible region, Antarctica, where several nations, including the United States, established permanent research facilities. The continent itself was set aside as a research laboratory for any nation via treaty concluded in December 1959. The U.S. site, McMurdo Station, would eventually host NASA research expeditions.

THE NASA METEOROLOGY PROGRAM

In addition to the atmospheric science research program NASA had adopted from the IGY effort, it gained an important space application that also came to include extensive meteorological research from ARPA: a weather satellite project known as TIROS (Television-Infrared Observations Satellite). ARPA, in turn, had acquired it from the army. NASA leaders had pursued transfer of TIROS because it would provide an early, highly visible success, thus helping to validate the new agency's existence.[43] But the public (and congressional) reaction to TIROS was so powerful that the agency expanded its efforts, becoming involved in a long-term collaboration with the Weather Bureau to advance weather satellite technology and to demonstrate the value of that technology by undertaking meteorological research.

TIROS was a moniker that accurately reflected the project's origin in a surveillance satellite program. In the late 1940s, at the RAND Corporation in Santa Monica, researchers had started thinking about the uses of space, despite the current technological inability to reach it. Their first report on the subject, "Preliminary Design of an Experimental World-Circling Spaceship," in May 1946, had discussed at a general level the potential economic and military value of space.[44] In May 1947, another report had analyzed the possibilities of space for military surveillance and intelligence collection. In 1950, RAND had pointed out that space surveillance could produce an unimaginably vast amount of data, and that planners would have to find ways of handling the torrent. The following year, RAND scientist William Kellogg had published a study of a meteorological satellite that helped inform the later IGY effort.[45]

In 1956, the U.S. Air Force had initiated a program known as WS-117L to develop a reconnaissance satellite. Lockheed won the air force's contract; RCA, the unsuccessful bidder, turned to the U.S. Army for funding. The army accepted RCA's proposal and initiated a reconnaissance satellite program of its own, named JANUS.[46] This was administered by ABMA. When ARPA was formed in 1958, JANUS became an ARPA project, although it remained under the administration of ABMA. At the same time, responsibility for reconnaissance satellites was removed from the army and JANUS had its television cameras detuned to a resolution that would serve for meteorology but not for intelligence gathering.[47] Its name was also changed to TIROS. Kellogg became chairman of the TIROS project's science advisory committee, helping define the vehicle's observing capabilities.[48]

When Dryden and Silverstein had begun organizing NASA in 1958, they had assigned Edgar Cortright to ARPA's Ad Hoc Committee on Meteorology, which oversaw TIROS. ARPA sponsored a two-day meeting on the subject of meteorological observations from space on 18 and 19 June 1958, after which Cortright summarized the technical possibilities. In addition to TIROS, the air force was proposing use of the WS-117L vehicle (ultimately known as Agena) for a WS-117W, a three-axis stabilized, 1360-kilogram polar-orbiting Orbital Meteorological System.[49] This WS-117W was to measure cloud cover, cloud definition, cloud layers and thickness, moisture content, ozone content, wind direction and velocity, albedo, the spectra of incoming radiation, heat balance, and lightning location—a tall order for a system that was to be operational by 1964.

The day before signing NASA's founding legislation, President Eisenhower had decided that meteorological satellite development should go to the new agency, and Cortright arranged for it to transfer on 13 April 1959. Cortright had also sought to bring researchers with relevant expertise into the agency, and he recruited William Stroud from Fort Monmouth to head the NASA meteorological program. Cortright, whose title was chief of advanced technology programs, assigned TIROS to the Goddard Space Flight Center.[50] He arranged for the creation of a Joint Meteorological Satellite Advisory Committee in May 1959 to coordinate meteorological satellite research and development efforts, composed of members from NASA, the Defense Department, and the Weather Bureau.[51] And finally, he hired Morris Tepper, a researcher at the Weather Bureau, to direct the agency's meteorology program at the headquarters level.

TIROS was initially a three-satellite project, and each satellite was to have three instruments. The largest would be the television system for transmitting visible light images to the ground. The television system consisted of two lenses, one wide-angle lens imaging a square area about 1287 kilometers on a side, and a narrow-angle lens that photographed a much smaller area inside the same region. The second instrument intended for it was an improved version of Stroud's infrared sensor that had flown on Vanguard 2. It could detect five different wavelength bands, allowing it to sense reflected solar radiation, water vapor absorption in the atmosphere, outgoing long-wave infrared radiation, and visible light. The third instrument was a lightweight version of Suomi's heat balance device. The first TIROS, however, flew with only the television camera.

TIROS 1 was launched on 1 April 1960 into an equatorial orbit, from which it

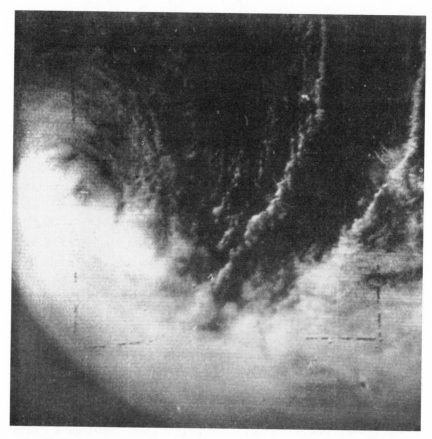

TIROS 1 image of tropical storm north of New Zealand, 10 April 1960. NOAA image spac0101, courtesy NOAA.

could photograph the Earth between 50 degrees north and south latitude. It began sending back images almost immediately, and over its 76-day life it transmitted 22,952 useable photographs. The images revealed structural features in clouds that were entirely unexpected. It provided the first photographs of oceanic storms, revealing a spiral-banded structure like that of hurricanes and photographing an unreported typhoon near New Zealand. TIROS also showed that mountain-wave cloud structures were much larger in scale than previously believed, extending from the Andes mountains across the entire width of South America and displaying short- and long-wave structure that was also unexpected.[52] Finally, it showed that unexpectedly rapid changes in the cloud patterns of vortexes occurred in the early phases of storm system formations. At the third meeting of SSB's Committee on the Meteorological Aspects of Satellites in June, Wexler told the group that the TIROS images had triggered a review of convective cell research to try to explain why cloud structure was so much more variable than previously believed and assumed clear organization over such different scales.[53]

The first TIROS also revealed that the Earth's magnetic and gravitational fields could affect spacecraft in low-Earth orbits in ways that could be used to help stabilize future spacecraft. TIROS 1, which had the shape of a short but wide cylinder, was spin-stabilized around its short axis to keep it from tumbling in orbit. Because of this, the satellite did not maintain a constant orientation with respect to the Earth below it; instead, its orientation changed continuously, with the cameras actually pointed at space more than 90 percent of the time. For the same reason, when they were pointed at Earth, the cameras also photographed the Earth at different angles. Using the photographs, William Bandeen and Warren Manger found that TIROS's orientation to the Earth was varying in a way that was unexpected.[54] It was experiencing precession, or wobbling, imposed by an outside source. The gravitational torque exerted on it by the Earth could not explain this.[55] Instead, it was the product of two forces. An interaction between the Earth's magnetic field and the magnetic field of the satellite exerted the primary force, while gravity imposed a lesser one. The magnetic field of a satellite could be varied deliberately by adding a loop of wire and controlling the magnitude of the electrical current within it, permitting a simple means of improving the stability of future satellites.

TIROS 1 was also enormously popular in the public arena and the more important political realm. While Van Allen's radiation belt discovery was probably more famous, TIROS provided results that were far more visible to average people (and politicians). Its photographs of the Earth's cloud cover placed a phenomenon of everyday life into a new context. The photographs of oceanic clouds suggested to anyone interested in the weather that TIROS could photograph storms at sea before they reached land, significantly lengthening storm warning times. The satellite received a four-column, front-page article in the *New York Times*, which also reproduced two of the satellite's first photographs.[56] Walter Sullivan, the paper's science writer, pointed out that TIROS presaged an "era when such vehicles [would] produce a constant stream of information of great, immediate value in our daily lives."[57] Other outlets followed the *Times'* lead, making TIROS and its instantly recognizable cloud photographs famous.

The Weather Bureau's Wexler, who had been involved in attempts to produce cloud photographs using rockets in the late 1940s, had foreseen the potential of meteorological satellites well before TIROS moved to NASA. Late in 1958, he

An early TIROS spacecraft being mated to its launch vehicle. NOAA image, courtesy NOAA.

had arranged for the establishment of a Meteorological Satellite Division in the Weather Bureau to begin looking at ways to link cloud-cover photographs to the standard weather maps that forecasters used, and he had also gotten NASA's assurance that while it would produce the early experimental weather satellites, it would not seek to make itself the operator of whatever permanent operational weather satellite system emerged. NASA also did not intend to carry out meteorological research using them, and when it adopted TIROS from ARPA in early 1959, it transferred $2 million to the Weather Bureau to fund its research into weather satellite operations and applications.[58]

Late in June 1960, SSB reviewed the satellite meteorology programs at NASA and the Weather Bureau. Summarizing the results for President Eisenhower's science advisor, Lloyd Berkner stated that the TIROS 1 had been "immediately useful," but there were significant challenges in interpreting the data. He continued that SSB believed that the Weather Bureau should have responsibility for the "basic design of the satellite observational systems, data analysis, and research," and therefore it should receive the funding for those efforts. NASA was currently paying for this, but it did not intend to support "exploitation of meteorological satellite data for either research or operational purposes subsequent to fiscal year 1961."[59] Berkner was endorsing an arrangement already made between Francis Reichelderfer, chief of the Weather Bureau, and NASA Deputy Administrator Dryden during April and May.[60]

The next step toward formalizing the relationship between the two agencies came in September, when Administrator Glennan invited Frederick Mueller, the secretary of commerce, and Elwood Quesada, administrator of the Federal Aviation Agency, to a lunch meeting to discuss the operational meteorological satellite program that Wexler and Reichelderfer sought. Glennan wrote that the three needed to clarify the responsibilities for carrying out satellite development, systems integration, and data processing so each agency could prepare its funding requests to the Budget Bureau.[61] At their 10 October meeting, they agreed to establish a new Interagency Meteorological Satellite Planning Committee that was to be chaired by a NASA official and that was to produce a development plan. Wexler protested, however, that NASA should have no role in planning operation of the system, and instead he wanted the system planning to be carried out within the existing National Coordinating Committee for Aviation Meteorology, which the Weather Bureau controlled.[62]

Reichelderfer took the issue to Glennan again in late November, gaining Glennan's agreement to use the existing National Coordinating Committee, with the addition of NASA members and a Panel on Operational Meteorological Satellites that would do the planning.[63] Reichelderfer also received assurance that NASA's leaders would tell their staffs to stay out of the operation of *operational* satellites. This was probably necessary because no one in NASA's meteorological satellite office, now run by Morris Tepper, or in NASA's science program more generally, believed that the weather satellite technology was really ready to be declared operational.[64] Nor did they believe that the Weather Bureau had created the capability to effectively process and disseminate the vast flood of routine data that even the very simple TIROS would produce. Nonetheless, political pressure to produce an operational weather satellite system quickly led to the formulation of a plan for a National Operational Meteorological System.

The agreement that evolved was built around a preexisting program at NASA Goddard to develop an experimental, fully stabilized polar-orbiting weather satellite that would follow the TIROS series. It was to have a much different set of instruments providing data that research-oriented meteorologists believed had great potential over the long term for providing a better understanding of the lower atmosphere's processes, but that no one as yet really knew how to make use of. Called Nimbus, it had been the result of advocacy by Cortright and William Stroud for research into new instrumentation. TIROS's small size and spin-stabilization limited its utility for carrying out instrument research, and initially Cortright had wanted to build a very large, multidisciplinary research satellite based on the WS-117L. Stroud, however, had argued that integration of the large, complex payload represented a huge risk. A simpler satellite, larger than TIROS but a good deal smaller than the Agena vehicle, yet still having a three-axis stabilization system that could keep its instruments aimed toward Earth continuously, made more sense.[65] Cortright had accepted Stroud's argument, and he successfully advocated this mid-sized Nimbus satellite program for the fiscal year 1960 budget.

Nimbus was not originally intended to be an operational weather satellite; instead, it was to be a research tool toward an eventual operational satellite. There were a good many ideas floating around the new space science community about what might be done to improve weather prediction with space technologies. Cloud pictures were only somewhat useful; they clearly improved hurricane warning, but they would not help much with forecasting outside the coastal regions. They also could not help advance the Weather Bureau's numerical prediction effort. To produce hemispheric and global forecasts, the models needed wind and temperature data. Radio astronomers had devised techniques to infer atmospheric and surface temperatures from the electromagnetic emissions of the other planets, and a handful of meteorologists thought this form of remote sensing might be useful for their research as well. Others believed that satellite-tracked balloons could provide the global-scale datasets needed by numerical prediction researchers. The Nimbus program was aimed at finding out which of many possible techniques would work out.

Nonetheless, in early 1961, the Panel on Operational Meteorological Satellites recommended basing the National Operational Meteorological Satellites system on Nimbus. The Weather Bureau's Harry Wexler and David S. Johnson had initially made this proposal to Morris Tepper at a meeting the previous November, and the deal was concluded easily due to a longstanding friendship between Reichelderfer and Hugh Dryden.[66] NASA agreed to have the first Nimbus ready for launch in 1962, and the TIROS series would be extended until then. The Weather Bureau would receive the appropriations for both TIROS and Nimbus, ensuring its control of the program, and transfer the money to NASA. President John F. Kennedy, in his special message to Congress of 25 May 1961 requesting the increased funding necessary for the Apollo project to reach the Moon, also requested additional funding for the Weather Bureau to pay for the operational meteorological satellite system. Congress approved the money that October, and in January 1962 NASA and the Department of Commerce, to which the Weather Bureau belonged, signed an interagency agreement to carry it out.[67] The resulting Nimbus Operational System (NOMS) was projected to cost about $60 million per year to operate.

The agreement between NASA and the Department of Commerce did not last eighteen months. There were several overlapping reasons for its failure, all of which were sufficient to justify breaking it. The first was simply that the technology was not ready for operational use, and it needed a good deal more research and development. The second, of course, was that NASA was a research-oriented agency, even within the Office of Applications that was host to the meteorological satellite program. Stroud had not wanted Nimbus to be an operational satellite, since its operational nature would substantially reduce its value for instrument research. Third, the technology's non-readiness led to substantial cost overruns that were beyond what the Weather Bureau could justify to the Budget Bureau. Fourth, the Defense Department decided that Nimbus was not what it wanted and it actively advocated for the Weather Bureau's defection. And fifth, both Wexler and Reichelderfer left the Weather Bureau, Wexler dying unexpectedly in August 1962. The two men that replaced them as leading meteorological satellite advocates, J. Herbert Holloman, assistant secretary of commerce for science and technology, and S. Fred Singer, who became director of the National Weather Satellite Center in June 1962, wanted a significantly different operational system than Nimbus. They sought to remove NASA from the weather satellite business completely so that they could pursue their own agenda.

The Department of Commerce began its efforts to break the agreement in April 1963 in a meeting between Holloman and Robert Seamans, the associate administrator of NASA. The projected cost of the Nimbus satellite had doubled at this point, and the first launch had been delayed to 1964 due to the need to replace the original solar cells with ones more resistant to radiation damage.[68] Singer's National Weather Satellite Center also believed that the lifetime of the satellites would be too short to justify the cost. Holloman and Seamans agreed to review the program over the next few months. Singer hired an outside contractor, the Aerospace Corporation, to produce an analysis of the program. Delivered in early September, the Aerospace Corporation's analysis told the Department of Commerce what it wanted to hear: the highly complex Nimbus satellite would have a lifespan of only about three months, leading to a program cost of $80 to $100 million per year, vice the $56 million that the Weather Bureau had anticipated. The value of the data received was not worth this cost, in Holloman's judgment. In the requirements he had forwarded to Seamans in June, Holloman placed a reasonable value on weather satellite data of $26.7 million per year.[69]

In its own analysis of the flap, the Budget Bureau's staff pointed to two real issues. There was a technical issue at stake, but it was not the projected life of the satellite or the reasonable value of the data. Holloman and Singer wanted to adopt gravity stabilization for their operational satellite, a technique being developed within the Defense Department. Nimbus was to use active thrusters to maintain its orientation, while gravity stabilization utilized the gravity and magnetic torques discovered by TIROS 1 to achieve the same effect. Thrusters needed fuel that would run out eventually, while the gravity gradient system would not. They also sought nuclear power to bring about a three-year lifespan, which solar power could not (yet) achieve. There was also, the Budget Bureau's analyst pointed out, a powerful "ad hominem" issue. Nimbus had been "plagued by bad feeling, bad interagency communications, and charges of bad management" for the past two years.[70] This reflected both strong and incompatible personalities on all sides as well as the divergent goals of the NASA program manager and his Weather Bureau counterpart.

It also reflected active efforts within the Defense Department to undermine NASA. The Department of Defense had been forced to give up its space projects by White House fiat in 1958, leading to the creation of NASA, and had never accepted the justice of that decision. It sought control over its own space destiny. Despite the joint Department of Defense–NASA–Weather Bureau agreement to place all operational weather satellites in the Weather Bureau, it had established its own operational weather satellite program in 1961. It wanted imaging satellites in a different orbit than the Weather Bureau did, preferring a polar orbit that provided early morning imagery to the noon orbit that the Weather Bureau sought in order to more effectively schedule reconnaissance flights and airborne refueling operations. This clandestine project had many different names over its lifetime, but it is generally known as the Defense Meteorological Satellite Program (DMSP). Its first satellite was a shrunken version of TIROS, utilized the magnetic-loop stabilization suggested by TIROS 1 and tested late in 1960 aboard TIROS 2, and could be placed into polar orbit using an existing booster.

More important to the Weather Bureau, its cameras were remounted to point out the satellite's side, not its base. By changing the orientation of the spin axis, the side-mounting would permit the cameras to photograph the Earth on each spin cycle, while the original TIROS's cameras spent 90 percent of each orbit pointed at space. This meant, National Reconnaissance Office historian Cargill Hall eventually wrote, 100 percent coverage of the Northern Hemisphere each day above 60 degrees latitude and 55 percent at the equator.[71] The first successful launch of this wheel-mode TIROS was 23 August 1962. The Department of Defense then actively recruited Singer, telling him about the classified DMSP program via a series of briefings. Singer recognized the short-term cost advantage of adopting the already-developed wheel-mode TIROS, and getting out of Nimbus had the benefit of permitting him to pursue the nuclear-powered, long-lived, gravity-stabilized satellite that he believed should be the ultimate operational system.[72]

On 12 September, Holloman informed the Budget Bureau by telephone that he intended to end the Commerce Department's participation in Nimbus in favor of the wheel-mode TIROS as an interim system. He also stated he planned to seek outright cancellation of Nimbus, in which, the Budget Bureau's official recorded, further investment was not "justified because the cost-performance potentialities (with heavy emphasis on cost) of the next generation of surveillance satellites now offers a so much greater potential as to justify initiating a new development program and taking the additional delay." The Budget officer also recorded cautioning Holloman not to place too much stock in any promise of an operational satellite that had not been matured to the point of "giving real confidence that multiple spacecraft procurements were justified."[73] That was good advice, as the Interim Operational System that emerged out of the Commerce Department's rebellion was still interim in 1970.

On 27 September 1963, Holloman drafted a letter to NASA's deputy administrator canceling the Department of Commerce's participation in Nimbus, although by this time he had backed away from seeking Nimbus's outright cancellation. He instead wished to immediately adopt an interim system "based on TIROS technology," the DMSP satellite he could not mention in an unclassified letter, and establish a new program "to meet the coordinated meteorological requirements and leading to a spacecraft lifetime such that the system operating costs are commensurate with its meteorological value." In formally telling the Budget Bureau its intentions on 2 October, the Commerce Department wrote that it expected to save $180 million between fiscal year 1964 and fiscal year 1968. A prototype of the new operational system could be initiated in fiscal year 1964, and in fiscal year 1968 the new operational system could be procured with an annual operating cost of $36 million, vice the $58 million previously allocated and the $80 million expected for Nimbus.[74]

On 3 October, NASA's Seamans responded with a memo warning that the Department of Commerce's action, while clearly within its rights, would "defer the date at which a fully operational meteorological satellite could be available." He also stated that NASA would continue the Nimbus program regardless, at least through the launch of the first two vehicles and probably the spare as well. They would provide the in-space test of the sensors for whatever operational system emerged, and they would also provide "unique and important observations needed for research."[75]

A *Wall Street Journal* article leaking news of the Commerce Department's revolt forced a relatively rapid settlement of the dispute. On 4 October, the two agencies gave the broad outlines of the agreement that would emerge, and a meeting between the budget director, the secretary of commerce, and the NASA administrator, the divorce was finalized. The Department of Commerce would get its interim TIROS system, and Nimbus would continue through the first two launches on NASA funds. The two agencies would study an operational system that would cost no more than $40 million per year to procure and operate. The meeting also caused the Budget Bureau staff to comment that no compelling justification for proceeding with development of an operational system in fiscal year 1965 was given by the Department of Commerce at the meeting. The Budget Bureau's staff was also unconvinced by the Commerce Department's basic argument that its not-yet-designed Operational Meteorological Satellite would be less expensive than Nimbus. Finally, the anonymous drafter also opined that Nimbus would provide the basis for whatever system the Department of Commerce got anyway, and that the actual technical disagreements were minor.[76]

On 2 January 1964, NASA and the Department of Commerce reached a new agreement adopting the DMSP's wheel-mode TIROS as the TIROS Operational System (TOS). The first of these was launched as TIROS 9 in January 1965; slightly more than a year later, a larger version carrying a navy-developed direct readout system that provided instant, lower-resolution images to inexpensive ground stations was launched as Environmental Science Services Agency (ESSA)

1, reflecting a name change for the Commerce Department's Weather Bureau. This series of TIROS-based satellites continued through 1970, alternating between the original high-resolution and the newer direct readout satellite so that one of each was always in orbit.[77] In 1970, the first of the Improved TIROS Operational System (ITOS) satellites replaced the TOS satellites. As the first three-axis stabilized operational weather satellites, they reflected a delay of several years over Nimbus, whose second "90 day satellite" flew in May 1966 and operated until January 1969.

NASA's Nimbus program, freed from the operational mission its leaders had never wanted, carried out the instrument research and meteorological science that had been their goal and is the subject of chapter 2. As the Budget Bureau had suspected, Nimbus-originated instruments formed the basis of ITOS. The Weather Bureau, on the other hand, never received the funds necessary to make the vast flood of images it received from its weather satellites fully useful. At an intellectual level, both NASA and the Weather Bureau had known in 1958 that data handling would be the bane of satellite research, but neither managed to come to grips with the problem. And because the Weather Bureau had rushed to declare TIROS and its successor operational, it could not justify to the Budget Bureau money to carry out *research* in data processing systems. Hence in her 1991 dissertation, ex-meteorologist Margaret Ellen Courain could cite Lee M. Mace, the Weather Bureau researcher responsible for turning the TIROS and ESSA imagery into useful products, as believing that forecasters did not really accept the imagery for fifteen years—the cloud photographs did not present information in a form they could use.[78] The operational satellite series provided an effective storm patrol during the 1960s, permitting early warning of approaching oceanic storms, but they did not assist routine forecasting.

The IGY's satellite program nevertheless set off a slow-motion transformation of atmospheric science. As the NASA meteorology program proceeded over the next decade, it developed new observing technologies that finally permitted demonstration of the generation circulation of the Earth's atmosphere, Harry Wexler's long-sought goal. It also eventually enabled the emergence of a global meteorology.

CHAPTER TWO

Developing
Satellite Meteorology

It cannot be said at this time that we really understand
how to use these new and unprecedented kinds of observations, and the prime motive is therefore exploration.
— *William W. Kellogg, July 1958*

The numerical weather prediction models developed during the 1950s at Princeton's Institute of Advanced Studies (IAS) and at the U.S. Air Force's Cambridge
Research Laboratory were part of the meteorological goal of long-range forecasting. They could forecast the future state of the atmosphere from a known initial
state over limited periods of time.[1] Defining that initial state, however, was a significant challenge. Poor data, or simple lack of data, led to poor forecasts. The
weather balloon network used to feed the models real-world data covered only
about 10 percent of the troposphere. The network did not extend into the Southern Hemisphere, into the tropics, or over the oceans. Without data from these
areas, neither human nor machine forecasters could produce accurate (*skillful* is
the term used in the profession) forecasts extending beyond two days, and less
than that south of the equator. Satellites promised global data. In the early 1960s,
researchers hoped that the availability of global data would eventually enable the
production of skillful month-long forecasts.

In any case, numerical meteorologists could not improve their models of the
atmosphere without global datasets. Charney's colleagues at Princeton had been

able to rely on the U.S. Weather Bureau's historical data for North America to build their regional models, and painstaking assembly of Northern Hemisphere data with the help of European allies permitted the construction of Northern Hemisphere models during the late 1950s. But because there was essentially no Southern Hemisphere data to collect, the performance of global models could not be checked against data. Generation of global meteorological data was the task NASA and the Weather Bureau pursued.

The information that numerical prediction models needed about the atmosphere was not the presence or absence of clouds, as the Television-Infrared Observation Satellite (TIROS) series of satellites provided, but its temperature and wind profiles. These determined atmospheric motion, and this was what numerical meteorologists sought from the computer. Two types of weather balloons provided these datasets in the Northern Hemisphere: radiosondes, which measured temperature, pressure, and humidity during their ascents; and rawinsondes, which were tracked by radar to derive wind velocities. No one knew how to measure these quantities from space in 1960, but there were lots of ideas. Verner Suomi's heat budget experiment had taken advantage of the fact that infrared wavelengths between 8 and 10 microns were known from laboratory experiments not to be strongly absorbed by the atmosphere. Stroud's five-channel radiometer had similarly utilized known atmospheric transmission characteristics. Writing to the head of the Advanced Research Projects Agency's (ARPA) meteorological satellite effort in 1958, RAND's Will Kellogg had listed and explained a variety of potential atmospheric emissions that might permit derivation of important quantities. Ozone emitted radiation around 9.6 microns, for example, which he concluded might permit determination of stratospheric temperature and circulation.[2] There were many possibilities awaiting an appropriate set of technologies.

In addition to these remote measurement techniques, satellites offered a means to expand balloon measurement techniques. Balloons were short lived and were released every twelve hours by national meteorological services; these, of course, operated extensively on land but not on the oceans. Because of this, they could not provide information about the oceanic atmosphere, leaving vast (but meteorologically important) data voids. The poorer nations of the world did not support meteorological networks, leaving them devoid of forecasting entirely. Satellites could solve these problems too. Polar-orbiting satellites could locate balloons once every twelve hours, ideally, receive data from the balloons' instruments, and relay it to a ground station. Geosynchronous satellites could do this nearly continuously. If long-lived balloons could be developed to make a global network of them economically feasible, satellite-tracked balloons launched from a handful of ground stations could provide wind and temperature measurements over most of the globe.

Between 1960 and 1975, NASA's Meteorological Programs Office, in collaboration with the National Oceanic and Atmospheric Administration's (NOAA) Environmental Satellite Service, the University of Wisconsin, and the National Center for Atmospheric Research (NCAR), developed and tested these two sets of technologies. This effort relied heavily on trials with the new satellite instruments using aircraft and balloon flights that were, in essence, inexpensive proof of method tests. These also occasionally revealed unexpected instrument capabilities, suggesting potential new uses for them. By the early 1970s, these new measurement technologies seemed capable of producing global meteorological datasets, permitting the construction and validation of global numerical prediction models.

IMAGING THE GLOBAL ATMOSPHERE

The first major goal of the NASA meteorological satellite program was the improvement of satellite imaging capabilities. On behalf of the Weather Bureau, it sought higher-resolution daylight imaging, new capability to produce imagery of the Earth's night side, and, to serve the real time needs of the Defense Department, an imaging system of lower resolution that could continuously transmit to less sophisticated ground stations than the primary satellite command and data stations. Finally, the agency sought imagery from geosynchronous orbit, which would permit more frequent imaging of most of the Earth's surface than the polar orbiters did.

Nimbus 1, launched 28 August 1964 into a polar orbit intended to cross the equator at the same time each day (a sun-synchronous orbit), carried three types of imaging equipment: an Advanced Vidicon Camera System (AVCS), the Automatic Picture Transmission (APT) system desired by the Defense Department, and a High Resolution Infrared Radiometer (HRIR). The vidicon and APT systems were both television-based systems that operated in the visible light portion of the electromagnetic spectrum. They differed in image quality and in the length of time required to transmit an image. The Nimbus 1 APT produced an 800-line scan image in 8 seconds, and transmitted it via a television signal, stretched over 200 seconds. Each image was about 1660 kilometers square, with a resolution of about 3 kilometers at the point directly below the satellite (referred to as nadir viewing). The AVCS on Nimbus 1 had three cameras that stored their images on a tape recorder for playback on ground station command. The three were arranged with one pointing directly downward with the other two angled 35 degrees to either side to produce a wider image. They also produced an 800-line scan, but the resolution was three times that of the APT camera, or about 1 kilometer. The tape recorder could store two orbits' worth of images, which could be transmitted to one of the two ground stations in four minutes.

The HRIR provided the satellite's nighttime capability. It was a single-channel scanning radiometer operating between 3.5 and 4.1 microns with a scan spot of about 6 kilometers in diameter. Its analog output was tape-recorded and played back to the ground station, which prepared two outputs: a data tape containing the raw radiance data and a film-strip-like photo facsimile. The photo facsimile provided the nighttime cloud imagery that the Weather Bureau sought for forecasting purposes, but the raw data proved more interesting to researchers. The sensor's data represented the radiance temperatures of surfaces within its field of view, and it could measure radiance temperatures between 210 and 330 degrees K. (The Earth's average is 280 degrees K.) In theory, at any rate, it could therefore provide the temperatures of cloud tops as well as that of land and oceanic surfaces.[3] This would be very useful data. The temperatures of cloud tops could be related directly to their altitudes, a meteorologically useful quantity. Global ocean surface temperatures would also be very useful to meteorologists, as well as to oceanographers. And land surface temperatures would supplement, though hardly replace, the measurements provided by meteorological ground stations.

One major challenge the satellite program's researchers faced in achieving these measurements was that the sensor produced an average temperature of the scene that it scanned. If the scene were free of cloud-free ocean or land, then the resulting temperature would relatively accurately represent that of the surface. Similarly, if the scene beneath the scan spot were entirely overcast, the cloud-top temperature would be accurate. If the scene consisted of both clouds and land or water surface, however, the sensor would give an incorrect result. Such cloud-contaminated scenes were both inevitable and a big problem. Human researchers could compare the photographic and data outputs to select cloud-free scenes for evaluation, but the manpower requirements for this made it operationally infeasible. Finding ways to improve sensor resolution to increase the number of cloud-free scenes and use data processing equipment to screen out cloud-contaminated ones automatically took several more years and a good deal of research.

Nimbus 1 did not operate for very long. After twenty-six days, its power system failed. The solar array paddles froze in a single position, which the project engineers blamed on the bearing lubrication leaking out into space. Without the ability to track the Sun as the satellite orbited, the solar arrays could not produce enough electricity to keep the satellite functional and it died. During its short life, however, it accomplished most of its goals. It produced more than 27,000 images and it demonstrated that the HRIR scanner could produce useable nighttime images. The HRIR's data was also the subject of numerous scientific publications delineating its benefits and drawbacks, including a lively series of articles on the cloud height measurement challenge.[4] Finally, its control system achieved a pointing accuracy of better than 1 degree, which permitted its images to be accurately related to the Earth's surface in the absence of clear terrain features. Its primary failure was in not achieving its desired lifespan.

NASA essentially repeated the Nimbus 1 mission in May 1966, with the launch of Nimbus 2. This vehicle carried the same instrumentation, with the addition of a copy of the TIROS five-channel infrared radiometer, renamed the Medium Resolution Infrared Radiometer (MRIR) to distinguish it from the HRIR. The MRIR was included to help define ways to evade the cloud-contamination problem. One channel duplicated the HRIR's channel at lower resolution. Others sensed the 6.9-micron water vapor emission band, permitting mapping of the water vapor distribution of the atmosphere; the 15-micron carbon dioxide band, permitting analysis of the distribution of carbon dioxide in the troposphere; the broadband 5- to 30-micron band for heat budget studies; and a visible light channel for measuring reflected solar radiation. Unfortunately, although Nimbus 2 functioned for 978 days, it suffered repeated tape recorder failures that undermined its science mission. The medium-resolution infrared imager shared the telemetry system's tape recorder, which failed in July, three months into the mission, rendering it useless. The mission engineers were able to shift the telemetry data onto the high-resolution system's recorder, reducing its useable scientific data, but that recorder failed in November. The AVCS recorder had failed in September, leaving only the APT system functioning for the remainder of the satellite's lifespan.

Despite the short lives of its radiometer instruments, Nimbus 2 succeeded in demonstrating several important capabilities. A group of scientists at Goddard used the Nimbus 2 HRIR dataset to demonstrate that, while the sensor could not give reliable sea surface temperatures in the daytime due to reflected sunlight, its nighttime data revealed the outlines of the major ocean currents and provided details on the temperature structure of the ocean surface.[5] The MRIR's data demonstrated that water vapor distribution in the atmosphere could be mapped from space, and it showed that, as most atmospheric scientists had expected, carbon dioxide was essentially uniformly distributed in the troposphere. The carbon dioxide result was particularly important to the meteorological program's future plans, which postulated using the radiances of carbon dioxide molecules to derive the temperature at different altitudes in the atmosphere. Non-uniform distribution would make that impossible. Finally, it showed that television cameras were not necessary for the production of "cloud pictures" for forecasting—as Bill Stroud had expected during the International Geophysical Year (IGY) program, visible and infrared channel radiometers could do the job just as well once engineers achieved a desirable resolution. After Nimbus 4, no further television-based systems flew in the meteorological satellite research and development program, and they were also deleted from the operational satellite program after 1972.

GEOSYNCHRONOUS IMAGING

When Ed Cortright had first proposed an intensive meteorological satellite development effort in 1959, it had included a second major component, Aeros. Aeros was to be a meteorological satellite in an orbit high enough that it would have the same angular velocity as the Earth's surface, permitting it to remain essentially stationary with respect to the surface. Known variously as synchronous, geosynchronous, and geostationary satellites, these offered two tantalizing possibilities in the early space age. First, as novelist Arthur C. Clarke had pointed out, they could serve as telecommunications relay stations, offering high capacity and global coverage with only a small number of vehicles. To meteorologists, and particularly the U.S. Weather Bureau, which operated a continental-scale telegraph network in order to acquire timely weather data, geosynchronous telecommunications satellites promised faster (and maybe less expensive) data distribution. More directly, cameras and/or imaging radiometers on such satellites would be able to generate multiple pictures of an entire hemisphere of the Earth every hour. The polar orbiters' primary drawback was that their twelve-hour orbits did not permit detection and tracking of short-lived severe storms—thunderstorms, tornados, and the like—because such storms rarely lasted more than a few hours. An imaging geosynchronous satellite could photograph the entire lifecycle of these storms, at the very least helping research meteorologists understand their evolution.

Aeros itself, however, was never approved or funded. Instead, the meteorological satellite program piggybacked the first demonstrations of geosynchronous imaging's potential on a pair of experimental satellites developed under the Office of Application's Advanced Technology Satellite (ATS) program. The ATS satellites were derivatives of Syncom, a geosynchronous communications satellite developed by Hughes Aircraft for NASA and first orbited in July 1963.[6] The ATS program's goal was demonstration of advanced satellite technologies in geosynchronous orbit, including stabilization and control systems, data processing, and telecommunications capabilities. Of the five planned missions, the first and third were to be spin-stabilized like the TIROS satellites, while the rest were demonstrations of gravity gradient three-axis stabilization schemes.

William Stroud, head of Goddard's Aeronomy and Meteorology branch, had taken the decision to place imaging experiments aboard some of the ATS satellites after receiving the results of three study contracts with Republic Aviation, RCA, and Hughes Aircraft. Republic had been tasked with the study of a complete geosynchronous meteorological satellite system, while RCA had been asked to investigate the feasibility of putting a TIROS into a highly eccentric orbit that would briefly reach geosynchronous altitude. This they intended as an experiment to help determine the optimum resolution for a geosynchronous satellite camera. Finally, they had tasked Hughes with examining the potential for placing a camera on the advanced version of Syncom being developed via the ATS project.

At a briefing to Goddard Center Director Harry Goett in November 1963, Stroud had argued that based on these studies, the most effective approach to getting a geosynchronous imaging capability was to act as an experimenter in the ATS program. His initial plan was to propose placing two AVCS units on each of the spin-stabilized ATS launches, and to do the same for the later gravity-gradient stabilized satellites. Because the spin-stabilized ATS satellites rotated at a hundred revolutions per minute, the images from this experiment would be smeared without some kind of electronic image motion compensation, and development of suitable circuitry was part of the effort. Stroud also proposed equipping some of the ATS satellites with dedicated communication circuits for relaying weather data and for broadcasting weather forecast maps.[7]

Over the next several months, however, the effort to develop motion compensation circuitry to permit putting AVCS cameras on the spin-stabilized ATS satellites foundered. The task proved too difficult for the relatively short time horizon before the first ATS launch scheduled for late 1966. Inability to resolve the motion compensation problem caused Goett to recommend not flying cameras on the ATS program's spinners after a design review in December 1964.[8] They would have to wait for the later gravity-gradient missions.

Three months before, however, Suomi and a colleague had submitted an informal proposal for a storm patrol experiment on one of the spin-stabilized ATS satellites.[9] They had devised an imaging system that employed the satellite's hundred rotations per minute spin as part of its scan motion, leading to its common

name, the spin-scan camera. Unlike the early TIROS cameras, their system utilized line scanning, imaging a narrow (approximately 10 kilometers wide) horizontal swath of the Earth on each rotation of the satellite. A motorized mirror
would provide vertical scanning of the Earth in a thousand discrete steps. Imaging
of the full Earth disk would take ten minutes, and they proposed transmitting the
data to Earth in real time and reassembling the image on the ground to eliminate
the data storage problems plaguing Nimbus. Most important, their system did not
require the troublesome motion compensation circuits that plagued the in-house
Goddard effort.

The Weather Bureau's new director, Robert M. White, liked Suomi's proposal.
It would get the much-desired geosynchronous capability quickly and inexpensively, as the ATS program was already funded and only the camera needed development. As he explained in a memo to David S. Johnson, however, it threatened
an already-approved effort at Goddard to launch a TIROS into a highly eccentric
orbit that would take it up to geosynchronous altitude for a few hours. This was
an experiment to determine the optimum resolution for a future satellite camera,
and the program office was concerned that leapfrogging this TIROS K mission
with an ATS would result in the loss of approved resources to something else.[10]
Further, the amount of time available between the proposal and the date at which
the spin-scan camera had to be ready for the integration and testing phases prior
to launch was too short. Typically instruments underwent extensive testing to
demonstrate their space-worthiness prior to being integrated into a satellite, but
there was barely a year left before the instrument had to be ready for integration
if it were to make the first ATS launch. White told Johnson that at a meeting of
the joint Satellite Program Review Board meeting at the end of January 1965
Suomi had contended that his camera could be ready by the November deadline;
NASA's representatives had not believed him.

White suggested to Johnson that he should press for a quick resolution of the
controversy by approaching an appropriate person at NASA. Ultimately, Homer
Newell's intercession proved necessary to get the spin-scan camera on the first
ATS flight. The camera was built by a Hughes subsidiary, Santa Barbara Research,
under a contract to the University of Wisconsin's Space Science and Engineering
Center, which had been founded by Suomi and Robert Parent, with grants from
NASA, the Environmental Sciences Services Administration (ESSA), and the
State of Wisconsin in 1965. A substantial engineering effort by the company,
aided by parallel delays in the satellite development effort related to thermal regu-
lation, permitted it to meet a somewhat later launch date in August 1966.

The spin-scan camera worked as expected, producing spectacular full-disk images of the Earth. It posed significant challenges for researchers, however. Spin-stabilized satellites precessed, or wobbled, in orbit slightly and the images had to be carefully (and manually) aligned to notable features on the ground. To trace the formation of storms or other cloud structures, researchers aligned a sequence of images on a light table and then copied them to movie film. The film loops then allowed qualitative study of storm evolution. The process was labor-intensive and expensive, limiting its operational usefulness. But it provided information about storm origins that had never been available before, making it attractive to a few researchers—especially if more efficient ways of producing the movie loops could be generated.

The ATS 1 camera was followed into space by a color version aboard ATS 3 in November 1967. This camera produced the first color images of the full Earth ever made, and its images became the subject of a number of scientific investigations. Plate 1 shows Earth on 18 November, from a position over the Atlantic. At ESSA's request, NASA positioned ATS 3 over the American Midwest during the spring of 1968 for a Tornado Watch Experiment. The satellite caught the genesis of several severe storms, and meteorologist Tetsuya Fujita at the University of Chicago turned its photographs into film loops for study.[11] Others combined ATS 3 imagery of a tornado event on 23 April with rawinsonde data to examine the role of convective warming in storm evolution.[12] The National Hurricane Center in Miami conducted a similar study during that year's hurricane season. This experiment caught the complete lifecycle of Hurricane Gladys, the first hurricane to have its entire lifecycle observed.[13]

The following year, ATS 3 supported a large field experiment near Barbados known as the Barbados Oceanographic and Meteorological Experiment (BOMEX). Conceived by Jule Charney, who had become interested in improving numerical models' performance in the tropics, and led by Joachim Kuettner, a German meteorologist who had worked with von Braun's group in Huntsville on the Mercury program, BOMEX was designed to try to measure the energy flow between ocean and atmosphere. The experiment consisted of four special observing periods carried out by a 500-kilometer-square network of radar-equipped ships, research aircraft flights, and satellite data collection. NASA parked ATS 3 over the observing area to provide cloud imagery to compare to data collected by the surface network. In the experiment's final days, the surface network and ATS 3 captured the formation of a small-scale, rapidly developing disturbance that peaked in three hours. Several groups of researchers used the ground and satellite observations to analyze the disturbance's formation; Suomi used the datasets to establish a correspondence between cloud reflectance in the visible and microwave radar regions of the electromagnetic spectrum.[14]

The following year, Robert H. Simpson, director of the National Hurricane Center, wrote that ATS 3's imagery had allowed meteorologists to describe the dominant circulation modes of a hurricane. This had been impossible before, because observations in the tropics were so scattered that no one had been able to assemble a coherent concept of hurricane evolution. The movie loops had permitted tracking of entire storms, determination of wind speeds (via cloud tracking), and derivation of vorticity and wind shear. Wind shear derivation permitted identification of formation spots for storms, because hurricanes could not develop in regions of high wind shears. Further, the imagery clearly showed that hurricanes tended to develop in low-pressure troughs that extended from the Central Atlantic to the Caribbean. Prior to the availability of the geosynchronous imagery, he concluded, "we were not able even to determine that this trough had day-to-day continuity, much less describe it in great detail on a day-by-day basis."[15]

The successes of the two ATS flights triggered an initially rapid movement toward an operational prototype of a dedicated geosynchronous weather satellite. Known as SMS, short for Synchronous Meteorological Satellite, NASA formally proposed this program in its fiscal year 1970 budget. Two of these SMS satellites would precede the first NOAA-funded operational satellite, GOES (Geosynchronous Operational Environmental Satellite). The Space Science and Engineering Center (SSEC) and Santa Barbara Research altered their spin-scan camera for the SMS flights to give it an infrared capability, replacing its television-like photographic apparatus with an imaging radiometer possessing eight visible light and three infrared channels. They also added a telescope that would permit high-resolution imaging of smaller portions of the Earth, so researchers could study storm formation in more detail. The revised instrument became known as the Visible and Infra-Red Spin-Scan Radiometer (VISSR, pronounced "visor") and first flew aboard SMS 1 in May 1974. The long gap between ATS 3 and SMS 1 resulted from the selection of a different spacecraft contractor, Philco-Ford Astronautics, which had difficulties engineering the vehicle. It proved not to matter, however, as ATS 1's and ATS 3's cameras functioned for more than a decade.

By 1970, therefore, the Meteorological Satellite Program Office had met the Weather Bureau's demands of the late 1950s for cloud imagery from space with three different series of satellites, TIROS/TOS, Nimbus, and ATS. Instrument technologies devised by researchers at the Goddard Space Flight Center and at the University of Wisconsin had decisively demonstrated the superiority of radiometers over television-based systems for imaging, and television was discarded during the following decade. Scientists had begun using the satellite images for research. Imaging was only one component of the global observing system NASA and the Weather Bureau sought during the decade, however. The agencies also sought the ability to produce temperature and wind profiles of the atmosphere to feed numerical weather prediction models.

MEASURING ATMOSPHERIC QUANTITIES

One of the two principal challenges meteorologists interested in global forecasting faced was obtaining wind profiles of the global atmosphere, particularly in the Southern Hemisphere, where the ratio of water to land was much higher than in the Northern Hemisphere and the equatorial belt. The numerical prediction programs needed wind data in order to predict the movement of the high and low pressure fronts that were the basis of weather forecasts, and balloons were the primary method of gaining wind measurements (aircraft reports were also widely used along major air routes). The now-traditional radiosonde and rawinsonde balloons suffered from limited geographic coverage and from lifespans measured in hours, however, making them poorly suited to the goal of global wind measurements.

In 1959, Vincent Lally, a meteorologist who specialized in balloon measurements, postulated that two thousand superpressure, or constant-volume, balloons, equipped with thin-film electronics and relaying their data via satellite, could provide the global temperature, pressure, and wind measurements that numerical weather prediction required. These Global Horizontal Sounding Technique (GHOST) balloons could have lifetimes of more than sixty days. Superpressure balloons, by their nature, remained at a single altitude without the need for ballast, and they could be placed at a variety of altitudes to provide data at several different levels in the atmosphere. Lally called these satellite satellites, reflecting their status as an auxiliary to the space-based platform that served as tracking system and data relay.[16]

One of Lloyd Berkner's final efforts as chairman of the National Academy of Science's Committee on Atmospheric Sciences had been production of a ten-year plan for atmospheric research in the United States. One of its recommendations had been the establishment of a national laboratory for atmospheric research, and in 1960 a consortium of research universities submitted a proposal to the National Science Foundation (NSF) to build and operate this facility. The NSF accepted their proposal, and the universities created an independent nonprofit corporation, the University Corporation for Atmospheric Research, to run it.[17] The facility itself was known universally as NCAR, and went into operation in 1961 in temporary quarters on the University of Colorado's campus in Boulder. Its founding director was Walter Orr Roberts, a solar physicist. One of Roberts's first actions had been to create a scientific ballooning facility. There was great scientific interest in balloons, in astronomy in particular—Princeton astrophysicist Martin Schwarzchild had flown a large solar telescope into the stratosphere during the late 1950s and planned a still larger one for the 1960s—and a balloon facility would help interest scientists in other disciplines in NCAR. Their interest and support, Roberts hoped, would help secure NCAR's future.

Roberts hired Lally away from Teledynamics, a corporation that had held a contract from the air force to develop a world weather forecasting system (Project 433L in the air force's jargon), to run the balloon facility. The series of decisions that had led to weather satellites being placed in a civilian agency had also led to placement of global forecasting hopes in the civilian National Weather Service, however, and Teledynamics' contract was eventually cancelled. Lally moved to NCAR in 1961, and the new balloon facility, located in Palestine, Texas, to avoid heavily traveled air routes, had opened in 1963.[18] After getting the facility into operation, and some successful early tests of his constant-level balloon system from Japan, Lally withdrew from running the balloon facility to concentrate on the development of his balloon system.

The first complete test of his GHOST system took place in 1966. Launched from Christchurch, New Zealand, in March, the balloons carried a solar-powered transmitter whose signals were tracked by an international network of ground stations. The GHOST balloons performed well at high altitudes, with one balloon making seventeen circumnavigations of the Earth in 192 days and others flying up to six months. At lower altitudes, lifespans were much shorter, averaging ten days. Icing was the principal culprit. Supercooled droplets in clouds crystallized on the thin balloon surfaces, and the extra weight brought the balloons down. Surface treatments for the balloons might resolve this, Lally argued in a *Science* article, but if not higher-altitude balloons could measure lower-level winds using a tethered instrument package.[19]

With the basic balloon technology demonstrated, Lally started working with researchers at the Goddard Space Flight Center on a means to track the balloons from space.[20] The system they devised was called the Interrogation, Recording, and Location System (IRLS). The balloon's payload consisted of a solar array to power the electronics, a low-power satellite transmitter, and the scientific instruments. Each balloon would be programmed with its own code; on each orbit, operators at the satellite's ground station would program IRLS to query specific balloons. The initial flight version had a 20-kilobyte memory module that permitted it to interrogate twenty platforms per orbit. The satellite determined the platform's location by the radio signal's round-trip time based on its own position; in the absence of modern satellite navigation systems, this was one of the few available options. The research group at Goddard, led by principal investigator Charles Cote, included buoys and other platform types in this first set of experiments to determine if the system could adequately track drifting oceanic buoys and land vehicles. The first satellite experiments were carried out with only two balloon-borne platforms.[21]

Nimbus 3 carried the first edition of the IRLS into space. This proved able to locate surface platforms to within about 2 kilometers, although with a directional ambiguity caused by its use of range-only measurement. A platform could be anywhere on the circumference of a range circle drawn around the satellite's position, and the system therefore required additional information to provide a positive location. The balloons carried a sun-angle sensor that provided a line of bearing. This line intersected the range circle at two points, and therefore a balloon could be in either position. Most of the time, only one of these positions would make sense meteorologically, but under certain circumstances this directional ambiguity would result in missing an interesting atmospheric phenomenon. The project team at Goddard had arranged for testing with a pair of Lally's GHOST balloons, with oceanic buoys, and with wildlife biologists for animal tracking, reflecting an effort to demonstrate the flexibility of satellite tracking technologies. The balloon testing was reasonably successful given the known constraints of the system, and it was flown again in an improved form on the subsequent Nimbus 4 mission.

The improved IRLS still contained the location ambiguity of the original, but this version could track a useful number of platforms. The science team had proposed using the IRLS/GHOST balloon system to carry out an investigation of the equatorial stratosphere's circulation.[22] In preparation, Lally's GHOST team had carried out a series of test launches from Ascension Island during 1968 and 1969, tracking them intermittently using ground stations.[23] This set of tests led to modification of the balloons, and during 1970 a science team led by Richard J. Reed of the University of Washington carried out launchings of twenty-six balloons to altitudes of 20 and 24 kilometers . Of these, eight balloons at 20 kilometers and three balloons at 24 kilometers were successfully tracked by IRLS for more than one month. The resulting data demonstrated the existence of long-period waves in the stratosphere.[24]

NASA ceased supporting the Southern Hemisphere constant-level balloon program after Nimbus 4, however, because NOAA had agreed to let the French Centre National d'Études Spatiales (CNES) provide a tracking system for its operational satellites (thus saving NOAA money). The French system, known as EOLE, used balloons that were essentially the same as Lally's, but the satellite-based tracking system operated on a different principle that offered the ability to track larger numbers of balloons. The balloon-based transmitters required were also simpler than those for IRLS, making the EOLE system potentially less expensive to use operationally. France launched a prototype EOLE satellite in 1968, and the full EOLE experiment took place in late 1971. After NASA's Wallops Island facility placed the satellite into orbit with a Scout rocket 16 August 1971, the Laboratoire de Météorologie Dynamique conducted launches of 480 constant level balloons from three sites in the Republic of Argentina. These flights were to 200 millibars, in the stratosphere, and represented the first large-scale experimental verification of theoretical models of stratospheric circulation.[25]

DEVELOPMENT OF RADIOMETRY

Meteorologists sought the balloon tracking system as a means of measuring wind and temperature in the atmosphere. But the GHOST satellite-balloon systems were only one possible approach, although it was one that deployed a technology very familiar to meteorologists. The primary alternative was radiometric sounding, which would rely on the atmosphere's emissions of energy. In a famous 1959 paper, Lewis D. Kaplan had argued that an infrared spectrometer capable of sensing the emissions of carbon dioxide might be able to produce a temperature and moisture profile of the atmosphere.[26] Earlier, other researchers had pointed out that measurement of the radiance intensity of carbon dioxide could be transformed mathematically into a temperature reading; Kaplan expanded this reasoning to contend that one could also infer a vertical temperature profile from the spectral distribution of the carbon dioxide emissions. Due to the way the atmosphere transferred radiation, measurements in the center of the carbon dioxide band could only come from the top of the atmosphere, while measurements from the wings of the spectral band would come from deeper down.

This radiance data could then be converted into temperatures using a mathematical inversion model. The model had to be based on detailed empirical

knowledge of the atmosphere, including its composition and its normal temperature and pressure profiles. Kaplan was also very clear in this paper that the instrument had to be designed for a specific inversion model. It wouldn't do to build an
instrument out of what happened to be available and then try to figure out what
its data meant later. Theory had to precede hardware, or the very complexity of
the atmosphere's radiation transmission processes would defeat meteorologists'
attempts to decipher the data.

Kaplan, who moved to the Jet Propulsion Laboratory (JPL) in 1961, had placed
his first instrument on the Mariner 2 mission to Venus, but his article had set in
motion a great deal of research into the mathematical foundation of a successful
instrument for future Nimbus missions. At the Weather Bureau, David Wark took
Kaplan's method, simplified it by limiting the problem to derivation of stratospheric temperature structure (where the cloud contamination problem would
be rare), and developed a model for a six-channel spectrometer sensing specific
intervals in the 15-micron carbon dioxide band.[27] Perhaps their most important
early finding was that reduction of the data required that the instrument have an
onboard calibration source to permit accurate, nearly continuous evaluation of
the measurement errors. Otherwise, errors destabilized the calculations very
quickly, rendering the results useless.

NASA and the Weather Bureau sponsored construction of a breadboard prototype of an instrument based on this model by the Barnes Engineering Company. This prototype was subjected to ground testing by researchers at the National
Weather Satellite Center to, as one co-experimenter put it, "obtain a qualitative
estimate over of the prevailing atmospheric temperatures over a limited distance,
and to show how variations in this temperature could be detected by the spectrometer."[28] This first test set, carried out during May and July 1962, was simply to
determine whether the instrument and its associated model would be able to
detect temperature changes at a distance. In qualitative terms, the experiment was
successful, producing results that were at least reasonable. Quantitatively, however, the results were poor. The datasets were contaminated by significant random noise, and limitations of the very simple instrument required sampling the
four channels' data sequentially, not simultaneously.

The instrument team also carried out a set of ground experiments using a commercial eight-channel spectrometer to investigate the possibility that additional
channels might improve accuracy. Some of the additional channels overlapped
infrared wavelengths that atmospheric water vapor absorbs, causing the measured
radiances in these channels to be much weaker than they would be in a dry atmosphere. This effect had to be removed from the calculation. In their paper describ-

ing the experiment, the researchers used local radiosonde data to remove the water vapor effect while acknowledging that this was self-defeating.[29] If one needed a water vapor profile from radiosondes to resolve a satellite instrument's temperature profile, the satellite instrument served no purpose—the radiosonde also provided a temperature profile. And one could not extract a water vapor profile from the satellite instrument and then use that profile to calculate the temperature profile. Further, they found that the additional carbon dioxide channels did not significantly improve the results. They concluded, in essence, that a different distribution of the instrument's channels was necessary. The future satellite instrument needed channels selected to provide a water vapor profile independent of the carbon dioxide radiance channels.

Wark's experimental team was satisfied enough with the qualitative success of the ground tests to sponsor a second-generation instrument for balloon testing. This instrument contained six carbon dioxide channels and a window channel to permit measurement of surface and cloud-top temperatures. It was equipped with a blackbody cooled by liquid nitrogen to simulate the space look that was the obvious calibration choice—the temperature of space was both known and unchanging—and was redesigned to be flyable on a high-altitude balloon. They had three copies made by Barnes Engineering, and these were tested in altitude chambers to evaluate their response to the altitude and pressure changes they would experience during a balloon ascent from the ground to 3 kilometers.

The team carried out the first balloon ascent with the instrument on 11 September 1964 at NCAR's balloon facility in Palestine, Texas. In addition to the experimental temperature profile instrument, the balloon carried cameras to photograph the clouds and terrain beneath it, several different temperature instruments to help identify atmospheric temperature effects on the instrument, and broadband radiometers to measure the entire incoming and outgoing radiative flux from the atmosphere. During the flight, the chase team used additional instruments on the ground and in a pair of aircraft to collect more data to compare to the balloon instrument's. The Palestine flight lasted seven hours and produced satisfactory data from all but one carbon dioxide channel, and the instrument was recovered with only minor damage. The following spring, the team performed a second flight at Sioux Falls, hoping to evaluate the instrument's performance at profiling a drier, colder polar air mass. Interference from a radio beacon on the balloon gondola, however, generated so much noise in the resulting readings that the data could not be resolved.

Writing about the results of this three-year-long series of experiments in 1967, Wark's team remarked that while neither flight experiment was flawless, they were each successful in that they revealed "hidden possibilities of failure of a satellite instrument and therefore contributed strongly to anticipated success" in the future satellite instrument.[30] The five useable channels of data from the first flight resolved to temperature curves that were very similar to, but not exactly like, those produced by local radiosonde measurement. Despite the similarity, however, Wark and his colleagues noted that the derived profiles did not meet their scientific standards and were not competitive with the local radiosonde results.

They traced the flaws to the empirical information they had used as an input to the calculations. The inversion equations needed to convert radiances to temperatures were nonlinear, and these generally could not be solved in a way that produced a unique solution. Instead, multiple solutions were almost always possible for a given set of data. But most of these solutions would be physically implausible. A solution that indicated a temperature of 1000 degrees C in the Earth's troposphere would clearly not be correct. One obvious way to constrain the range of possible solutions to ones that were physically reasonable was to use a database of past temperatures, and the team had constructed an average summer profile from ninety August and September radiosonde flights taken during preceding years to use in transforming the instrument's data. They concluded that the atmosphere on the day of the flight had been significantly different near the tropopause than this average profile they had used. The inversion mathematics forced the derived temperature closer to the average profile than was warranted by the atmospheric conditions on the day of the test, producing the difference between the radiosonde and derived temperatures. The team concluded that in constructing the empirical profiles necessary to resolve the inversion equations, researchers had to be careful to choose soundings that reflected the full range of the atmosphere's variability. Otherwise they would not get accurate results.

A space-borne version of the Weather Bureau's infrared temperature sounder went into orbit onboard Nimbus 3 on 14 April 1969. This flight provided the first demonstrations that space-based sensors could produce relatively accurate temperature profiles of the Earth's atmosphere. NASA and the Weather Bureau had arranged for two weeks of special observations to evaluate the performance of the Satellite InfraRed Sounder (SIRS), launching radiosondes during satellite overpasses to provide in situ temperatures to compare with the instrument's derived temperatures. The historic first sounding, taken the day of the launch, was over Kingston, Jamaica, and produced a very satisfactory match. Conditions at Kingston were clear, providing an easy test for the new technology, but soundings taken under other conditions were also relatively good. In writing up their results for *Science*, principal investigator David Wark remarked that the main limitation in deriving accurate temperatures came from accurately determining the boundary condition, the surface temperature or the cloud-top temperature.[31] Improved measurement in the window region of the atmosphere might help alleviate this by giving the temperature of the cloud tops.

Nimbus 4, launched 8 April 1970, carried aloft a set of incremental advancements to the Weather Bureau's SIRS instrument. This version of SIRS employed a mirror to permit scanning across the satellite's track, increasing its coverage. It also had additional infrared channels chosen to allow inference of both temperature and water vapor profiles and a window channel that permitted determination of surface or cloud-top temperature, an addition necessary to overcome the challenge of establishing the boundary conditions for the inversion equations. This version of SIRS was transitional, however, in the sense that its replacement was already in development. By the time it reached space, the Weather Bureau had developed a new cloud-clearance methodology that needed a somewhat different instrument.

During 1967 and 1968, William L. Smith, who had joined the Weather Bureau's satellite center in 1966 from the University of Wisconsin, and Harold Woolf, who had come from MIT in 1963, had developed a method for removing the effects of clouds from the soundings. It required interpolation between the radiances of two adjacent scenes in the relevant carbon dioxide and window channels to derive an equivalent clear-scene radiance.[32] That equivalent radiance could then be used in the inversion equations to produce a temperature. Their method drove the development of a new sensor scheduled for the Nimbus 5 flight. Known as the Infrared Temperature Profile Radiometer, this had two new window channels and generated a matrix of scan spots each 32 kilometers wide. The scan spots provided the series of independent scenes necessary for the decontamination process at the cost of some additional complexity in the instrument.

The Weather Bureau's approach to remote sensing of the atmosphere was not the only possible radiometric approach available. The Weather Bureau instruments were essentially diffraction grating spectrometers that measured specific, narrow wavelength bands. This reduced the amount of data transmitted by the instrument and therefore the amount of data processing required to resolve it, but at the cost of eliminating detail from the derived profiles. In essence, it produced an average temperature of a thick slab of atmosphere. This made it fundamentally different from the temperature set produced by a radiosonde, which provided a vertical, and nearly continuous, set of point measurements. Another approach, taken by Rudolf Hanel of the Goddard Space Flight Center, was to sample the entire carbon dioxide radiance band using a Michelson interferometer.[33] Converting its output required a different mathematical technique using Fourier transforms, and the large amount of data presented an enormous data processing task. The advantage of an instrument based on this technique was that it preserved the fine vertical structure of the atmosphere and therefore promised the most accurate profiles if its data processing challenges could be overcome. Hanel's IRIS flew on Nimbus 3 and Nimbus 4, demonstrating that the basic technique was sound. One could produce temperature profiles from an interferometer. These flights also demonstrated that, at least from the standpoint of operational utility, the interferometer approach was undesirable. The data processing burden

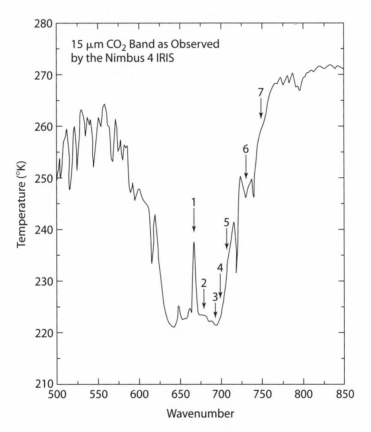

A portion of the Earth's infrared spectrum from the IRIS instrument on Nimbus 4. The image also indicates (with numbered arrows) the segments of the spectrum sampled by William L. Smith's SIRS instrument. From: W. L. Smith, "Satellite Techniques for Observing the Temperature Structure of the Atmosphere," *Bulletin of the American Meteorological Society*, 53:11, November 1972, p. 1076.

imposed by the instrument was beyond the computing capacity available to the Weather Bureau, and the Nimbus program dropped the interferometer approach from subsequent flights.

One interesting result of Hanel's IRIS flight was the first observational determination of the greenhouse effect of the Earth's atmosphere. It was very well accepted by the early 1970s that the Earth was about 33 degrees C warmer than it would be in the absence of the atmosphere, but this knowledge was based upon theoretical calculation and on laboratory measurements of the radiative characteristics of various trace gases. But the effect had never been measured in the real atmosphere. IRIS, whose spectral range covered the entire infrared spectrum, was able to make the necessary measurements, revealing the details of the infrared spectrum of the atmosphere. Hanel published his results in 1972.[34] IRIS also went to Mars in 1971.

During the late 1960s, finally, JPL's Moustafa Chahine began working on a new temperature retrieval method that would not require the statistical database that the Weather Bureau's did to constrain the range of possible temperature solutions. Such a database would obviously not be available for the other planets, and Chahine wanted a method that would allow temperature sounding of any planetary atmosphere. His so-called physical relaxation technique was first used to invert data from the Pioneer 10 mission to Jupiter. The temperature instrument on this flight was a four-channel infrared spectrometer, and the results, while poor compared to radiosonde methods on Earth, were the first such measurements for one of the gas-giant outer planets.[35] More channels with better accuracy would improve the results, and Chahine was able to get funding for a balloon-based instrument for Earth sensing from NASA after this demonstration.

He also began working with Lewis Kaplan, Jule Charney, and John Shaw to develop a new cloud clearance scheme based on the use of additional shortwave infrared channels to help distinguish cloudy scenes from clear ones. NSF funded modification of the balloon instrument to use the new method, which was tested in a series of airplane flights near Port Charles, Louisiana, in 1975. This eighteen-channel spectrometer demonstrated the ability to provide temperatures accurate to within 1 degree K, even in the presence of high cloud fractions, suggesting better atmospheric soundings could be had from a 1980s generation of space-borne instruments.[36]

MICROWAVE REMOTE SENSING

Both SIRS and IRIS utilized the infrared spectrum, an approach that suffered from the reality that cloud cover almost completely absorbed infrared radiation. Hence neither instrument could produce temperature profiles below clouds; instead, they read the cloud-top temperature. This would result in errors in numerical forecasts using the data. In the case of a persistent overcast, these errors would grow very rapidly as successive soundings failed to produce data. Another approach was available, however. The atmosphere was also largely transparent to microwave radiation, which would penetrate most types of clouds, including all of the known persistent types. Atmospheric oxygen radiated in the microwave region, and this permitted determination of atmospheric temperature in the same way carbon dioxide infrared radiances did. The technologies employed to make microwave instruments derived directly from astrophysics, where microwave remote sensing had been flourishing since the early 1950s.

The first microwave temperature instrument, however, went to Venus. In 1962, NASA's JPL launched its first successful planetary spacecraft, Mariner 2, equipped with a two-channel microwave radiometer. This instrument had been chosen to resolve a controversy over Venus's surface temperature, which according to Earth-based radiotelescopes seemed to be more than 500 degrees C. This was controversial because Venus was not close enough to the Sun to have a temperature this high without a far stronger greenhouse effect than Earth possessed, but which few astronomers thought was possible. A strong enough greenhouse effect seemed to require an atmosphere several hundred times as rich as Earth's in carbon dioxide, or else a gas that was a much stronger absorber in the infrared spectrum. Yet the Mariner radiometer confirmed that the high temperature was real, leading to many efforts over the next decade to try to explain how it was possible. It also helped spark interest in using microwave radiometers to sense Earth.

At a NASA-sponsored symposium on passive microwave sensing techniques held in July 1962, Michael L. Meeks of MIT's Lincoln Laboratory, which specialized in microwave technologies, had led an extensive discussion of the oxygen spectrum and its potential relationship to atmospheric temperature. Attendees included Morton Stoller, NASA's director of applications; William Stroud, Rudolf Hanel, and William Bandeen from Goddard Space Flight Center; Lewis Kaplan; and Vern Suomi. Meeks, along with Harvard College Observatory's A. E. Lilley, had carried out laboratory experiments designed to characterize the emission and absorption characteristics of oxygen at atmospheric and a selection of lower pressures.[37] A series of thirty-six emission lines could be distinguished in oxygen's spectra from above the atmosphere, in theory, at least, which could be used for temperature profiling.

Victor Chung, an MIT graduate student working under Alan Barrett, presented another possible use for microwave sensing at this meeting. He had investigated techniques for sensing water vapor concentration and precipitable water. This relied upon emissions at 22.2 gigahertz, known as the water vapor resonance line.[38] W. E. Vivian, of the Conductran Corporation, presented a method of using microwave emissions to determine sea state, even in overcast conditions. The fact of these water vapor emissions caused radio astronomer Frank Drake, of the National Radio Astronomy Observatory, to comment that it probably explained why the microwave radiometers used on radio telescopes had three to five times less useful sensitivity than they did in theory—they were being contaminated by atmospheric emissions. He explained that at the Green Bank Observatory, Torleiv Orhaug had undertaken research to characterize water vapor emission impact on the telescope's sensitivity, finding that it was almost always controlled by atmospheric water vapor, and that rainstorms appeared clearly in the telescope's data. This, he concluded, while highly annoying to radio astronomers, should make "all meteorologists joyful."[39]

During the next several years, several of Alan Barrett's graduate students at MIT carried out various investigations of microwave techniques. William B. Lenior completed a PhD thesis on microwave temperature profiling.[40] He received a NASA grant to carry out balloon experiments with a microwave radiometer; in an initial series of flights out of NCAR's Palestine, Texas, balloon facility, he received disappointing results. He found that the observed radiance intensities did not match those predicted by theory. But he later found that he had not accurately accounted for the antenna's performance. In a second set of balloon experiments carried out during July 1965, he verified that the oxygen emissions did behave in accordance with theory.[41] Also under a NASA grant, graduate student David Staelin carried out solar extinction experiments using the 8.5-meter antenna at the Lincoln Laboratory during 1964, showing that passive microwave sensing could provide water vapor profiles and possibly total liquid water content.[42]

Staelin recalled many years later that Goddard's William Nordberg was the driving force behind the Nimbus series' microwave instruments. From this research area, two types of instrument gradually emerged. Goddard's instrument, for which Thomas J. Wilheit, Jr. was principal investigator, was a mapping radiometer. Known as the Electrically Scanned Microwave Radiometer (ESMR), this instrument's center frequency was 19.35 gigahertz. It was first test-flown onboard NASA's Convair 990 research aircraft during May and June 1967. In fourteen flights, twelve over water, Goddard's research team found the instrument could, as they expected, determine surface temperatures, map ice fields and the sea/land boundary, and detect areas of heavy rainfall.[43] Only thick cumulous clouds affected the instrument, showing up as regions of high brightness and offering a potential means of deriving the liquid content of the clouds. Surface emissivity affected the temperatures generated from the instrument's data, but for over-water use this was essentially irrelevant as the water surface had a constant, known emissivity. Hence the instrument could derive sea surface temperatures even in the presence of most clouds. It could also measure the spatial extent of sea ice.

The other type of microwave device generated in this research program was the much-desired temperature profile instrument. MIT's David Staelin was principal investigator for what was eventually named the Nimbus E Microwave Spectrometer (NEMS). This was a five-channel radiometer, with two water vapor channels intended to derive water vapor and liquid water content of oceanic clouds as well as sea surface temperatures, and three oxygen channels designed to produce temperature profiles of the stratosphere. NEMS flew on Nimbus 5 in 1972, demonstrating that the basic principles it was based on were correct, and it was followed in 1975 by a scanning version, the Scanning Microwave Spectrometer (SCAMS), on Nimbus 6.

The principal disadvantage of the microwave instrument was that its horizontal resolution was much poorer than that of the infrared instrument, a drawback that called for retaining both. Its vertical resolution was also poor. A disadvantage that emerged in the testing program was that while its performance in the upper troposphere and stratosphere was equal to or better than that of the infrared instrument, its performance in the lower troposphere was inferior. Hence the microwave approach by itself was not seen as a complete solution to the sounding problem. Instead, the weather satellite program adopted both infrared and microwave sounders.

The operational sounding instrument that emerged from the NASA-NOAA program was named the TIROS Operational Vertical Sounder (TOVS), retaining the famous TIROS name while no longer having anything in common with the TIROS series of vehicles. TOVS consisted of three units: Bill Smith's High Resolution Infrared Spectrometer (HIRS), which was essentially the same as his Infrared Temperature Profile Radiometer with additional channels; the Stratospheric Sounding Unit, a version of the United Kingdom's Selective Chopper Radiometer; and the Microwave Sounding Unit, Staelin's instrument. The three instruments flew individually on Nimbus 6, launched in 1975, which served as a functional prototype for TOVS. The Nimbus 6 mission also helped to work bugs out of the retrieval algorithms, cloud clearance schemes, and data processing system for the operational series of satellites that replaced the Improved TIROS Operational System (ITOS) series beginning in 1977.

At the same time that NASA and NOAA were finalizing the design of the sounders for the operational weather satellites intended for the late 1970s, NASA researchers at the centers were beginning to explore still more new techniques for making atmospheric measurements. Some of these were means of making traditional meteorological measurements more accurately, while others were intended to make new measurements. The Goddard Space Flight Center built and flew two new Earth Radiation Budget instruments on Nimbus 6 and 7, trying to improve upon Suomi's simple instrument.[44] Also at Goddard, Donald Heath began developing and flying instruments designed to measure stratospheric ozone production and loss, flying his first instrument on Nimbus 4 in 1970. Finally, at Langley Research Center, James D. Lawrence began assembling a research group to do atmospheric remote sensing using various kinds of lasers. Lasers promised to be able to measure aerosols, cloud particles, water vapor, and certain trace gases in the atmosphere at much higher resolution than other techniques. Lasers were new devices, however, and two decades would elapse before they reached space.[45]

NASA, NOAA, and SSEC, finally, embarked on a joint effort to develop a version of the spin-scan camera used on the geosynchronous satellites that would provide temperature soundings. The value of adding sounding capabilities to the geosynchronous satellites was primarily to improve severe storm forecasting. The short time horizons of midwestern storms meant that the twelve-hour orbits of the polar orbiters would rarely detect them. The geosynchronous satellites, however, could produce a sounding every thirty minutes with an appropriate instrument, eventually leading to improved storm forecasting and better warning times. Known as the VAS instrument, short for "Visual/infrared spin-scan radiometer Atmospheric Sounder," this instrument first flew on GOES 4, launched in 1980, and was the basis for an extensive severe storms research program during that decade.[46]

By the mid-1970s, then, NASA and its collaborators had developed a set of instruments for producing global meteorological data using both radiometric and balloon methods. To some degree, the instruments' capabilities overlapped. GHOST, ITPR, and SCAMS all were able to produce temperature profiles in the atmosphere. They were not equal, to be certain. GHOST proved to be limited to higher altitudes, preventing it from generating wind and temperature data from the lower atmosphere, which remained necessary data for the forecast models. Each of the radiometric sensors had specific weaknesses that other sensors were chosen to complement. The new instruments were hardly perfect, possessing calibration problems, large errors, difficulties in the presence of clouds, and some reliability problems. Yet they provided the first source of daily global meteorological data ever realized.

The remote sensing technologies that were the basis of the meteorological satellite instruments came originally from astrophysics, where practitioners had developed them for examining the atmospheres of other planets. Because NASA was the *space* agency and during the 1960s carried out extensive planetary science as well (see chapter 4), it had been a natural home for planetary astronomers and astrophysicists. In turn, their ideas and technologies had influenced the development of meteorological capabilities. This fertilization would recur as NASA moved into atmospheric chemistry during the 1970s, as new demands were placed on the agency.

Finally, the joint NASA–Weather Bureau success at producing technologies for global datasets enabled the execution of an international atmospheric research program during the 1970s intended to realize the forecasting gains made possible by these new sensing technologies. Known as GARP (Global Atmospheric Research Program), an idea of Jule Charney's that came to fruition after Lally's GHOST balloon system provided the first inkling that global data might be possible in 1966, this program marked the achievement of a truly global meteorology.

CHAPTER THREE

Constructing a Global Meteorology

Does the flap of a butterfly's wings in Brazil set off a tornado in Texas?

—*Edward Lorenz, 1972*

In early 1961, President John F. Kennedy's science advisor, MIT physicist Jerome Wiesner, had asked the National Academy of Sciences' Committee on Atmospheric Science to propose a ten-year program for the profession. The report, drafted by meteorologist Sverre Petterssen, called for establishment of a set of new international institutions to further expand meteorology's reach.[1] An International Atmospheric Science Program should carry out scientific research on the global atmosphere, an International Meteorological Service Program should provide global-scale forecasting, and a World Weather Watch should sustain a global atmospheric observation system. These recommendations formed one basis of UN Resolution 1721, "International Co-Operation in the Peaceful Uses of Outer Space."[2]

In 1950, the United Nations had established a World Meteorological Organization (WMO), and this was the entity tasked with implementing the meteorological portions of Resolution 1721.[3] The WMO staff generated a report recommending the establishment of the World Weather Watch's observation capability within the WMO, while arguing that the atmospheric research program was properly the domain of the International Council of Scientific Unions (ICSU). Resolution 1802, adopted in December 1962, had accepted this arrangement.

The principal architects of the WMO's report were the Soviet Union's academician V. A. Bugaev and the Weather Bureau's chief of research, Harry Wexler. Wexler had been personally involved in postwar meteorology's first great advancement, the construction of workable numerical weather prediction models during the late 1940s and through the 1950s, and was an important promoter of scientific internationalism via his International Geophysical Year (IGY) efforts.[4] He was also, of course, a supporter of satellite-based meteorology, funding Vern Suomi's first satellite instrument for the IGY, establishing the Weather Bureau's satellite center, and, until his untimely death in 1962, serving as a highly respected advocate of satellite meteorology. He had believed that the union of these two great new technologies of numerical prediction and satellite data could produce truly global forecasts that might eventually be accurate to periods of a month.

NASA's role in the Global Atmospheric Research Program (GARP) that evolved during the late 1960s was as a provider of new technologies and support for large-scale field experiments. GARP's principal purpose was to provide quality-controlled global meteorological datasets for use in improving future numerical prediction models, and therefore the goals of the program were unobtainable without space-based measurements and telecommunications. But GARP required field experiments, too, to provide ground truth for the space-based measurements and to collect data that could not yet be gained from satellites. In keeping with the longstanding geophysical tradition of field science, these expeditions were international in organization and very large in scope. GARP's final field experiment, the First GARP Global Experiment (FGGE) of 1979, finally achieved Charney's dream of a global, quality-controlled, extensive meteorological dataset.

FORMING GARP

Science writer James Gleick, in his history of chaos science, remarked that GARP was founded in "years of unreal optimism about weather forecasts." There was, he continued, "an idea that human society would free itself from weather's turmoil and become its master instead of its victim."[5] The linkage of satellite observation to numerical forecasting would, in this view, permit not only month-long forecasts but eventually weather control. John von Neumann had believed that weather control could eventually result from this research area, and he was hardly alone. Robert M. White, head of the Weather Bureau and eventually the National Oceanic and Atmospheric Administration (NOAA), then believed weather control was within reach. Nobel Laureate Irving Langmuir did too, spending many years on cloud seeding research. NASA's Homer Newell was routinely asked about the relevance of NASA's meteorological research for weather control during his annual budget testimony before Congress, suggesting the importance of the issue to the agency's funders.

The program that became GARP started as an American initiative. Acting in its capacity as U.S. representative to the ICSU, the National Academy of Sciences had asked Jule Charney to serve as the organizer of the American proposal for WMO's research program. Charney's plan was based on three principles: the atmosphere was a single system such that disturbances in one area propagated around the world in four to five days; a new approach to observational techniques based on both satellite-derived quantitative data and satellite-relayed in situ data was necessary to improve prediction; finally, high-speed digital computers were capable of coping with the torrent of data such satellites could provide.[6] Its observing system was based on Vincent Lally's constant-level satellite-balloon system, at this point still an unknown quantity.

In March 1964, Morris Tepper, Jule Charney, and Philip Thompson visited colleagues in London, Paris, Geneva, and Brussels to explain Charney's proposal and seek their reactions. Here they found great enthusiasm for Charney's plan, with the only expressed concern being the balloon subprogram. The French government's space research establishment, Centre National d'Études Spatiale (CNES), had already gotten approval for a satellite-balloon tracking experiment that was originally to be launched in 1967 from Algeria.[7] During negotiations over who would pay for the eventual operational global observing system, the French government had agreed to fund the satellite-balloon subsystem, and EOLE was the result.

The balloon program, however, raised two potential challenges. First, the European scientists all emphasized that the potential impact of the balloons on aircraft needed to be investigated. Jet aircraft operated at altitudes near those the balloons needed to be at to provide data on the desired wind velocities, and the balloon payloads needed to be designed so that they would not damage aircraft. The second was that the balloons probably could not fly over the Soviet Union. During the 1950s, the United States had flown intelligence cameras and radiation detectors over the USSR, causing a diplomatic fiasco that resulted in complete ban on balloon overflights. This meant that the Global Horizontal Sounding Technique (GHOST) and EOLE balloon flights would have to be confined to the Southern Hemisphere. This restriction would significantly impair the utility of the balloon system for the operational global observing system unless the USSR could be recruited into the effort.

Three months after the three American delegates' visit to Europe, ICSU agreed to form a Committee on Atmospheric Sciences to plan the global experiment. At its first meeting in Geneva, held during February 1965, this committee agreed that the research program should be directed at understanding the general circulation of the troposphere and lower stratosphere and should contain two elements.[8] In a theoretical element, the program should develop dynamical, that is, numerical, models of the general circulation of the atmosphere that included radiation, momentum, and moisture movement on local and regional scales as well as the global scale. Second, the program should specify the observational needs of global atmospheric research, including the technological capabilities of its sensors and its telecommunications system, and carry out full-scale observation programs over time-limited periods.

Prior to the second meeting of the ICSU's Committee on Atmospheric Sciences in April 1966, Charney had assembled the formal American proposal for the research program. Titled *The Feasibility of a Global Observation and Analysis Experiment*, Charney's proposal became known as the Blue Book for the color of its cover.[9] It also divided the research program into two problem areas: exploitation of space and data processing technologies to provide global observations, and improvement of scientific understanding of turbulent transport of matter and energy in the atmosphere. ICSU's committee, in turn, proposed carrying out the global experiment in 1972. The year would be "designated as a twelve-month period for an intensive, international, observational study and analysis of the global circulation in the troposphere and lower stratosphere." In preparation, researchers would carry out a series of other investigations. Tropical circulation was poorly understood, and as the tropics were where most of the Sun's radiation reached Earth, a tropical subprogram was essential. Energy exchange between the atmosphere and land and ocean surfaces was also a poorly understood process, but one essential to accurate long-range weather prediction, and an observational program to determine the dynamics of these energy flows was vital. Finally, design studies of a global observing system that could meet the scientific needs of the research program had to be carried out.

By the end of the year, however, it was already clear to the committee's members that their chosen date was highly unrealistic. Neither the satellite-based temperature profile instrument nor the balloon-tracking system would reach space before 1969 due to the loss of Nimbus B; even if they worked as expected, their project scientists would need several more years to understand their capabilities and limitations. Subsequent, improved instruments would not be available until 1974 or 1975. No one expected that the first pair of operational geosynchronous satellites would be available until those later dates either—and it was absurd to believe in 1966, when satellites routinely failed in a few months, that the spin-scan cameras on the two Advanced Technology Satellite (ATS) satellites would still be sending back pictures in 1972. Furthermore, the subprograms themselves were going to require a good deal of effort. Design studies for the global observing system required numerical simulation on large computers whose time was expensive and often difficult to acquire. The tropical subprogram would involve an international flotilla of ships that had to be loaned by national governments, a complex, time-consuming process. For all these reasons, the global experiment had to be postponed to 1976.

In early March 1967, at the third meeting of the Committee on Atmospheric Sciences, the scientist-delegates began to discuss the details of what they now called GARP. NASA's Morris Tepper, chairman of the Committee on Space Research's meteorological subcommittee, had established three panels to look at different aspects of the future observing system at the previous meeting, and at this meeting the chairmen of these panels presented their findings. UCLA meteorologist Yale Mintz, who specialized in numerical modeling, told the committee that what modelers needed was a set of global observations of the atmosphere extending over a few months, up to a year. Such a dataset would provide a detailed, global snapshot of the atmosphere that model researchers could use to initialize global prediction models and a set of real-world results to compare to the models' output forecasts. This was the only way the models could be improved. William Nordberg presented the status of temperature profile instruments for satellites. And J. E. Blamont, from France's Centre National de la Recherché Scientifique, presented the status of the satellite-balloon research. Preliminary experiments with the National Center for Atmospheric Research's (NCAR) GHOST and the French EOLE balloon systems, had shown mixed results. Flights at high altitudes had gone relatively well, with some of the balloons surviving more than two hundred days. But lower-altitude balloons (500 millibars and below) tended to ice up. This panel concluded that "there [was] little likelihood of the availability of a global balloon-satellite observing system by 1972."[10]

The outcome of this meeting was a set of recommendations on the structure, timing, and contents of the proposed global program. Completion dates for the major field experiment should be moved to 1972–73 and for the final global experiment to 1975–76. The group asserted that a large-scale tropical observation subprogram should be the primary field experiment, to be carried out in the

1972–73 period. Finally, they recommended that the somewhat unwieldy committee structure be replaced by a special joint scientific committee that could provide a unified front for GARP and that could carry it out relatively unhindered by the three organizations that supported it (ICSU, WMO, and the International Union of Geodesy and Geophysics [IUGG]).[11] Their recommendations were accepted by the three sponsoring organizations later that year, and the new committee became the GARP Joint Organizing Committee.[12]

This third meeting left the details of the global experiment unplanned, however, and the Committee on Atmospheric Science's chairman, Bert Bolin of the Stockholm Meteorological Institute, another veteran of Charney's numerical group at Princeton, arranged for a Study Conference to be held in Stockholm in early July 1967 to complete them. He invited specialists in all the different subfields of meteorology that the global research program had to address — boundary layer flux, air-sea interaction, convective processes, meso-scale phenomena, atmospheric radiation — and have them work with the numerical modelers to define the program.[13] At this conference, the global program took its (mostly) final form. The conference ratified the importance of the tropical subprogram and pushed its date back to 1974, when they hoped better satellite instruments and the satellite-balloon system would finally be available, although the group retained the 1975–76 date for the global experiment.

The two-year American budget formulation cycle ensured that nothing much happened to get GARP going until 1969, however. In the words of NOAA's Robert White, NOAA's GARP office had to "mark time" while waiting for funds to come through. This had the fortunate result that GARP offices at NASA and NOAA received their go-aheads just as one element of the future observing system, the infrared sounder, got its first successful space-borne test. The successful retrieval of tropospheric temperatures by the Nimbus 3's Satellite InfraRed Sounder (SIRS) instrument team served as an additional stimulus to American GARP efforts. Morris Tepper took the retrievals up to a meeting with Robert Jastrow, Jule Charney, and Milton Halem in Jastrow's office at the Goddard Institute for Space Studies (GISS) in New York, where the data convinced the three men that it was finally time to start carrying out the detailed design studies that would result in the eventual GARP experimental observation system.[14] The Nimbus 5 and 6 launches with improved temperature sounders and balloon tracking systems were already in the development pipeline, and these were scheduled to be in orbit by the time all the rest of the infrastructure necessary to carry out the research program was in place.

After Tepper's meeting at GISS, he established a planning committee to map out a GARP strategy for NASA. He obtained permission to establish a GARP project office at Goddard Space Flight Center in Maryland, with Harry Press as the project manager and Robert Jastrow the project scientist. At the 1969 meeting of the Joint Organizing Committee, the structure of GARP was finalized, and NASA became responsible for specific tasks within it. GARP would consist of the large tropical experiment known as GARP Atlantic Tropical Experiment (GATE); a Data Systems Test (DST) that would carry out an evaluation of the global observing system; and the FGGE, which would produce its first global datasets. In the United States, the National Academy of Sciences was tasked with handling the planning and Academy president Philip Handler appointed Jule Charney to chair the U.S. Committee for GARP. NOAA became lead agency, with NASA responsible for the hardware development for the global observing system and for the DST. The space agency was also responsible for carrying out simulation studies necessary to support the detailed planning for the DST and the FGGE.[15] Finally, NOAA contracted the planning of the tropical field experiment to NCAR in Colorado.

SIMULATION STUDIES

A crucial component of GARP planning were simulation studies carried out by GISS, the Geophysical Fluid Dynamics Laboratory (GFDL) at Princeton, and NCAR. Using numerical prediction models, these studies addressed two important questions: the optimum configuration of the future global observing system, and the realistic time horizon of predictions using it. The first question would affect the technologies chosen for the global observing system, and how much building and operating it would cost. The second was aimed at understanding what GARP actually had the potential to achieve. Despite the enthusiasm for month-long forecasts, it was not at all clear by late in the decade that this was even theoretically possible.

After seeing the temperature soundings from Nimbus 3's SIRS instrument in April 1969, Jule Charney had asked Robert Jastrow and Milton Halem at GISS to collaborate on a study to determine whether in situ wind measurements were really necessary for the proposed global observation system. GISS had been founded by Jastrow at Columbia University in New York in May 1961. GISS served as a center for theoretical modeling and data analysis studies, which Jastrow had believed NASA needed for its science program. The university location would foster better links with the scientific community, and much of GISS's early work had been in the development of atmosphere models of Venus and Mars.

By this time, Charney had grown disenchanted with the constant-level balloon system. The balloons' short lives at low altitudes made a balloon-based observing system expensive to maintain at all the different altitudes the numerical models needed. He had also had a thought that the models might not actually need wind measurements in any case. At a numerical simulation conference in Tokyo in 1968, Charney had postulated that since wind in the real atmosphere derived from temperature differences, one might be able to simulate this process in the model by continuously inserting temperatures while the model was running.[16] This historical temperature data would, he thought, permit the model to generate wind fields in the lower atmosphere accurately without any need for an actual wind measurement. Winds from the upper atmosphere, necessary to provide a check on the calculations, could be obtained from either the constant-level drifting balloon system that both the United States and France were working on or from another of Vern Suomi's ideas, wind vectors derived by tracking clouds using geosynchronous satellite cloud imagery.

This was the thesis that Charney wanted GISS to evaluate. GISS's Halem obtained a copy of the Mintz-Arakawa two-level general circulation model from Yale Mintz at UCLA to run the experiments with. Using GISS's IBM 360-95 computer, he, Jastrow, and Charney performed simulation experiments to investigate Charney's idea. In a first set of experiments, they sought to determine an optimum period between temperature insertions. Insertion too frequently created spurious gravity waves in the model atmosphere, and they established twelve hours as the optimum period. Then they simulated the results that two potential observing systems might give. Temperature profiles generated by a barebones observing system consisting of two Nimbus satellites orbiting twelve hours apart, and nothing else, produced winds of useable accuracy, but only if their temperature errors were 0.25 degrees C or less. This was far better than what the SIRS sounder obtained. Simulations of a more robust observing system consisting of the two Nimbuses, upper troposphere and stratospheric winds from satellite-tracked balloons, and surface pressures from satellite-monitored buoys provided much more satisfactory results.[17]

In their resulting article, the three men merely concluded that their simplified model had only shown the possibility that historical temperature data insertion could result in accurate wind fields. Other researchers needed to do much more experimentation with more sophisticated models to check and refine this conclusion. Halem recalls that the paper was nonetheless greeted with a great deal of skepticism.[18] A lot of researchers were surprised that one could insert temperature data at all during the model run without causing spurious oscillations. The numerical prediction models were initial state models. Operators fed them observational data at the beginning of a run and then left the model alone to calculate the desired length of time; the models were not designed to be updated. In fact, the tendency of global circulation models to destabilize when fed new data, or sometimes simply reject the real-world data and continue using their internally calculated results, turned out to be a very difficult challenge for researchers in numerical modeling. Eventually, Charney's "direct insertion" methodology fell out of favor and was replaced by a more complex, but more effective, methodology called "four-dimensional assimilation."

The article also served as a preliminary study of what a global observing system would have to consist of to produce the desired outcome of GARP, thirty-day global weather predictions. GISS undertook more studies during the second half of 1969 to further help define the GARP observing system, leading to considerable unease in the profession about the achievability of their goals. Halem and Jastrow found that with the error limits set by GARP planners of 3 meters per second for winds, 1 degree C for temperatures, and 3 millibars for surface pressure, they could achieve skillful predictions of only three to four days. Reducing the wind error alone to 1 meter per second could increase predictability to eight days, but to reach two weeks, the upper altitude wind observations had to have errors of less than 0.5 meter per second, the temperature soundings less than 0.5 degree C, and the surface pressure 0.5 millibar.[19] These errors were far beyond the state of the technical art. As Halem put it thirty-three years later, "with GARP error limits, we wouldn't be able to make monthly forecasts. And that disturbed people."[20] Indeed, these numerical experiments showed that two-week forecasts were impossible within either the proposed GARP error limits or those imposed by the state of observation technologies.

Turning their attention to studies of what might be achievable within the limitations of near-term satellite technology, Jastrow, Halem, and their team at GISS found that the GARP Observing System probably would not need wind information from balloons or surface pressures from buoys to meet its requirements, except in the tropics. Indeed, through further simulations with the Mintz-Arakawa model, they determined that GARP error limits for winds and surface pressures were too generous and actually destructive of forecast accuracy. Based on the 1 to 2 degrees C error that the Nimbus 3 SIRS instrument was achieving, Jastrow and Halem reported, insertion of historical temperature profiles during the model runs produced wind and pressure fields that were more accurate than GARP error specifications for winds and pressures. Adding wind and pressure data at GARP error limitations produced worse forecasts than using the temperature data alone.

They concluded that GARP wind and pressure error specifications should be tightened to 1.5 meters per second and 2 millibars, respectively, and that if observations confirmed these simulation results, the global observing system would not need ongoing measurement of wind velocities by the satellite-balloon system.[21]

The GISS team also carried out simulations directed at other aspects of the observing system. At Suomi's request, they analyzed the potential utility of vertical temperature profile instruments like those on the polar orbiters on geosynchronous satellites.[22] First, they examined the impact of geosynchronous sounding, without corresponding polar-orbiting satellites. Jastow and Halem reported that the geosynchronous sounders alone would result in inferior forecasts. This was due to their inability to provide soundings above 60 degrees latitude, which the satellites could not see from their equatorial orbits. The poles were crucial to wind determination, and without polar soundings the wind errors grew very rapidly. When added to the soundings provided by two polar orbiters, however, geosynchronous satellite soundings resulted in a substantial reduction in wind error. More important, the simulation studies showed that the geosynchronous sounding data could substitute for the loss of one polar-orbiting satellite's sounder, preventing a reduction in forecast skill and providing a valuable backup capacity. Hence, temperature sounders on geosynchronous satellites would be a useful addition to the global observing system.

Finally, the GISS team studied the need for a side-scanning capability in the satellite sounder. The SIRS instrument on Nimbus 3 was fixed and only observed the atmosphere directly below the satellite (nadir viewing), which meant that it did not provide complete global coverage in each orbit. Instead, it sounded a relatively narrow swath of the Earth and only repeated each swath every few days. This, Jastrow and Halem reported, was insufficient for a two-polar orbiter observing system. It would not provide temperatures of a large enough portion of the atmosphere to accurately control the wind and pressure fields in the circulation model. The satellite instrument needed to scan at least 30 degrees to either side of the satellite's track to ensure nearly complete coverage of the Earth on every orbit. They also found that four polar-orbiting satellites without side-scanning instruments would achieve the same results, but with obviously greater costs.[23] In planning for GARP, a side-scanning capability had already been assumed for the polar orbiters, however, and this result ratified GARP planners' intentions.

Other simulation studies were also carried out for GARP. Akira Kasahara at NCAR and Joseph Smagorinsky, director of GFDL at Princeton, conducted simulation experiments aimed at determining the overall predictability of the atmosphere.[24] This was a controversial subject, as earlier experiments by MIT meteorologist Edward N. Lorenz had indicated that GARP's goal of long-range weather forecasting was impossible. He had found that using the same initial data, his simplified model would output the same results for the first few days of a forecast and then start to diverge. Eventually, the forecasts from successive runs bore no relationship to each other at all. Early on, he recognized that when he had entered the initial state data into the model, he had rounded the numbers to fewer decimal places than the computer used, introducing a small error—much smaller than the measurement errors of real instruments. This small error then grew as the computer worked through its iterative calculations. The growth and propagation of error had been enough to produce the forecast divergence he had witnessed. Lorenz expanded this result in a 1963 paper to argue that his results suggested that the weather was so sensitive to initial conditions that the meteorological profession's dreams of monthly forecasts would never be realized.[25] One could never measure the initial state of the atmosphere precisely enough to accomplish it.

Lorenz's paper was well-known, but highly controversial, in the small community of numerical prediction researchers. While he had explained something that they had seen happen with their own models for more than a decade, his explanation was one that challenged the foundation of their beliefs. Numerical researchers like Charney, Smagorinsky, and Philip Thomas had been trained in the new physics-based meteorology, and the physics community prided itself on its ability to achieve prediction. Physicists were all trained to believe that once one understood the mathematics underlying a phenomenon, one could predict its future states accurately, forever. Yet this belief was based upon an assumption that errors would remain small throughout a calculation.

From a big-picture perspective, Lorenz's argument was that this would not be true for nonlinear phenomena. Errors would inevitably grow as the calculations progressed, eventually overwhelming the original data. Hence, Lorenz's argument was one that physicists, and even most meteorologists, did not accept easily. It was a denial of a central tenet of their science. The simulation experiments carried out between 1969 and 1971 at GISS, NCAR, and GFDL, however, served as further confirmation that Lorenz was correct. The inability of the GARP observing system to achieve prediction lengths beyond a week reflected the inherent sensitivity of the model to the data it was fed. Simulation studies of predictability continued for several more years, but the principal remaining question was whether these simulation studies were adequate reflections of reality. This could not be known until the GARP observing system was built and its data used to confirm this disturbing limit to predictability.

Finally, the observing system simulation experiments led to the downgrading of the constant-level balloon system's priority. By the Joint Organizing Committee's 1971 meeting in Toronto, it was already clear that a satellite-balloon system covering the entire Southern Hemisphere was unnecessary. The ability to use temperatures in place of winds at most altitudes had reduced the need for a wind-finding system to a single altitude that was referred to as a reference level, which would provide a check on the model calculations. But simulation studies had demonstrated that a balloon system would not produce the data meteorologists wanted. In the simulations, the balloons tended to cluster in certain areas, leaving other areas without coverage. This clustering tendency reduced the usefulness of the system to operational forecasting models, which needed the data to be relatively uniform in spatial distribution. EOLE therefore became an experiment to determine whether the simulations were accurate representations of the (still relatively unknown) Southern Hemisphere's general circulation when it was finally carried out in 1973, and the Southern Hemisphere reference level need was filled by a drifting buoy system.

The remaining need for a balloon system seemed to be for determination of winds in the tropics. The substitution of temperatures for winds in the tropics did not work because the equations that inferred the winds from temperatures, in Vince Lally's words, "blew up." They required a non-zero Coriolis force acting on the air masses being measured, and at the equator the Coriolis force was zero. Hence, prediction models routinely had large errors in their tropical winds that propagated into the mid-latitudes, and an equatorial observing system in addition to the satellite sounders seemed necessary. In three sets of studies, NCAR, GISS, and GFDL researchers examined the question of what kind of system was really necessary.

There were two possibilities for a tropical wind system: a variation of the constant-level balloon system Lally named Mother GHOST (formally the Carrier Balloon System), and the use of winds derived from tracking clouds via imagery from the geosynchronous satellites. The Mother GHOST was a large GHOST balloon that remained at 200 millibars while releasing dropsondes on command relayed to it via geosynchronous satellite. The dropsondes would provide a vertical wind profile while allowing the Mother GHOST to stay at an altitude high enough for a relatively long life. The alternative, cloud-tracked winds, depended upon successful development of a way to produce them quickly and inexpensively. Suomi's Space Science and Engineering Center (SSEC) was working on a system that replaced the film-loop-based method with a computer-based semi-automated one known as WINDCO. It would only provide winds at two altitudes, however, simply because the people doing the altitude assignment could only usefully distinguish between high- and low-altitude clouds.

The studies carried out by the three simulation study centers produced mixed results at first, with GISS finding that there was no improvement in wind errors through use of either system based on studies with the two-level Mintz-Arakawa model. Using Smagorinsky's nine-level model, the GFDL staff had found that the data did improve forecasting. Further studies by all three organizations indicated that the difference resulted from the vertical resolutions of the models, with the more realistic nine-level models consistently showing improved forecasts from use of the tropical wind data.[26] The Mother GHOST system was the only one that resulted in tropical wind errors within the error limits that the Joint Organizing Committee had specified; the cloud-tracked winds improved the wind errors at all latitudes, but still resulted in wind errors in the tropics that were greater than desired. Hence, Mother GHOST was approved for the FGGE. For the earlier tropical experiment, a constant-level balloon system proposed jointly by NCAR and SSEC, the Tropical Wind, Energetics, and Reference Level Experiment (TWERLE) was selected to provide a check on the cloud-tracked winds, as Mother GHOST would not be available in time.

By 1973, then, the global observing system for GARP consisted of a permanent system composed of two polar-orbiting satellites provided by the United States and five geosynchronous satellites providing complete, and to some degree overlapping, coverage of the Earth between 60N and 60S latitudes. Two of these were to be provided by the United States, with one additional satellite each from Japan, the Soviet Union, and the European Space Agency. These satellites would form the space portion of the World Weather Watch system during and after GARP. A series of special observing systems would complement the permanent system during the global experiment's intensive observation periods: a Southern Hemisphere drifting buoy system to define the surface reference level that in the Northern Hemisphere was provided by ground stations and weather ships, and Mother GHOST to provide tropical wind profiles. These studies had also shown that neither monthly nor two-week forecasts would result from the technologies planned for the 1970s. Instead, they would probably achieve forecasts of four to five days—the same forecast length already available from the conventional surface data.

At GISS, Jastrow had responded to all the interest in extending forecast length by hiring meteorologist Richard Somerville to lead an effort to improve the Mintz-Arakawa model. The Mintz-Arakawa model had an extremely efficient scheme for calculating the motion of air around the world and Somerville kept that dynamical core, but his group began reworking most of the rest of the model processes. They expanded the number of atmosphere levels from two to nine, to improve the vertical fidelity of the model, and wrote a new radiative transfer code. Jastrow was still interested in longer-range forecasts, and pushed them to make a model that could make "farmer's forecasts," as he called them. These would be seasonal forecasts, aimed at helping farmers choose the best crops for the next growing season. The new model was completed in 1974, and the group first used it to explore some old questions, including whether day-to-day solar variability affected the weather.[27]

GATE AND THE DSTS

The first major GARP experiment was the tropical experiment that had been part of the original GARP proposal. Originally named TROMEX, for Tropical Meteorological Experiment, and intended for the equatorial Pacific, it evolved through several iterations into GATE. Carried out simultaneously was a set of DSTs intended to identify problems with the observing system's hardware and software. Planners for both GATE and the DSTs understood that properly handling the data stream from the observing systems was the largest challenge facing them. The scientific purpose of GARP experiments was collection of high-quality datasets for use in future research; the experiments would fail if the data was unusable. Similarly, if the data could not be processed in near real time, it would not be useful to the global operational forecasting system GARP was to demonstrate for the future World Weather Watch. The GATE and DST experiences proved enlightening.

One of the central questions in synoptic-scale meteorology during the 1960s was how energy moved from the tropics, which received the majority of the Earth's overall solar input (insolation), into the mid-latitudes and then to the poles. At its highest level, the process was well understood. Solar radiation passed through the atmosphere and was absorbed by the oceans, heating the surface. Evaporation from the surface then carried the energy into the atmosphere. When this water vapor condensed into rain, this energy, technically called the latent heat of evaporation, was released, heating the surrounding air. Because the atmosphere is mostly transparent to the incoming sunlight, this is the primary mechanism for transfer of solar energy into the atmosphere. What meteorologists did not know were the details of how this happened—how much energy was received, on average, by each kilometer of ocean, how much water evaporated and recondensed, and most important for their global forecasting ambitions, how the small-scale convective systems that resulted from this process affected the synoptic-scale motions of the atmosphere they wanted to predict. This last question was the central scientific objective of GATE.

The first Television-Infrared Observations Satellite (TIROS) satellite had presented meteorologists with some intriguing data from the remote tropical oceans. Its cloud photographs had revealed the existence of very short-lived, small (i.e., 30 to 100 kilometers), but intense convective systems. They had been labeled *cloud clusters* for their appearance in these images. The clusters formed and dissipated within twelve hours, usually appearing and disappearing between two orbits of the satellite. These cloud clusters were an obvious mechanism for energy transfer, and they had caught the interest of Jule Charney and Vern Suomi, among many others. Because study of them offered the potential to provide important insights into the dynamics of the energy transfer process, they became GATE's central phenomenon.

In late October 1968, Suomi hosted a study group in Madison to speculate about the potential linkage of these cloud clusters to larger-scale phenomena, prepare a preliminary tropical cloud climatology from the prior year's satellite imagery, and formulate recommendations on how to construct a ship-based observation network that would permit study of the full life-cycle of the clusters. They produced a recommendation that the experiment be conducted around the Marshall Islands using an overlapping set of ship networks, with a meso-scale array to study the clusters themselves and a second array covering ten times the first's area to permit linking the meso-scale data with synoptic-scale phenomenon.[28] GATE would also require the satellite-based global observing system in order to complete the series of linkages from the small scale to the global. In particular, the experiment depended upon the geosynchronous imaging satellites, whose ability to produce images of the same area in relatively rapid succession would be crucial, and it needed a tropical balloon system to better define winds. These recommendations became the basis of the formal National Academy of Science's proposal to the U.S. government regarding GATE.[29]

The study group at Madison had chosen the Marshall Islands because the cloud clusters were quite common, occurring every four to five days, and because the region was logistically feasible. The chain had a large number of small, uninhabited islands that could be used as observing stations, thus reducing the number of ships required, and there were airfields available at Kwajalein and Eniwetok islands for logistical support and to host aircraft-based experimenters. However, these were military installations and that led to the shift of the tropical experiment to the Atlantic. The tropical experiment, while principally planned by the United

States, was an *international* experiment. There would be a substantial number of Soviet ships involved—as it turned out, the Soviet Union provided more ships to GATE than any other nation—and hosting them at the American naval station at Kwajalein for the experiment period of nearly a year was infeasible. The experiment area was therefore replanned into the equatorial Atlantic, with an enlarged ship array as a substitute for fewer island-based observing stations.

The Nixon administration's March 1970 approval of the National Academy of Science's plan for American participation in GARP permitted detailed international negotiations over funding as well as ship, aircraft, and satellite availability to go ahead.[30] By late 1971, the experiment had taken on its final form. There were to be three special observing periods during the summer of 1974, each three weeks long. A network of thirty-eight ships, organized into three nested arrays straddling the equator, would stretch from the east coast of South America to the west coast of Africa.[31] The ship array was be supplemented by thirteen aircraft, whose missions would be planned daily based upon the next day's forecast and adjusted on the fly through use of imagery from a geosynchronous imaging satellite. Initially, the experiment was to be run from the British Meteorological Office's facilities in Bracknell, England, but after a diplomatic row caused by the expulsion of Soviet diplomats on charges of espionage, Bracknell became politically unacceptable.[32] Senegal offered use of a new facility at Dakar's airport for the experiment's headquarters, and this became home to the experiment's Scientific Management Group. The Dakar site had the advantage of putting the management group in the same location as the aircraft, facilitating mission planning, at the cost of requiring construction of a ground station for receipt of the satellite imagery.

NASA's primary contribution to GATE was a set of satellites for the global observing system and experiments related to those satellites. NASA Goddard had scheduled the launch of the prototype of NOAA's Geosynchronous Operational Environmental Satellite (GOES), SMS 1, for mid-1973, and this would be parked over the experiment area to provide the overhead imagery the planners considered crucial. It would also provide the experiment data link support and NOAA's Weather FAX service to permit rapid distribution of forecast maps. Goddard was also responsible for providing Nimbus 6 to the experiment. In many respects, this was a functional prototype for NOAA's next generation of polar-orbiting weather satellites. It carried the prototypes of Bill Smith's High Resolution Infrared Spectrometer (HIRS), David Staelin's Scanning Microwave Spectrometer (SCAMS), and John T. Houghton's Pressure Modulated Spectrometer. These three instruments were to be functionally unified into the TIROS Operational Vertical Sounder on TIROS N, the actual hardware prototype of the new operational

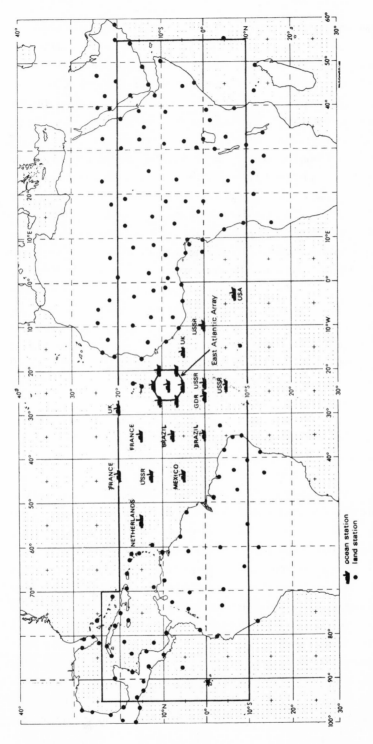

GATE ship array (*center*), with GATE land-based observation locations marked on continental areas. From Joachim P. Kuettner, "General Description and Central Program of GATE," *Bulletin of the American Meteorological Society*, 55:7 (July 1974), p. 713.

satellite due for launch in the late 1970s. Nimbus 6 also carried the tracking and communications equipment for the TWERLE experiment, which was an evolution of the Interrogation, Recording, and Location System (IRLS) system demonstrated on earlier satellites. NASA also provided other support to GATE, including its tracking ship USNS Vanguard, Ames Research Center's Convair 990, and data processing facilities at both Goddard and GISS. But the satellites were the centerpiece of NASA's GATE effort.

The satellites also turned out to be its primary challenge. SMS 1 suffered development problems related to Goddard's choice of a contractor that lacked the necessary expertise and resources to complete it on schedule, and its launch date slid from mid-1973 to June 1974. It reached orbit barely in time for the experiment, whose ships left port 15 June for the first observing period. Its delay caused a great deal of concern on the U.S. Committee for GARP, because it was the only geosynchronous satellite to have infrared imaging capability and the Scientific Management Group intended to use the nighttime cloud imagery in its aircraft mission planning. The agency was less fortunate with Nimbus F, which did not launch until well after GATE. Its delay was a product of instrument development problems. The contractor for the HIRS instrument had not been able to deliver it in time. Hence, the satellite observing system that ultimately supported GATE consisted of the still partly operational Nimbus 5, SMS 1, the still-functioning ATS 3, and two operational NOAA polar-orbiting satellites, NOAA 2 and NOAA 3. These carried the earlier version of Smith's temperature profile instrument, the Vertical Temperature Profile Radiometer (VTPR), and the Advanced Very High Resolution Radiometer (AVHRR).

In addition to these changes in the satellite observing system, there were other changes in NASA's support for GATE during the three years between acceptance of the 1971 plan and its conduct in 1974. NCAR's Paul Julian and Robert Steinberg from the Lewis Research Center in Cleveland had conceived a way to acquire wind data from commercial airliners equipped with inertial navigation systems.[33] These sensed wind motions automatically, and Julian and Steinberg's idea was to equip the navigation systems with recorders and pick up the resulting tapes when the aircraft landed. These would only provide wind data for a very narrow set of altitudes, and only along airline flight routes. But several routes overflew the GATE array, and NASA and NOAA arranged contracts with several airlines flying those routes to install recorders and collect the data tapes. They also equipped a U.S. Air Force C-141 that had a daily route over the array for the experiment.[34]

GATE's three field phases took place between June and September 1974. By this time, GATE had evolved to include five hundred different experiments organized into a Central Program and five subprograms: a synoptic-scale program, a convective-scale subprogram, a boundary-layer subprogram, a radiation subprogram, and an oceanographic subprogram.[35] The Central Program's objectives were to examine the interaction between smaller-scale tropical weather phenomena and the general circulation and to improve numerical modeling and prediction methods. Each of the subprograms supported the Central Program in some way. The synoptic-scale subprogram supported it through description of synoptic-scale disturbances within the experiment region and by providing the datasets for numerical models. The convection subprogram included the cloud cluster investigation that had been important to GATE's foundation as well as a budget experiment vital to understanding scale interaction. The boundary-layer subprogram included surface flux measurements needed for the convective studies and for efforts to parameterize convective processes. The radiation program focused on radiative heating and cooling rates and processes, also necessary quantities for parameterization efforts, while the oceanographic subprogram was aimed at ocean-atmosphere forcings.[36]

In his early comments after GATE field phase, experiment director Joachim Kuettner, a veteran of the NASA Mercury program and a meteorologist who had specialized in mountain-induced waves in the atmosphere, wrote that "it [was] common experience that no field project achieves 100% of its goal." Atmospheric scientists had to work in the "laboratory" of the Earth's atmosphere, and between its vagaries and those of the machinery of the observing systems, they were generally lucky to get a majority of the observations they sought. In GATE's case, the availability of real-time and near-real-time data relayed via satellite had allowed the Scientific Management Group to identify parts of the observation system that were not performing as expected and either fix them or compensate for them — repositioning ships with unreliable wind-finding equipment from more to less important areas, replacing ships with mechanical difficulties, or reassigning aircraft missions. He estimated that GATE accomplished about 80 percent of the observations intended for it, with the most disappointing results coming from the conventional surface stations of the World Weather Watch.[37]

Originally, GATE and the DSTs had been scheduled to coincide. GATE's requirement for data from conventional, special, and space-borne observing systems had made it an obvious opportunity for verification of the data transmission and processing system that was to be the basis of a future American global forecasting system. The DST project office at NASA Goddard had generated a set of four tests to be carried out on individual pieces of the observing system, such as processing of geosynchronous imagery to extract wind velocities from cloud motions, between 1972 and the beginning of GATE. During GATE, a fifth test encompassing all of the observing systems would be carried out, followed by a sixth, and final, test during the 1974–75 winter—one could not be certain that the observing systems functioned well in all conditions without testing them in the best and worst seasons, after all. The delayed launches of SMS 1 and Nimbus F, both intended to be functional prototypes of the operational GOES and TIROS N series satellites, however, forced the postponement of the DST series. Hence the two full-up tests were carried out for sixty-day periods in August–October 1975 and January–March 1976.[38]

The results of the final two DSTs were somewhat disturbing. Portions of the tests went very well. The TWERLE experiment, for example, carried out launches of 393 instrumented, constant-level balloons to 150 millibars altitude in the tropical Southern Hemisphere between October 1975 and January 1976. The RAMS tracking system performed essentially as expected, giving a location accurate to within about 3 kilometers, and demonstrating a form of clustering that had not been seen with the mid-latitude EOLE experiment. A substantial number of balloons clustered in the Gulf of Guinea region, which project scientist Paul Julian interpreted as verifying analyses from conventional data that suggested large-scale, long-duration divergence. TWERLE's RAMS system also successfully tracked and received data from drifting buoys, an important part of the future global observation system.[39]

But the DSTs also demonstrated that any future global observation system needed to have much greater quality control in its data processing path. The quality control challenge derived from the use of cloud-tracking to derive wind velocities and from the well-known cloud clearance difficulties with the infrared temperature profile instruments. During the early 1970s, Suomi had developed a method of partially automating the cloud-tracking process that allowed derivation of wind velocities from geosynchronous satellite imagery. Initially called WINDCO, and later McIdas, the system utilized a midi-computer as a workstation, a remote mainframe for processing, and a television-quality monitor that was linked via a lookup table to the much-higher-resolution image data. This permitted operators to work with the high-resolution data without actually having to display it—video displays with sufficient resolution to display the 120-megabyte satellite images did not exist.

One of the principal difficulties of tracking clouds in successive images had been that the satellite's own motion was imperfect, and it had to be subtracted from the overall cloud motion. WINDCO did this using two corrections, one based on the satellite's known motion from the satellite tracking system, and one based on registration of the images to an obvious landmark on Earth. The WINDCO operator chose a landmark visible in successive images with a light pen, and the computer made the necessary corrections to its model of the satellite's motion. The operator could then select clouds for tracking and produce the wind set. Suomi liked to call this method of combining human intelligence with automation man-interactive computing, and this was the genesis of the system's final name, the Man-Computer Interactive Data Access System (McIdas). McIdas, demonstrated for NASA and NOAA officials in April 1972, became the means of production of the cloud-tracked wind sets for GATE and for the DSTs in 1974–75.[40] SSEC was assigned responsibility for receiving and archiving the satellite imagery during these experiments and producing four sets of wind data each day.

But when GISS and the National Meteorological Center tried to use the wind sets in their experimental global models during 1976 and 1977, they found the data had damaging errors. The principal flaw was in the operator's assignment of altitude to the clouds being tracked, and thus to the resulting wind vector. The datasets were supposed to contain wind vectors at two levels, and the operators had not been able to reliably distinguish between upper- and lower-level clouds. Although the number of errors was actually small in relation to the size of the dataset, the erroneous winds had large impacts on the resulting forecasts.

Similarly, the temperature profiles derived by the National Environmental Satellite Service from Smith's infrared temperature sounder on Nimbus 6 contained substantial errors. Some of the errors derived from the cloud clearance problem, while others occurred under specific meteorological conditions. The automated inversion method that Smith and Hal Woolf had developed proved not to give accurate results under all conditions. The erroneous temperatures caused forecast errors just as the wind errors had. Writing for the record in November 1977 about the temperature sounding challenge, one member of the U.S. Committee for GARP commented that the committee had concluded that the data processing methods used during the FGGE should not be the same as those used during the DSTs. A special effort to check the quality of the sounding retrievals during the experiment was going to be necessary.[41]

Most disturbing of all to GARP participants, however, was the National Meteorological Center's assessment that the satellite temperature profiles did not significantly improve forecast skill in the Northern Hemisphere.[42] This determination was independent of the problem of erroneous soundings; even when they were weeded out and discarded, the National Meteorological Center's forecast model produced essentially the same results as it did when given only the conventional radiosonde data. In the Southern Hemisphere, where almost no conventional data existed, the satellite data improved forecasts significantly, bringing Southern Hemisphere forecasts to nearly the same level of skill as those in the Northern Hemisphere. This ratified the results of the Observing Systems Simulation Experiments, which had suggested that this generation of satellite sounders would only produce skillful four- to five-day forecasts. But GARP planners, most of whom represented Northern Hemisphere nations, had expected that satellite data would extend Northern Hemisphere forecasts beyond the four to five days possible with conventional data. Instead, the DSTs had suggested that the satellite soundings produced no benefit to their nations at all.

The results of the DSTs were troubling, but their purpose, after all, had been to evaluate the functioning of the data processing system prior to GARP's primary goal, the FGGE (also known internationally as the Global Weather Experiment). In that sense, the DSTs had been very successful. While GARP planners could not address the disappointing predictability outcome prior to the FGGE—this, they understood, required a new generation of satellite sensors, new prediction models, or both—they could fix the quality control problem. In January 1978, the First GARP Global Experiment Advisory Committee met in Madison to work out how NASA, NOAA, and SSEC would deal with it. At this meeting, NOAA's Bill Smith argued that the satellite sensors already provided most of the data necessary to produce better outcomes.[43] Assignment of altitudes to the clouds tracked to produce winds could be done more accurately if the McIdas operators had access to the cloud-top temperature data provided by the window channel since cloud temperature was directly related to altitude. Similarly, McIdas operators could cross-check temperature soundings with the National Meteorological Center's analysis charts, and with cloud imagery to evaluate them if these were made available to the operator. In this way, trained meteorologists' subjective judgment could be used to check the performance of the automated cloud clearance process.

While having to use humans to inspect the results of the automated sounding results slowed the process somewhat, only a small fraction of the twenty thousand soundings generated by TIROS N's HIRS instrument each day would require human intervention. A set of automated filters that compared soundings to nearby radiosondes already threw out obviously bad measurements. Another set of filters that compared the soundings from the infrared instrument to those from the microwave instrument was under development at the National Meteorological Center. Because the two instruments had significantly different resolution, instead of automatically discarding soundings that differed, these filters would flag them for inspection. Under certain meteorological conditions, one would expect the large-area microwave sounding to differ from the higher-resolution infrared result without either of them being incorrect. Meteorologists could identify these and validate or reject the sounding. This was another area McIdas's ability to display multiple data sources graphically would help speed the process.

As a result of this meeting, the committee recommended that SSEC develop and implement McIdas software to enable its use for these quality control measures. In a "special effort," the FGGE datasets would be checked by meteorologists using McIdas; later in the year, SSEC proposed for and won this task. It also built a McIdas system for NOAA's National Environmental Satellite Service for operational use. As a result of these recommendations, the Goddard GARP project office restructured the data flow path for the FGGE, linking SSEC's McIdas terminal to a mainframe computer at GISS for the data processing and to the National Meteorological Center so that forecast analyses could be imported directly into McIdas.[44]

Finally, recognition that full automation of the retrieval process would not produce data of sufficient quality caused the National Environmental Satellite Service's director, David Johnson, to adopt Suomi's man-interactive data processing concept for the operational system post-FGGE. McIdas overcame one of the central problems of the early space age: the ability to produce overwhelming amounts of data without a corresponding capacity to analyze it all. McIdas's graphical display of large datasets maximized its human operators' capabilities while preserving the digital nature of the data. While it did not reduce the data torrent, McIdas put the data in a form more meaningful to humans, allowing its operators to apply knowledge and judgment in evaluating the data.

THE FIRST (AND LAST) GARP GLOBAL EXPERIMENT

The FGGE, renamed the Global Weather Experiment (GWE) as it became clear to the GARP Joint Organizing Committee that it would not be followed by further global experiments, was finally carried out in two special observing periods, January–March 1979 and May–July 1979. Initially planned to encompass an entire year of detailed observation, the GWE was reduced to two sixty-day intensive observation periods by a combination of lack of funds and lack of interest. The purpose of the experiment was the production of datasets for use in improving numerical forecast models; originally, GARP planners' beliefs in the possibility of month-long forecast had generated the need for larger, longer-term datasets. By 1979, GARP's founders no longer believed that thirty-day forecasts were possible, and they could not justify the cost of a full year's detailed, quality-controlled data-sets. Their apparent inability to provide a great leap in forecast length or skill had reduced politicians' interest in the program, with a consequent reduction in financial support. The two sixty-day periods would provide enough meteorological diversity for model improvement.

The GWE also served as an operational test of the prototype of the new American operational polar-orbiting satellite, TIROS N, and this was the source of GWE's delay from its originally planned year of 1976 to 1979. NASA's TIROS N effort had started in 1971 but had soon run into troubles. In this case, the troubles were not primarily technological. In 1972, the Office of Management and Budget (OMB) had embarked on one of its occasional "streamline the government" initiatives and put TIROS N on hold while it investigated whether the nation should continue to maintain two separate polar-orbiting weather satellite programs, the Defense Department's Defense Meteorological Satellite Program (DMSP) series and the NOAA series.

OMB's preference was to eliminate the civilian program. This had sparked a meeting of the U.S. Committee for GARP to discuss the issue, and the committee decided to advocate in favor of the civilian program.[45] Vern Suomi and Richard Reed recruited National Academy of Sciences president Philip Handler into the effort, and Handler raised the subject with the director of the White House Office of Science and Technology Policy, Russell Drew, resulting in a meeting on 15 October 1973. Suomi followed that meeting up with a letter delineating the scientific requirements imposed on the satellite sensors by the GWE experiments: the sounder's accuracy needed to approach 1 degree C, which required use of both infrared and microwave sounders. Because France was providing the balloon and drifting buoy tracking and data system for the GWE, the United States also had a commitment to provide space and an appropriate interface on the satellite for it. These were the vital requirements for the GWE, Suomi had argued, and whatever system OMB chose needed to support them.[46]

To a degree, OMB relented after the scientists' intercession. The two polar orbiter programs remained separate, but NASA was required to use the Defense Department's satellite and modify it to take the instruments it and NOAA had developed over the previous decade. The name TIROS N remained attached to the project, however, and the delay pushed its launch back to October 1978. The first NOAA-funded operational version of the satellite, NOAA 6, followed it into orbit in June 1979.

One more satellite was supposed to join the GWE constellation, Seasat. Developed at the Jet Propulsion Laboratory (JPL), Seasat carried a radar altimeter to precisely measure the height of the ocean surface. Sea surface height varied with wind, current, and temperature, and this measurement was of interest to physical oceanographers. Seasat also carried a scatterometer, permitting it to indirectly measure wind velocity at the ocean surface. If this experiment worked out, the sea surface wind measurement would provide a replacement for the lower-altitude variants of the constant-level balloon system that had been too short-lived to be of use.[47] This would be particularly useful for tropical forecasting, where winds could not be inferred from temperature histories by the prediction models accurately.

Seasat was launched on time, but failed in orbit after 106 days. It returned enough data to demonstrate that the scatterometer and surface height functions produced good results, but did not function long enough to be used during the GWE. It took NASA many years to repeat the experiment because it could not reach agreement with the U.S. Navy to help fund it. Instead, NASA eventually arranged a joint effort with France known as TOPEX/Poseidon to replace the surface height measurement, and with Japan to replace the scatterometer. Neither of these flew before 1990, however. Seasat's loss reduced the completeness of GARP datasets, meaning, for example, that surface wind measurements in the Monsoon Experiment's area would be less than desired. Further, of course, its data would not be available for operational forecasting, which NASA had hoped to achieve.

During the GWE, the satellite network was supplemented by all the special observing systems developed during the preceding decade: dropsondes, the Southern Hemisphere drifting buoy system, and an equatorial constant-level balloon system much like TWERLE. There had been a great deal of concern about these during the preceding few years, because the experiment's funding nations did not provide enough money to carry them all out. Vincent Lally's Carrier Balloon System had been intended as the source of vertical wind profiles for comparison to model-produced wind fields, but its high cost, combined with doubts about the utility of its data—the constant-level balloons' tendency to cluster meant large data gaps—had resulted in its cancellation in favor of aircraft-based dropsondes. France cancelled its funding of the ARGOS tracking system, and the system was salvaged by a donation from the Shah of Iran. The Soviet Union was unable to meet its commitment of a geosynchronous meteorological satellite, and NASA, NOAA, and the National Science Foundation (NSF) had to scrounge for the funds necessary to revive the now-mothballed ATS 3 for the duration of the experiment, and constructed a ground station in Europe to serve it.

Despite all of these departures from the original plans, the observing systems performed as expected during the two intensive observing periods, as did the revised data processing procedures. NOAA P-3 and C-130, and USAF WC-135 and C-141 aircraft flew dropsonde missions from Hickam Air Force Base and Acapulco International Airport, Howard Air Force Base, Ascension Island, and Diego Garcia to provide tropical wind measurements. Boeing 747 airliners equipped with the Lewis Research Center's automatic data reporting system submitted 240,000 observations via the geosynchronous satellites. NOAA provided and operated sixty-four drifting buoys in the Southern Hemisphere for the reference-level experiment, supplementing a larger international flotilla.

Hence, all of the preparation that had gone into getting ready for FGGE paid off in an experiment that was essentially anticlimactic. The datasets were prepared and archived between 1979 and 1981 and became the basis of future research on prediction models. Because the special observing periods were carried out in conjunction with another large-scale field experiment in the Indian Ocean, the Monsoon Experiment, the data was also in high demand for research into the lifecycle of these annual events. This was precisely what GARP's founders had hoped for, and the smoothness of the FGGE and the quality of its data after all the delays and disappointments they had experienced reflected both their hard work and their ability to adapt when things did not quite work out the way they had hoped.

WHAT HATH GARP (AND NASA) WROUGHT?

GARP had been founded to advance two sets of technologies, numerical models and satellite instruments, and to use them to enhance human knowledge of atmospheric processes and to increase forecast lengths. It's fair, then, to ask what it actually achieved. Writing in 1991, the European Center For Medium Range Forecasting's Lennart Bengtsson credited the post-GWE global observing system, the development of data assimilation schemes for non-synoptic data, and improved numerical models with having achieved five-day forecasts in the mid-latitudes with the same level of skill as one-day Northern Hemisphere forecasts had had in the early 1950s.[48] While this was far from the two-week forecasts Charney and many others had hoped for in the mid-1960s—and the monthly forecasts predicted earlier in the decade—it was consonant with the results of the Observing System Simulation Experiments and with the DST results. The convergence of these different experiments on the same number produced, over time, a great deal of confidence within the meteorological profession that the models represented the atmosphere's large-scale processes relatively accurately.

Bengtsson also addressed a highly controversial subject: whether better models or the addition of satellite data had resulted in these gains. Several examinations of this subject had resulted in the conclusion that improvements in the models had produced most of the gains.[49] When run using only the FGGE satellite data, the prediction models' forecasts degraded about a day faster than when run using only the conventional data. The model researchers believed that because the satellite measurements were volumetric averages and tended to wash out the fine vertical structure of the atmosphere, they resulted in analyses that consistently underestimated the energy available in the atmosphere, leading to earlier forecast degradation. Satellite advocates believed that the assimilation schemes for the data were at fault, because they treated the satellite data as if it were radiosonde data delivered at non-synoptic times. Instead, the satellite data was a fundamentally different kind of measurement, and assimilation schemes designed for it would show better results.[50]

In giving the Royal Meteorological Society's Symons Memorial Lecture in 1990, NOAA's Bill Smith contended that both of these arguments had validity. But pointing to a surprising recent study that showed rapid improvement in forecast skill between the end of the GWE and 1986, followed by a plateau of skill in subsequent years, Smith argued that improvements in model physics and assimilation schemes following the GWE had achieved all that was possible with the current, late 1970s generation of satellite instruments. He conceded that their poor vertical resolution was responsible for the lack of significant positive impact the satellite data had on Northern Hemisphere forecasting. Future improvements in forecast length and skill required a new generation of satellite instruments with much greater vertical resolution.[51]

Smith believed that future instruments could produce radiosonde-like data, and in 1979 had proposed an interferometer-based instrument known as HRIS — the High-Resolution Interferometer Spectrometer.[52] A Michelson interferometer like Hanel's InfraRed Inferometer Spectrometer (IRIS), but higher in spectral resolution, this technique would not produce the volumetric averages that Smith's spectrometer-type earlier instruments did. Instead, it would produce a continuous atmospheric profile, exactly like a radiosonde. His proposal was funded by NASA and NOAA, and began to fly on NASA's ER 2 in 1986. It took a second form as a satellite instrument called CrIS, the Crosstrack Infrared Sounder, which was designed to supplant the infrared sounding unit on the TIROS N series in the late 1980s.

But no new instruments flew before the turn of the century, reflecting GARP's principal shortcoming — it did not live long enough to carry out its full program.

The FGGE had not been intended to be the last GARP global experiment; instead, it was to have provided the data necessary to design better models (which did happen) and, aided by those new models, to build and fly improved satellite sensors (which did not). The reason GARP did not complete its program, Smith had not bothered to tell his audience of practitioners, was that NASA had restructured its satellite instrument development programs in ways that unintentionally led to a two-decade-long hiatus in new instruments for the polar orbiters.

The first generation of NASA's leadership retired during the late 1970s, and the new leadership did not think well of the approach the old Meteorological Programs Office had taken toward instrument development. Shelby Tilford, who became head of NASA's Upper Atmosphere Research Program (UARP) in 1977, and later head of its overall Earth sciences program, recalls that the conflict had been over the relationship between instrument developers and model developers. Within NOAA, the National Environmental Satellite Data and Information Service (NESDIS) designed instruments and operated the satellites, while the National Weather Service developed models. The conflict had come to a point when NOAA/NESDIS had sent over to NASA requirements for a next-generation sensor and the model developers at the National Weather Service had refused to verify them. Indeed, they took a position of rejecting the value of satellite data entirely. Because the satellite data did not produce better forecasts than the radiosondes, the Weather Service only employed the satellite data from the Southern Hemisphere and used the radiosonde data in the Northern Hemisphere. Tilford saw little sense in continuing to spend money on a program to develop sensors whose data would not be used. So NASA and NOAA leaders agreed to end the Operational Satellite Improvement Program in 1982.[53]

They did not, however, intend meteorological satellite instrumentation development to end. Instead, the two agencies agreed to incorporate new instrument development into NASA's atmospheric sciences program, with the agency supporting both the instruments and the models needed to use them. This would prove the value of satellite remote sensing to science as GARP studies had not, and perhaps eventually permit overcoming the resistance to satellite data that had grown in the National Weather Service. The new instruments could be transitioned to NOAA after NASA had demonstrated their capabilities. Yet the election of Ronald Reagan in 1980 undermined this plan. Reagan's budget officials believed that the government should not provide operational functions, such as the meteorological satellite systems, and sought to privatize them. While this effort failed, in the process, the administration cut both NASA and NOAA budgets substantially. This left NOAA without the ability to finance even incremental improvements to the weather satellite series, and the instrument generation of 1978, with only minor updates, continued to fly through the end of the century.[54]

In pursuit of GARP, and driven by their own ambitions to remake meteorology into a global science, NASA, NOAA, NCAR, and SSEC produced a technological legacy of sophisticated global models and instruments to feed them that had been dreams when they had started in 1960, and in a few cases that had not been thought of at the time. They made mainstream the use of simulation studies, graphical display of data, and remote sensing. Yet they did not achieve what they had set out to do: provide a revolutionary increase in forecast length.

Instead, what GARP and its myriad supporting studies accomplished was a powerful demonstration of the unpredictable nature of the atmosphere. In the process, it also provided a large, expensive case study of the limits of the belief system postwar scientists had gained from physics. The ability to predict phenomena, the ultimate test of a physical theorem's correctness, was not fully applicable to meteorology. Instead, regardless of the quality or completeness of meteorologists' understanding of atmospheric processes, one could not predict the weather into an indefinite future. The weather, and the atmospheric processes that produced it, had a strong element of chaos at its root.

This issue of limits to predictability, and the directly related idea that tiny, even immeasurable, changes can have global effects, was profoundly disturbing. It was quickly adopted in the public sphere, achieving a cultural resonance as the "butterfly effect," the name given it by Ed Lorenz in a 1972 lecture—a butterfly flapping its wings on one side of the world could change the weather on the other.[55] But within the scientific community, this exploding of the belief that once one fully understood a phenomenon one could make accurate long-range predictions left in its wake a community that had to rebuild itself around the notion of *uncertainty*. One could make probabilistic forecasts with sufficient understanding of a complex phenomenon, but the deterministic, long-range predictions the postwar meteorology community had thought was in its reach was impossible.

The Joint Organizing Committee of GARP voted itself out of existence in 1980. Or, rather, voted to transform itself into the Joint Scientific Committee of the World Climate Research Program. GARP's goal had been twofold, to improve weather forecasting and to investigate the physical processes of climate; having done what seemed possible with the weather, its leaders turned to Earth's climate. Interest in climate research had grown throughout the 1970s, in part due to

NASA's planetary studies and in part due to increasing evidence that humans had attained the ability to change climate by altering the chemistry of Earth's atmosphere. NASA would never play as large a role in the World Climate Research Program (WCRP) as it had in GARP. But its role in climate science would far surpass its role in weather forecasting. One of Robert Jastrow's young hires at GISS, radiative transfer specialist James E. Hansen, would become one of the foremost climate specialists in the world during the 1980s, and one of the most controversial.

CHAPTER FOUR

Planetary
Atmospheres

Yale Mintz painted this immense canvas of applying
numerical models to every atmosphere in the solar
system. It was a beautiful vision.

—*Conway Leovy, February 2006*

During the long years of the Global Atmospheric Research Program (GARP),
NASA's space science organization had been busily exploring what space scientists call the terrestrial planets, Mercury, Venus, and Mars. These three, unlike the
giant outer planets, are made of rock and metal, in roughly the same amounts as
Earth. At the beginning of the space age, there was a great deal of expectation
within the scientific community that at least the nearer two, Venus and Mars,
would be climatically like Earth. Indeed, most scientists of the late 1950s assumed
that an annual "wave of darkening" across the face of Mars represented the bloom
of plant life of some form in a Martian spring.[1]

The 1960s proved shocking to the scientific community, as the Mariner spacecraft built by the Jet Propulsion Laboratory (JPL) proved that these worlds were
radically unlike Earth. Despite having developed out of the same material, and in
Earth's and Venus's cases, with only a small difference in orbital distance and thus
solar intensity, they had each evolved in very different directions. Only Earth had
apparent life; by the late 1970s, the others were judged not merely dead worlds but
ones incapable of hosting any kind of life. This forced the community to think

very hard about the relationship between atmospheric chemistry and climate. This work on planetary climates caused NASA scientists to become involved in a brewing debate over whether humans were changing Earth's climate.

James R. Fleming links the modern theory of greenhouse warming to the work of Guy S. Callendar, who demonstrated a gradual increase of carbon dioxide in the Earth's atmosphere and made important new measurements of the infrared spectra of a number of gases, including water vapor and carbon dioxide, and Gilbert Plass, who made new measurements of carbon dioxide's spectral characteristics and built the first modern infrared radiative transfer computer code. Their work, published between 1939 and 1958, revived and put into its modern form the hypothesis that human emissions of carbon dioxide might cause the Earth to warm.[2] Spencer Weart traces the subsequent evolution of global warming theory through the work of Roger Revelle, Hans Suess, Charles Keeling, and more recent hypotheses of abrupt climate change.[3]

With the exception of Plass, whose work spanned Earth and planetary atmospheres, these scientists were all Earth scientists. They were interested in the climate dynamics of Earth. But NASA, as the space agency, had little direct interest in fostering Earth science during the 1960s. Its meteorological satellite program was an applications program aimed at fostering better weather forecasting by the Weather Bureau. It was not aimed at making NASA itself an Earth science powerhouse. But NASA gradually became one during the 1970s and 1980s, with its institutional interests centered around the very large question of global climate change.

NASA, however, came to an institutional interest in climate studies via a somewhat different route than had Earth scientists. Astronomers were also interested in the atmospheres and climates of other planets, and in fact the Weather Bureau's Harry Wexler had let a contract to Seymour Hess shortly after World War II to use the Lowell Observatory's telescopes to try to understand the general circulations of the Mars and Venus atmospheres. Wexler hoped these would provide clues to the general circulation of Earth's atmosphere, which could not yet be seen from a suitably large distance to make large-scale sense of. This effort didn't succeed, as even the largest telescopes could not see enough detail.[4]

It was from NASA's interest in fostering planetary astronomy that it reached an interest in studying Earth's climate. By 1970, it was already clear that humans were altering the chemistry of Earth's atmosphere. And the late 1970s nearly saw the end of American planetary science. This would help bring the space agency back to Earth.

THE VENUSIAN ATMOSPHERE

Two years after the first Television-Infrared Observations Satellite (TIROS) weather satellite went into Earth orbit, NASA's JPL had succeeded in sending a spacecraft to visit Earth's twin planet, Venus. Of the nine major planets in the solar system, Venus was known to be the closest in mass to the Earth, and it had long been suspected of having conditions on its surface similar to those of Earth's Carboniferous period (345–280 million years before present).[5] Very high temperatures and sea levels, and a nearly planetary-scale tropical climate, marked the Carboniferous. Life was so abundant and rich that most of the world's major coal beds were laid down during this period. Venus was expected by many to be a somewhat warmer version of this period, although there were a handful of indications that this wasn't true. As late as 1955, reputable scientists could still publish arguments that Venus was covered by a worldwide ocean.[6] This idea did not survive the very first mission to Venus, however. Venus turned out to be extraordinarily hot. By the end of the decade, it was believed instead to have been the victim of a runaway greenhouse effect that had left its surface the temperature of molten lead—even though the planet surface received less energy from the Sun than did Earth.

In a series of meetings held at California Institute of Technology in late 1960 and early 1961, the Space Science Board (SSB) of the National Academy of Sciences had discussed the state of knowledge of the atmospheres of Venus and Mars. Will Kellogg of RAND Corporation served as chairman, and a familiar group of astronomers, atmospheric scientists, and remote sensing specialists had gathered, Lewis Kaplan, Yale Mintz, and Carl Sagan among them. With the space age just beginning and the possibility of gaining a closer perspective on, and thus better measurements of, these planets, the collected scientists hoped to provide a scientific strategy for answering some of the outstanding questions about them. A key question about Venus regarded its surface temperature. In June 1956, scientist Cornell Mayer at the Naval Research Laboratory (NRL) had announced radiotelescope measurements of Venus that showed the planet radiating strongly in the microwave region.[7] They argued that this implied Venus's surface temperature averaged 600 degrees K, a clear refutation of the Earthlike Venus consensus.

In 1960, while still a student, Carl Sagan had made one of the first attempts at explaining how Venus could maintain a surface temperature of more than 600 degrees K. Using the known radiative characteristics of various gases, he constructed a model atmosphere for Venus composed primarily of carbon dioxide, including small amounts of water vapor to explain the Venusian atmosphere's apparent absorbitivity in the far infrared. To achieve the high temperatures, Sagan calculated that Venus's atmosphere needed to have a carbon dioxide abundance equivalent to 3 to 4 times the mass of Earth's entire atmosphere; assuming that the ratio between nitrogen (the dominant gas in Earth's atmosphere) and carbon dioxide remained the same, Venus's atmosphere would be about 300 times as dense as the Earth's. This was a radically different concept of Venus, and left it, in Sagan's words "a hot, dry, sandy, windy, cloudy, and probably lifeless planet."[8]

Sagan's model, however, had not been widely accepted within the scientific community. It was criticized, for example, because under the gas pressures represented within the mainstream scientific literature, carbon dioxide did not display sufficient infrared opacity. Sagan, however, had used values drawn from boiler and steam engineering literature representing much higher pressures, where carbon dioxide displayed strong "pressure broadening" of its infrared absorption lines. He argued that carbon dioxide could reach 99 percent infrared opacity under these conditions, sufficient to produce extreme surface temperatures.

Another model of the Venus atmosphere was presented at this meeting, the aeolosphere. In this model, the Sun's energy was deposited high in the atmosphere, and intense winds carried heat to the surface. The strong winds kept the surface permanently shrouded in dust. This dust also served as a strong infrared absorber, keeping the surface perpetually hot. In this model, virtually no light at all would reach the planet's surface, rendering it extraordinarily hot and dark.[9] Spacecraft measurements might be able to distinguish between these measurements by examining the microwave emissions of the planet, and JPL's Mariner R spacecraft, which was scheduled for launch in 1962, would carry a microwave spectrometer to test these hypotheses.

It was also possible that the surface was not as hot as NRL's observers had thought, although Kellogg and Sagan considered this lingering hope "for low temperatures and a habitable surface . . . rather dim."[10] It was possible that the microwave radiation that Mayer had detected was not from the planet's surface but from a highly energetic ionosphere. This hot atmosphere/cold surface model might permit the surface to remain at a more reasonable temperature.

The notion of a habitable Venus was dispelled by the earliest spacecraft-based measurements. In 1962, the United States achieved its first successful planetary mission with the Mariner 2 mission to Venus. This was a flyby mission, with the vehicle passing about 35,400 kilometers from Venus during mid-December. Mariner 2 carried both infrared and microwave radiometers for examining the planet. The microwave radiometer, whose experiment team included Alan Barrett from MIT, provided the telling surface measurement. The instrument had been de-

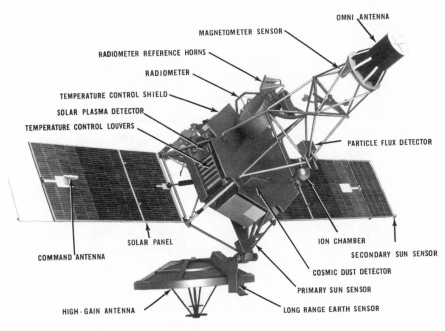

Mariner Venus 1962. JPL image P2009, courtesy NASA/JPL/Caltech.

signed to enable selection between the hot surface/cold atmosphere model and the cold surface/hot atmosphere model by scanning across the planet disk. If the microwave radiation intensity increased toward the limb of Venus, the thin crescent of atmosphere between the planet surface and space, this would indicate that the atmosphere was the source of the radiation, not the surface. Similarly, if the limb darkened, or showed less microwave intensity than the surface, then the source would be a hot surface. The Mariner data clearly indicated that the second case was true.[11] Venus had a very hot surface.

This left open the important question of why it did, and a number of theoretically inclined researchers tackled this question during the decade.[12] In 1969, for example, Andrew P. Ingersoll of the California Institute of Technology proposed that a water vapor–induced runaway greenhouse effect had occurred on Venus. On Earth, atmospheric water vapor remained in equilibrium with surfaces of liquid water because, given the current composition of the atmosphere, incident solar flux was insufficient to cause continually increasing evaporation. Instead, the colder upper atmosphere forced water that evaporated from the surface to precipitate back out, maintaining a stable, if delicate, balance. On Venus, where incident top-of-the-atmosphere solar flux was considerably higher than on Earth, this had not happened. In his calculations, water vapor outgassing from the young planet had never condensed as it largely had on Earth, producing a greenhouse effect that continually increased in intensity.[13] James Pollack, Carl Sagan's first graduate student and head of Ames Research Center's atmospheric modeling group, made similar calculations the same year.

Researchers at the Goddard Institute for Space Studies (GISS) in New York added carbon dioxide to the evolution of the Venusian greenhouse the following year. They started from an assumption that the amount of carbon dioxide in the Venusian atmosphere was approximately the same amount held in the Earth's crust in the form of carbonate minerals. The formation of these minerals, however, is temperature dependent and also requires the presence of water. In the absence of liquid water, which their model of Venus's climate evolution indicated had never been possible, nearly all of the carbon dioxide outgassed by the young planet would remain in the atmosphere instead of being deposited in rock. This would maintain the greenhouse effect as water vapor disintegrated in the upper atmosphere under solar bombardment and the hydrogen escaped into space. Over billions of years, hydrogen escape would gradually remove all of the water vapor from the atmosphere, leaving the dense carbon dioxide atmosphere behind.[14]

There were also alternative theses to the greenhouse model. One of these was Richard Goody's deep convective model. Goody had argued the majority of Venus's atmospheric energy was deposited in the high cloud layer on the sunward side of the planet and was carried to the planet's night side by large-scale zonal currents. There, of course, radiatively cooling air masses would descend, bring energy into the lower atmosphere, and eventually return to the day side. Many scientists believed that even the very thick carbon dioxide atmosphere of Venus could not have sufficient infrared opacity to generate Venus's high surface temperature through the classical greenhouse effect alone; very little solar energy was thought to reach the surface through the cloud layers, and hence this small amount would have to be entirely retained by the atmosphere to maintain the high surface temperature. That did not seem possible, and Goody's dynamical model was designed to overcome the limitations of Sagan's simple radiative model.[15]

These theoretical studies were also stimulated by the increasing tempo of spacecraft operations at Venus. In 1967, the American probe Mariner 5 had made a flyby of Venus. This spacecraft was not equipped for extensive atmospheric investigation, although it did confirm the very high surface temperature via radio occultation. More detailed investigation of the predicted extreme surface conditions was carried out by the Soviet Union with a series of atmospheric entry probes and landers beginning the same year. The first of these was Venera 4, which successfully entered the Venusian atmosphere on 18 October, the same day as the Mariner 5 flyby. It released thermometers, gas analyzers, a barometer, and an atmospheric density probe. These showed that the atmosphere was more than 90 percent carbon dioxide. The temperature and pressure reached 535 degrees K and 18 atmospheres before the probe failed.[16] Initially, the Soviet mission scientists had thought the probe had reached the surface; Sagan and Pollack later demonstrated that it had not, instead failing while still descending on its parachute. Yet it was judged a highly successful mission, and a nearly continuous series of Venera spacecraft followed: two more atmosphere entry probes in 1969, the first successful Venus lander in 1970, a second lander in 1972, and still more in 1975, 1978, 1981, and 1984. The last pair carried two French constant-level balloons as well, designed to float in the cloud system.[17]

The composition of the Venus clouds was finally figured out in 1972, based not on spacecraft data but on ground-based observations. The key measurement was of the index of refraction of the clouds' primary constituent. This was hardly a new quantity. Every optically transparent substance has an index of refraction, but as astronomer Ronald Schorn reports in his history of planetary astronomy, no one had thought to check the clouds' index against those of known substances. Two persons finally suggested the answer nearly simultaneously late in 1972: Godfrey T. Sill and Louise G. D. Young.[18] They proposed droplets of sulfuric acid as the most likely culprits. Sulfate aerosols were common products of volcanic eruptions on Earth, providing an obvious mechanism by which they might have been injected into the Venusian atmosphere. In 1974, James Pollack at the Ames Research Center obtained measurements of the near-infrared spectra of the clouds using a Learjet aircraft equipped with an infrared telescope. Comparison of these reflected spectra to laboratory spectra of strong solutions of sulfuric acid provided confirmation. The Venus cloud sheet was sulfuric acid.[19]

The Soviet 1972 lander provided important new data on both the atmosphere and surface. A photometer aboard Venera 8 demonstrated that sunlight equivalent to an overcast day on Earth reached the Venusian surface through the thick cloud mantle, making photography feasible. The amount of light was only 2 to 3 percent of the total received at the top of the atmosphere, but this suggested that the greenhouse model of the atmosphere was roughly valid. Further, the wind measurements made during the lander descent showed very high-velocity winds in the upper atmosphere that flowed from the sunlit side to the dark side. Onboard measurements demonstrated that the cloud layers—there appeared to be three layers in the Soviet data—extended far deeper into the atmosphere than previously believed. And, finally, the probe found that the atmosphere was essentially adiabatic from the surface to 50 kilometers.

Pollack's group at Ames used the availability of a new supercomputer at Ames, the ILLIAC IV, to construct a numerical model of Venus's general circulation during 1974 and 1975. This was based on Yale Mintz's general circulation model of the Earth's atmosphere modified with a new radiative transfer scheme based on one developed by Andrew Lacis and James Hansen at GISS. There were several striking results from their calculation. The first was that the cloud layer absorbed a substantial part of the incoming solar radiation, leading to substantial heating of the surrounding atmosphere on the sunward side of the planet (as predicted by Goody and others). Much more solar energy was deposited in their model cloud layers than at the surface, in opposition to the way energy is deposited on Earth. High-altitude zonal winds carried energy to the night side of the planet, as expected.

Another surprising finding was that the sulfate aerosol clouds were also strong absorbers in the thermal infrared spectrum, and therefore enhanced the overall planetary greenhouse effect. This appeared to resolve some problems with the greenhouse model caused by insufficient infrared opacity several critics had pointed out. Further, the cloud absorption in both the visible and infrared rendered the thick cloud layers essentially adiabatic and isothermal. And a shear zone, a region in which horizontal wind speeds changed dramatically, appeared at the base of the lowest model cloud layer. Both the adiabatic cloud region and the shear zone were consistent with the measurements made by the Venera probes.[20]

In 1975, NASA gained new start approval for a dual-spacecraft mission named Pioneer Venus. Based at Ames Research Center, this mission had two goals: radar mapping of the surface of Venus, and examination of the chemical and thermal characteristics of the atmosphere by a set of entry probes. The concept for the mission came from Richard Goody, who had recruited to his cause Donald Hunten, Nelson Spencer from the Goddard Space Flight Center, and Vern Suomi; in 1969, the four had prepared a mission plan titled A Venus Multiple Entry-Probe Direct-Impact Mission. It had not been well-received by NASA, as the four were quite critical of the way the Venus program was currently being carried out by the agency. Goody had appended a note to the report that argued that the agency's Venus missions were not being designed to investigate specific scientific questions, a comment virtually guaranteed to alienate NASA management (and that of JPL, which was responsible for the Mariner missions). Goody wanted measurements from within the Venusian atmosphere to compare to his circulation model, and NASA did not seem interested in providing them.[21]

But the leaders of both the Goddard Space Flight Center and Ames Research Center, who would compete to carry it out, found it in their interests to pursue the project. Their advocacy kept the concept alive, aided by an Ames demonstration in 1971 that a high-speed entry was possible with existing technologies. The Soviet Venera plans then motivated NASA leadership to take the proposal seriously, forming a Pioneer Venus Science Steering Group in January 1972. NASA sought new start approval from Congress for fiscal year 1974 in anticipation of a 1977 set of launches; the delay of approval to fiscal year 1975 set the effort back to a 1978 launch.[22]

The two Pioneer Venus spacecraft were known as the Orbiter and the Bus. The Orbiter's job was radar mapping of the planet's surface, although it also carried instruments designed to investigate the upper atmosphere, above the cloud layers. The key instruments needed to investigate the lower atmosphere and the Venusian greenhouse effect were housed by the Bus. It carried three small probes and a large probe, each of which had instruments to measure solar and infrared flux in the local atmosphere, temperature, pressure, and the optical properties of particulates. The large probe also carried a gas chromatograph, to measure the abundance of various atmospheric gases, and a neutral mass spectrometer. Finally, the Bus itself was designed to enter the atmosphere carrying a second neutral mass spectrometer. Only the large probe had a parachute to slow its descent. The small probes and the Bus free-fell during entry. The small probes had heat shields to permit them to descend all the way to the surface, while the Bus was expected to burn up at an altitude of 80 kilometers or so, above the cloud layers.

All five entry spacecraft entered the atmosphere on 9 December 1978. The large probe and all three small probes descended through the atmosphere, impacting the surface eventually at about 32 kilometers per hour. The instrumentation also functioned relatively well, although the important heat flux instruments failed (oddly simultaneously) at 12 kilometers altitude. They also displayed behavior that made their data unreliable, so the data could not be used to detail the energy fluxes within the atmosphere. However, the temperature, pressure, and wind profiles of the atmosphere made by the probes corroborated those made by the preceding Soviet landers, clearly showing very similar adiabatic conditions, shear zones, and cloud structure. The data was also generally in keeping with Pollack's model of the atmosphere. One major difference between the modeled atmosphere and the measured one, however, appeared in the convective stability of the lower atmosphere. Pollack's model had shown a relatively rapid overturning of the lower atmosphere, as had some others. The data from the probes showed the lower atmosphere to be convectively stable, with very low wind velocities. The extremely dense lower atmosphere transported heat so efficiently that it maintained uniform surface temperatures even without high velocities; in the absence of sharp surface temperature gradients, there was no mechanism to drive strong convection.[23]

By the end of the first two decades of space exploration, Venus had been transformed from a potentially habitable planet into an extraordinarily uninhabitable one. Most, and perhaps all, of its carbon dioxide had wound up in its atmosphere, unlike Earth's, where the majority had been sequestered away in sedimentary rock. The perpetual yellow haze that had been visible to Earth-bound astronomers had turned out to be a sulfuric acid cloud sheet that simultaneously reflected away most of the incoming solar energy and enhanced the infrared greenhouse effect that kept the planet so hot. Finally, the atmosphere's circulation appeared radically different, with enough energy being deposited within the cloud sheet to produce high-velocity zonal circulation that carried heat to the planet's perpetual dark side, while the deep atmosphere was relatively quiet. Venus had evolved along a far different route than had Earth, despite its similar size and overall composition.

THE MARTIAN ATMOSPHERE

Mars, even more than Venus, had been seen as a potential abode of life prior to the space age. In 1877, Italian astronomer Giovanni Virginio Schiaparelli had reported the existence of canali on Mars, leading to a series of speculations about liquid water on Mars and the existence of an old, and perhaps superior civilization, on the planet. This line of thought had reached its peak with the work of American astronomer Percival Lowell. Working largely during the 1890s, Lowell contended in both the scientific and the popular press that an "annual wave of darkening" on Mars represented the seasonal growth patterns of irrigated agriculture—and thus evidence of an inhabited, and civilized, planet.[24] While most of the scientific community had not accepted that the canali were evidence of life, they were seen as evidence of liquid water and therefore also of Earthlike conditions. Things turned out to be a bit more complicated than that.

In 1963, Hyron Spinrad, a recent PhD graduate of the University of California, Berkeley, completed a seminal study of the atmospheric radiation of Mars using the 100-inch reflecting telescope at Mount Wilson that undermined the case for liquid water on Mars. On the night of 12–13 April, Spinrad obtained a near-infrared spectrum of Mars that showed that there was, in fact, water in the Martian atmosphere, but the amount was tiny. Initially, he estimated the amount as having been 5 to 10 microns of precipitable water, to put it in conventional Earthly meteorological terms. This was a controversial finding, as the spectra appeared at the limit of detectability. They were also too faint to reproduce, so his plates could not be seen by others without traveling to California. Since other astronomers could not examine his evidence, it was easy for them to disbelieve his result.

Spinrad's plate also displayed absorption lines for carbon dioxide. This was entirely unexpected. The carbon dioxide absorption bands are relatively weak, and Spinrad's ability to detect them at all meant that there was a great deal of the gas in the Martian atmosphere—more than in Earth's, in fact. Finally, it occurred to Spinrad to wonder whether carbon dioxide was the primary gas in the Martian atmosphere, as it was in Venus's (but not, of course, Earth's). If that turned out to be the case, then his measurement suggested a surface pressure of only 35 millibars, well below the figure generally accepted at the time.[25] By comparison, the average sea level pressure on Earth is 1013 millibars; the lowest pressures at sea level occur in the middle of hurricanes and typhoons, but had never been seen below 870 millibars.

His findings were extremely important to NASA, whose engineers needed to know Martian atmospheric pressure and density accurately for the design of atmospheric entry and landing systems, so the head of NASA's astronomy program called a meeting in Washington to try to come up with a consensus answer. Instead, estimates ranged from 8 to 100 millibars—although the 8 millibars estimate was a half-joke made by Lewis Kaplan, based on the remote possibility that the Martian atmosphere was entirely carbon dioxide.[26]

The issue of Martian atmospheric pressure was largely settled in 1965, by the first spacecraft to reach Mars successfully. This was JPL's Mariner 4, which flew by the planet on 15 July 1965 at a distance of about 9600 kilometers. Mariner 4 carried a camera but no atmospheric sensors; nonetheless, it provided important information about the Martian atmosphere. The photographs sent back showed an unexpectedly crater-pocked, desert-like Mars, effectively discrediting the vision of a planet inhabited by advanced sentient life and finally debunking the Martian canali. Mars lacked even the flora expected to be the cause of the "wave of darkening." This Moonlike surface proved that Mars had a very thin atmosphere. A thick atmosphere would have burned up incoming meteorites. And the craters' existence also showed that Mars no longer had a significant hydrologic cycle. On Earth, craters are buried by erosional processes led by liquid water; on Mars, they clearly lasted for billions of years.

Mariner 4 image showing ancient, cratered terrain on Mars, taken in July 1965. The largest crater visible in the image is 151 kilometers in diameter; other, younger craters are superimposed on top of it. A ridge running from the lower left corner of the image cuts through the crater wall as well. JPL image 02979, courtesy NASA/JPL/Caltech.

But Mariner's science team also used the vehicle's radio to obtain a better measurement of atmospheric pressure. Shortly before launch, a JPL engineer proposed using the radio transmissions for an occultation experiment. As the spacecraft passed behind Mars from Earth's perspective, its radio signal would be altered by the Martian atmosphere in ways that were detectable from Earth and that would provide information on the atmosphere's temperature and pressure. His proposal caused some anguish for JPL and Mariner management, as it meant not transmitting imagery, the mission's primary objective, during the experiment. But they agreed to it, and the resulting findings clarified the status of the Martian atmosphere considerably. The atmospheric density was about 1 percent of Earth's, with a surface temperature of about 180 degrees K and a pressure between 4 and 7 millibars.[27] This was, roughly, the same as the estimate Lewis Kaplan had half-jokingly made several years earlier—and stunningly low.

Two more Mariner flyby missions visited Mars in 1969. Mariners 6 and 7, equipped with identical cameras but no atmospheric instruments, flew past the planet in July and August of that year, returning additional photographs of a dead, and heavily cratered, surface. At the same time, several different astronomy groups confirmed the 1963 detection of water vapor in the Martian atmosphere, again using infrared spectroscopy. They also were able to demonstrate that the water vapor content varied with latitude and season. This helped lead a JPL group under Crofton "Barney" Farmer to design a spacecraft-based instrument to provide more detailed measurements from Mars orbit; this became the Mars Atmospheric Water Detector and would fly on the 1976 Viking mission.[28]

The Mars exploration effort achieved its first orbiting spacecraft in 1971, with Mariner 9 (its twin, Mariner 8, made its home on the bottom of the Atlantic Ocean). As Mariner 9 approached the planet, ground-based astronomers detected a brilliant cloud that rapidly spread to cover the Noachis region. In two weeks, it covered the entire visible disk of the planet. The spacecraft went into orbit 13 November, and the images it returned showed a completely featureless Mars. The dust storm had covered the entire planet. Mariner 9 had been intended to map the planet photographically in preparation for a 1976 landing attempt by the Viking project; the vast dust storm seemed to make that impossible. The only visible features were Nix Olympica and three other peaks that protruded above the storm. This revealed to the camera team that they were the largest volcanoes in the solar system by far, dwarfing anything on Earth, but it did not help them accomplish their primary mission.[29]

Mariner 9 also carried a copy of Rudy Hanel's Infrared Interferometer Spectrometer (IRIS) instrument, developed originally for the Nimbus meteorological satellite series. Initially intended to produce atmospheric temperature profiles, it could also generate other useful information. Hanel's team used the data to determine the abundance and distribution of water vapor in the Martian atmosphere, finding 10 to 20 precipitable microns over most of the planet. They also found a seasonal change, in which water vapor disappeared from the south polar region and reappeared in the north polar region. And they prepared detailed analyses of the spectral characteristics of Martian carbon dioxide, which differed from Earthly characteristics slightly due to the very low pressure. IRIS could also provide surface temperature measurements through the extremely dry Martian atmosphere. Hanel's group at Goddard used the surface temperatures to help them construct a pressure map for Mars; they also were able to construct topo-graphic maps for several regions of Mars by using surface pressure measurements to estimate altitudes.[30]

The IRIS data also sparked a great deal of interest in the radiative properties of the suspended dust in the Martian atmosphere, particularly within Pollack's group at the Ames Research Center. The temperature profiles produced by the instru-ment showed that the dust had the effect of absorbing sunlight and heating the atmospheric layer around it, while simultaneously, the lack of energy reaching the Mars surface caused it to cool rapidly. The atmosphere adjacent to the surface therefore cooled as well. This had led to a temperature profile that was essentially flat between the dust layer and the surface—put in meteorological terms, the lapse rate was nearly zero. The very high daily surface temperature range (a prod-uct of the very dry, thin atmosphere) also narrowed, from about 75 degrees C to 35 degrees C. Finally, because the heat distribution in the atmosphere changed, the circulation also changed. A lofting effect attributed to solar heating of the dust-laden atmosphere resulted in the dust being transported as high as 50 kilo-meters.[31] On Earth, this altitude represented the boundary between the top of the stratosphere and the thermosphere, where even extreme volcanic events had never been known to deposit particulate matter. These data showed a Martian atmosphere whose behavior was extraordinarily strange, and very unlike Earth's. It also eventually led to the famous nuclear winter hypothesis.[32]

Mariner 9 is perhaps best known for its revelations about the Martian surface, however, not the atmosphere. After the planetary dust storm finally ended in late December, the imaging team began to produce the photographic map of the planet that was their primary goal. In the process, they found that the images from the previous flyby missions had been highly misleading. While Mars certainly had Moonlike cratering, it also had vast volcanoes, and still more interestingly, vast chasms and canyons.[33] To the geologists on the imaging science team, many of these looked like the natural drainage structures on Earth—extensive dendritic patterns, outwash plains, and deltas. There was no visible water to have made them, but the signs of past water were unmistakable, unless reasoning by Earth-analogy, which was all that was available to the geologists, was itself misleading.

The science team ruled out other substances relatively quickly. A surface pres-sure 5 times Earth's would have been necessary to allow liquid carbon dioxide to carve such channels, and Mars did not have it. While wind also causes erosion, Earth winds did not cut winding, tributary-laden channels the way water did. Imagery of the Tharsis region clearly showed braided channeling, a common feature of outwash plains on Earth. Further, many of the river channels appeared

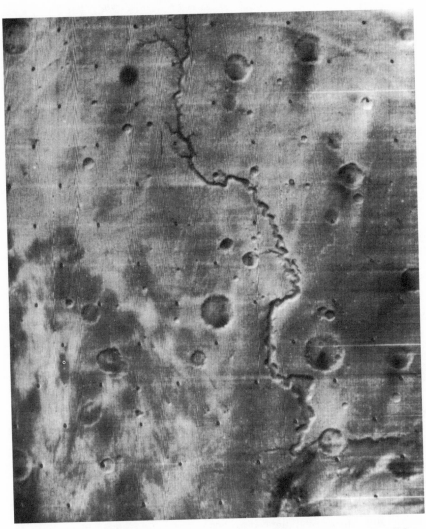

Mariner 9 photo showing what mission scientists thought was an ancient runoff feature. JPL image P12753.tif, courtesy NASA/JPL/Caltech.

to come from regions of chaotic terrain that seemed similar to glacial regions on
Earth, where subsurface heating had led to the collapse of permafrost-laden terrain. Hence the imaging teams concluded that Mars had once had extensive
surface water that had since gone missing.[34] The fascinating scientific questions
then became, what happened to it? And, how could Mars, with its thin wisp of an
atmosphere, have once had a climate capable of supporting liquid water on the
surface?

Mariner 9 ran out of maneuvering fuel and was shut down 27 October 1972, having radically altered the scientific community's beliefs about Mars. The science teams turned to evaluating the reams of data it had sent them. One important line of research that emerged was the effort to conceive of a Martian climate that could have hosted an oceanic surface. One of the first papers to appear on this came from Carl Sagan. In a 1972 paper, Sagan had argued that Mars had two stable climates, one warm, with a surface pressure approaching Earth's, and its current cold age. Then in early 1973, he and two of his students, Brian Toon and P. J. Gierasch, published a paper in *Science* that reported on a study they had done of the effects of orbital obliquity, solar luminosity, and polar cap albedo on Martian climate.[35]

Using a general circulation model of the Martian atmosphere that had been adapted from Yale Mintz's Earthly general circulation model, for the 1973 paper they examined poleward heat transport in the present atmosphere as a means of estimating the overall stability of the present climate. Then, assuming varying amounts of carbon dioxide in the polar caps, at the time thought to be primarily carbon dioxide with a lesser amount of water ice, they had evaluated the effect of larger amounts of atmospheric carbon dioxide on Martian surface temperatures. An all-carbon dioxide atmosphere of 40 millibar surface pressure seemed enough to permit water at the equator; if much larger amounts of carbon dioxide had been available in the remote past, as the giant Martian volcanoes suggested was possible, a water-enhanced greenhouse effect of 30 degrees K could have permitted average equatorial temperatures to exceed water's freezing point.

They then turned to an examination of what could have caused the transfer of sufficient carbon dioxide from the polar caps to the atmosphere and back, starting with orbital variations. Their circulation model indicated that a 15 percent increase in energy absorbed at one pole was necessary to create the necessary conditions. They turned to orbital mechanics for their answer. A Czech scientist, Mitrofan Milankovitch, had argued in a series of papers prior to World War II that Earth's ice ages had been driven by very slow, small changes in the Earth's orbit. His work had not received wide acceptance prior to the 1970s, however. This is partly because the four large glaciations believed to have happened during the past 20 million years were not enough—Milankovitch's analysis suggested that there should have been many more.[36] But Sagan and his colleagues, who were astronomers, not Earth scientists, chose to apply Milankovitch's reasoning to Mars.

The two major components of orbital climate forcing are distance from the Sun and obliquity (the wobble of the spin axis). Planetary orbits are not perfectly circular; instead, they are elliptical and vary slowly. Sagan's analysis of the Martian orbit found it too stable and circular for distance from the Sun to have changed enough to produce the necessary alterations in insolation. The observed obliquity, however, was quite sufficient to raise polar temperatures and produce the necessary outgassing of whatever carbon dioxide was present.[37] This would occur on a cycle of about a hundred thousand years. Hence Sagan's team concluded that "the atmospheric pressure on Mars has been both much larger and much smaller than present values during a considerable portion of Martian history."[38] Current models of solar luminosity changes suggested that the Sun was at its long-term minimum intensity; if correct, its gradual increase over the next several million years could also produce a warmer Mars. Finally, decreasing the albedos of the poles (making them darker and more absorptive) also appeared capable of producing such changes. In fact, their model was most sensitive to albedo change. Only a 4 percent reduction in polar albedo was necessary to produce the same effect as a 15 percent increase of insolation. There were, then, at least three ways to construct a warmer, wetter, and possibly living Mars that were physically realistic.

The three were arguing that only relatively small changes were necessary to bring about a radical shift in Martian climate. This was a direct attack on a long-held belief that the planets, once their catastrophic period of formation was over, were basically stable. Mars not only had changed since its formation, it would change again. Humans had the misfortune to have evolved when Mars (and Earth) were in cold modes (at least Sagan thought it a misfortune, as he would not get to witness a habitable Mars), but a warmer Mars would occur, eventually. Stability of planetary climates could no longer be assumed.

The final Mars mission of the decade was Viking, which consisted of two orbiter/lander pairs. The orbiters were modified Mariners, designed and built by JPL, while the landers were the responsibility of Langley Research Center, which also managed the overall program. The first landing had been scheduled for 4 July 1976, but was postponed when the chosen landing site proved unsuitable; the actual first landing was on 20 July, at a site on the Chryse plain. The landers drew the most attention by far, as they returned the first surface images of Mars and because they were equipped primarily to look for life in the Martian soil. This they did not find. Instead, they found no evidence of organic compounds in the soil. Very high levels of ultraviolet radiation reached the surface of Mars, and this appeared to have destroyed whatever organic matter had once existed. The Martian environment appeared to be self-sterilizing.[39]

The orbiter data, however, painted a somewhat more positive picture of Mars, at least for those willing to satisfy themselves with a Mars that might once have harbored a warmer, wetter past. Barney Farmer, a British infrared spectroscopist who had been hired to establish a spectroscopy lab at JPL in 1966, had designed an infrared-based instrument for the Viking orbiters called the Mars Atmospheric Water Detector. Over the four-year life of the first orbiter, this provided a detailed examination of the variations of water in the atmosphere. A clear pattern emerged from this data. The Martian atmosphere was nearly saturated over the summertime pole but dried almost completely during the winter season. Yet the relationship between the polar cap and water vapor content was not absolute—water vapor appeared to be generated outside the polar regions as well. Writing in the *Journal of Geophysical Research,* Farmer and his colleague Peter Doms argued that the data suggested water ice was bound within the surface material of Mars through all latitudes poleward of 40 degrees.[40] If true, Mars had vastly more water than had been supposed on the basis of the extent of the polar caps, and certainly more than the tiny amount in the atmosphere. But if the regolith planet-wide held water ice, then Martian oceans during the planet's warm periods became conceivable, and it would certainly explain where the water that had formed the surface features went. It was still there, frozen into the subsurface. To optimists like Sagan, oceans' worth of water meant a Mars that might well have had life in the past, and could have it again when the planet returned to its warmer self.

In the time span of a decade, then, Mars in the human imagination went from a living world to a Moonlike dead one, and thence finally to one that might once have had (and might again) have life. The possibility of life was critically dependent on the Martian climate, which had turned out to be currently inhospitable to life as 1970s scientists understood it to be, but good observational evidence suggested had not always been so. This effort to understand Mars drew scientists to make broad analogies to the planet they thought they understood better, Earth, and to construct rather speculative arguments on very limited data. Farmer's argument regarding the ice content of the Martian regolith, while physically plausible, was based on a chain of inferences, not direct measurement; he would not be shown to be correct until 2002, long after he had retired. Similarly, Sagan's model of Martian climate change was based on a good deal of calculation and extrapolation, and on very little data. Actual measurements that suggested that his vision of cyclic changes in the Martian climate might be correct did not appear until 2004, when new surface data from a pair of geological robot landers clearly showed layered sedimentary deposits. The timescale, however, remained unmeasured, and the consensus as of 2005, several years after Sagan's death, was that Mars's warm period had been confined to its first billion years.[41]

Sagan's willingness to engage in what were to many scientists improvable speculations made him highly controversial. But space science itself was increasingly controversial during the decade because of how it was being conducted. The effort to understand the Martian surface, and the related climate that could explain it, caused the Mariner and Viking science teams to engage in interdisciplinary research. Farmer, for example, was by training a spectroscopist originally interested in trying to measure the Sun's infrared spectrum. Water in the Earth's atmosphere made that difficult, and he had had to develop expertise in the infrared signature of water vapor to accomplish his measurements. This, of course, had made him an obvious person to make measurements of Martian water vapor. But he had gone far beyond that, establishing an argument regarding the distribution of water ice in the Martian subsurface.

This was problematic, because the scientific disciplines provide the social infrastructure for the sciences: physicists determined what methodologies were acceptable in physics, established standards of evidence for physics, edited the physics journals, and operated the all-important peer review system for physics, as did geologists for geology, astronomers for astronomy, and meteorologists for meteorology. While the boundaries were permeable to a degree, with physicists colonizing all of these other fields during the period, this was a one-way exchange—meteorologists did not switch to physics. Understanding the complexity of the terrestrial planets' atmospheres seemed to require collaborative research between members of very different disciplines. The mechanisms of science, particularly peer review, did not handle this well. The controversy remained subdued and relatively minor as long as the planetary scientists only studied other planets—no one else much cared what planetary scientists had to say about cold, dead Mars. But this changed when they began to turn their insights, and instruments, on the Earth.

RECKONING WITH THE EARTH

In 1972, Sagan and George Mullen had published a short paper in *Science* that encompassed the climate histories of both Earth and Mars. Their motivation was a thorny problem deriving from stellar astronomy and fusion physics. Specialists in both these fields believed the Sun had been substantially dimmer when it had first formed four and a half billion years ago. Known as the Faint Early Sun hypothesis, this meant that Earth should have been a frozen ball of ice for most of its history. But the geological evidence available in the 1970s was that it had never been one. This required explanation. By the end of the decade, the explanation had become highly controversial. The brilliant independent scientist James E. Lovelock extended Sagan's reasoning about the Earth's physical climate into an argument that life itself was responsible for the Earth's comfortable climate.

Sagan and Mullen began their analysis of the Earth's past climate with a discussion of the greenhouse effect that maintains Earth's current average surface temperature about 30 degrees K higher than it would be in the absence of an atmosphere. They then turned to the Faint Early Sun hypothesis, reviewing the various estimates of the Sun's evolution, which ranged from luminosity changes of 30 to 60 percent. They chose 30 percent to be conservative, and then ran the Sun backward to evaluate the effect on surface temperatures. Average temperatures in their simple radiative model dropped below the freezing point of seawater 2.3 billion years ago. The geologic evidence available in the 1970s was relatively clear, however, that the Earth was largely ice-free as far back as 3.2 billion years, and life clearly existed in the form of algal mats called stromatolites at 2.8 billion years. (In this paper, Sagan accepted some rather controversial microfossils as extending life back to 3.2 billion years as well.) There was thus a substantial conflict between what should have happened and what had actually happened, and their purpose was to explain it.

They believed that only a much stronger greenhouse effect than the current Earth's atmosphere provides was necessary to keep the Earth from freezing under the faint early Sun. The two therefore sought an atmospheric composition that was more in keeping with the Earth's climate history. They needed a gas that was a strong absorber in the mid-infrared, as the carbon dioxide bands were relatively saturated—they accepted as true Syukuro Manabe's calculation that doubling the carbon dioxide content of the Earth's atmosphere, in the absence of any other feedback processes, would result in a 2 degrees C increase in average surface temperature. This was not enough to counter a 30 percent dimmer Sun.

The molecule they found most suitable was ammonia. In the presence of oxygen, this gas is highly reactive and would not last the necessary billion or so years, but the early Earth's atmosphere had no oxygen. There was some evidence from oceanic clay minerals that ammonia had been a minor constituent of the Earth's atmosphere in its youth; there was also evidence from a very famous experiment by chemists Stanley Miller and Harold Urey that showed an ammonia-bearing atmosphere was necessary for the formation of amino acids, the basic constituents of living organisms. They thus postulated that Earth's early atmosphere was a mixture of carbon dioxide, water vapor, and ammonia, with additional minor greenhouse contributions from methane and hydrogen sulfide. Their model required that small amounts of ammonia remained in the atmosphere up to the Precambrian-Cambrian boundary (570 million years before present), but this they found plausible even with the evolution of small amounts of atmospheric oxygen somewhere between 1 and 2 billion years ago. The origin of photosynthetic life in the early Cambrian, and subsequent production of the modern oxygenated atmosphere, would then have resulted in the removal of the ammonia. They concluded that "the evolution of green plants could have significantly cooled off Earth."[42]

British chemist James E. Lovelock, one of the handful of scientists in the late twentieth century able to make a living as an independent consultant, then expanded on this argument to contend that life itself was responsible for Earth's comfortable climate. Lovelock had started serving as a consultant at JPL in the early 1960s, working with a group of scientists on the question of life detection on other planets, specifically Mars. In his popular book *Gaia: A New Look at Life on Earth*, he recounts that he wound up at odds with his research group fairly quickly. They wanted to look for signs of life in the soil (on Earth, of course, soil teems with living things), but Lovelock disagreed with that approach. He thought that the most obvious place to look for telltale signs of life was in the atmosphere. His key insight was that a planet with abundant life would have an atmosphere in a state of extreme chemical disequilibrium.[43] In other words, it would contain chemicals that would long ago have been stripped out by reactions with the solid surface or have been destroyed in reactions with other gases. This led him to reinterpret the history of Earth's atmosphere, and the climate that it regulates.

Lovelock began from an idea that the most general function of living organisms was to reduce entropy. Entropy, a thermodynamic quantity that had to continually increase generally, but could be reduced locally with the consumption of energy, cannot be directly measured. And in any case, living processes were not the only natural processes known to reduce entropy. Hence this was not a useful formulation of the problem. Instead, he started to look at living processes as factories, which reduce the entropy of the materials going into them while increasing the entropy of their surroundings via their waste products. Waste products were the key. Just as factories deposit some of their waste in the atmosphere, so did living things. Lovelock, early on working with Dian Hitchcock and later Lynn Margulis, started looking at the gaseous waste products produced by life: oxygen, carbon dioxide, nitrogen, ammonia, and methane.[44]

When Lovelock was writing, the Earth's atmosphere consisted of 21 percent oxygen, by volume, 78 percent nitrogen, with the remaining 1 percent made up of various trace gases. Carbon dioxide, for example, was 0.03 percent of the atmosphere's volume. Yet in terms of chemical equilibrium, this was highly improbable. Over the Earth's billions of years of existence, oxygen, a highly reactive gas, would have been extracted by chemical weathering of surface material and be, as it was on Venus and Mars, undetectable. Similarly, the most chemically stable form of nitrogen was in the form of nitrate ions in the oceans, not as a noble gas in the atmosphere. Hence in a chemically stable version of the Earth, the atmosphere would be mostly carbon dioxide, as were the atmospheres of Mars and Venus, and contain neither nitrogen nor oxygen.[45]

Lovelock was not the first to recognize the unstable nature of the Earth's atmosphere. Rather, he was building on a minority view in geochemistry. In the majority view, the Earth's unlikely atmosphere was explained as a product of planetary outgassing, with the oxygen provided by the photodissociation of water vapor in the upper atmosphere. The resulting hydrogen, as it had on Venus, would then escape into space, leaving oxygen free in the atmosphere. Yet this view did not comport with evidence available by the late 1960s regarding the dissociation rate of water in the upper atmosphere, or with the rates of consumption of oxygen in the weathering processes. Nor did it square with the interplanetary view. Lacking both a magnetic field and an ozone layer, Venus experienced much larger high-energy fluxes at the top of its atmosphere than Earth did, which would lead to a higher dissociation rate and more rapid hydrogen escape and oxygen production. And, of course, whatever water Venus had once had was gone. But there was no measurable residual oxygen. Hence the majority view no longer explained the available evidence.

Lovelock credits Swedish chemist Lars Gullen Sillen as being the first to question the control mechanism for oxygen, nitrogen, and carbon dioxide.[46] But Lovelock extended Sillen's point to argue that life itself maintained the relative abundances of these gases. Photosynthetic plants consumed carbon dioxide and released oxygen, while animal life consumed oxygen and released carbon dioxide. The trace amounts of methane in the atmosphere, about a billion tons, were already well-known to be a mostly biological byproduct.[47] The presence of these gases was, for Lovelock, the ultimate proof of life, and in a 1965 article for *Nature*, he set out his argument that the atmosphere was the place to search for Martian life.[48]

But in the process of thinking about how to find life, he began to reconceive the Earth as a single, self-regulating organism. After a 1969 presentation in Boston, he began working with Lynn Margulis, then Sagan's wife, to refine and flesh out the idea. They eventually published two important articles in 1973 and 1974 in which they described their hypothesis. They started out by telling a story about the Earth's climatic evolution that began with Sagan and Mullen's model of the Earth under the faint early Sun.[49]

Sagan and Mullen had argued that the early Earth had needed ammonia in its atmosphere to compensate for the much dimmer early Sun. Margulis and Lovelock looked at the problem of the faint early Sun somewhat differently, however, asking how the Earth's temperature had remained essentially constant during 3 billion years of gradual solar intensity increase. Their reading of the geologic evidence suggested that the Earth had not varied in globally averaged temperature by more than 10 degrees C in the past 3.5 billion years. During that time, solar luminosity had increased by 40 to 60 percent. Yet a virtually unchanging climate seemed highly unlikely given what was known about the solar system at the time. Following Sagan and Mullen, Manabe, and Rasool and De Bergh, they accepted that only a 10 percent change in solar luminosity was necessary to provoke either a runaway greenhouse effect and evaporation of the oceans, or alternatively bring about an iceball Earth. But that had not happened. Instead, despite the slowly warming Sun, the Earth's temperature had remained effectively constant. This is what Lovelock and Margulis sought to explain.

They began their story with the need for an early ammonia-laden atmosphere to keep Earth largely ice-free, and then presented evidence for the very early evolution of life. The 1960s and early 1970s had seen a number of important discoveries of microfossils in some of the most ancient rocks still accessible on the Earth's surface, extending the history of life back to about 3 billion years. More important, these discoveries had expanded the variety of early life. The discovered fossils were prokaryotes (blue-green algae), and these displayed what Margulis and Lovelock called "metabolic versatility." They existed via a wide variety of different metabolic processes, and as a result generated a variety of different waste products. They were also widespread, and certain types formed large structures—the stromatolites Sagan had mentioned. These were distributed globally by the late Precambrian, and had been in existence for about 1.2 billion years. One of the most common metabolic cycles for prokaryotes on the present Earth results in the production of oxygen from carbon dioxide; Margulis and Lovelock argued that these vast stromatolite beds were the likely source of the oxygen that gradually became present in the Earth's atmosphere during the Precambrian.[50]

The transition from an anoxic, ammonia and carbon dioxide atmosphere to the present oxygen-rich one should have destabilized the Earth's climate by dramatically reducing the atmosphere's greenhouse capacity, but it had not. This led them to argue for active control of climate. They used the metaphor of a planetary engineer, whose employer had assigned him a planet and directed him

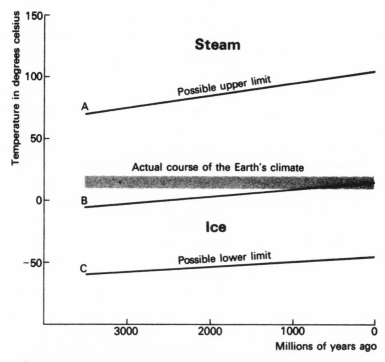

Fig. 1. The course of the Earth's average temperature since the beginning of life 3.5 aeons ago is all within the narrow bounds of the horizontal lines between 10° and 20°C. If our planetary temperature depended only on the abiological constraints set by the sun's output and the heat balance of the Earth's atmosphere and surface, then the conditions of either the upper or lower extremes, marked by the lines A and C, could have been reached. Had this happened, or even if a middle course were followed, line B, which passively goes with the sun's heat output, all life would have been eliminated.

Margulis and Lovelock's demonstration of Earth's climate stability.
Reprinted from James Lovelock, *Gaia: A New Look at Life on Earth*, revised edition (Oxford University Press, 1995), p. 20.

to maintain a specific set of temperature and acidity specifications for several billion years. Then they reviewed the tools available to the engineer for temperature control: control of the planet's radiation balance, its surface emissivity, the composition of its atmosphere, and the distribution of suspended particulates. As seemed to be the case with Mars, small changes in planetary albedo could effect sizeable changes in temperature. The engineer could change this by, for example, darkening the polar regions, something Harry Wexler had studied briefly during the 1950s. Similarly, organisms could impact albedo by changing their colors, by changing the color of the sediments they trapped and fixed, and even by altering the color of snow and ice. The same was true of surface emissivity. The global distribution of stromatolites would have allowed them to alter the Earth's overall emissivity through changes in their surface porosity and composition, for example.

Organisms also altered the chemical composition of the Earth's atmosphere, impacting its radiative qualities. Nearly all organisms either consumed or produced carbon dioxide. Ammonia, the gas Sagan and Mullen had proposed as maintaining the Earth's warmth under the faint early Sun, was also a "very active product of microbial metabolism." It was a waste product of many organisms, and was also consumed by nearly all bacteria and fungi. Hence, while the amount of it in the current atmosphere was vanishingly small, this was because virtually all of the billion or so tons produced each year by biologic processes were also being consumed. To Margulis and Lovelock, microbial consumption explained the near-disappearance of ammonia from the early atmosphere; ammonia-fixing microbes would have thrived on the young Earth, and as they drew down the atmospheric reservoir of ammonia would have been increasingly pressured into environments where they would be in contact with ammonia producers. This would have had a large radiative impact on the Earth's atmosphere, as under the faint early Sun removal of ammonia at too high a rate would have sent the Earth into an iceball mode from which it could not recover. Indeed, their reading of the Earth's chemical history suggested a crisis for its thermal equilibrium in the late Precambrian, but the geologic record did not seem to contain evidence of one.[51] They took this as evidence of active control of the climate by biologic actors, postulating that selection pressures on local populations produced a response to the cooling Earth that eventually counteracted it.

The need for an active control agent led the two to conceive of the Earth as a single organism they named Gaia, for the Earth goddess of the ancient Greeks (also known as Ge, from which derived the names for geology and geography).[52] Greek philosophy had been based on the notion of a balance of nature, which was essentially what their vision of a self-regulating Earth implied. This naming was the first source of controversy for their hypothesis, because it imbued the Earth with a quasi-religious mysticism that did not comport well with the belief systems of most of their scientist peers—particularly physical scientists. Their metaphor of a planetary engineer, while intended to help simplify the explanation, was not well-received either. It implied a conscious regulator, which was not what they were arguing. In the same article they deployed this metaphor in, they also explained that Neodarwinian mechanisms of selection were the means by which planetary control was maintained.[53] This did not protect them from vocal criticism by their biologist peers, who eventually forced them to reformulate the hypothesis with more care toward the details of the current evolutionary synthesis.

In their seminal 1974 *Icarus* article, Margulis and Lovelock commented that "probably a planet is either lifeless or it teems with life. We suspect that on a planetary scale sparse life is an unstable state implying recent birth or imminent death."[54] The combination of living processes and evolutionary ones was so powerful, in their view, that organisms could remake a planetary environment to facilitate their own spread. Hence over the eons of deep time, life would take over a planet, make it more suitable, and eventually be found everywhere. In this view of life, there were no marginal environments. Life would be found in any local environment of an inhabited planet—or nowhere. This did not bode well for NASA's dreams of finding life on Mars, or anywhere else in the solar system. If life existed at all off Gaia, it would be readily apparent from its impact on the composition of the Martian or Venusian atmospheres. Radiotelescopes, and telescope-aided infrared spectroscopy, were all one needed.

Stripping away the mysticism inherent in the Gaia label, the two were presenting a view of the Earth that could be grasped by systems engineers, a profession that specialized in (nonliving) feedback control systems. In his 1979 popular exegesis of the Gaia hypothesis, Lovelock devoted a chapter to cybernetic theory, the mathematical basis for feedback systems. For Earth scientists, Margulis and Lovelock were presenting a view of the world that required examination of complex, interlocking feedback loops. Some of these feedback loops, such as the hydrologic cycle that was of great interest to meteorologists, were primarily physical. At least in the 1970s, evapotranspiration was perceived as only a minor participant in the water cycle. Other obvious cycles, such as the carbon cycle, were both physical and biological. Understanding them required the very interdisciplinary research that the American scientific community did not consider serious science, and was not set up to foster. In fact, they expressed the hope that their hypothesis would change that particular scientific dynamic.[55]

Taken in a larger view, Lovelock and Margulis were arguing that the Earth's climate had been fundamentally altered by the evolution of life. Living things affected the chemistry of the atmosphere, altering its composition. Changing the atmosphere's chemistry affected its radiative characteristics, and over geologic time, these biogenic changes had produced the Earth's current comfortable climate. Life, they were arguing, had achieved the ability to make planetary-scale changes eons ago. This early life, however, had done so unwittingly. They did not weigh in on whether humans had achieved this ability as well, but in these papers did not need to. Lovelock, in fact, had already demonstrated that human emissions of chlorofluorocarbons (CFCs) had changed the composition of Earth's atmosphere, but he did not yet perceive the consequences. And Charles David Keeling, by the early 1970s, had conclusively demonstrated that human emissions of carbon dioxide were also changing the composition of the atmosphere. As carbon dioxide was a greenhouse gas, humans had clearly achieved the power to change the Earth's climate.

NASA Chief Scientist Homer Newell remarked in his 1980 memoir that space science had proven to be integrative.[56] Planetary science had drawn on many scientific disciplines to develop new knowledge about the other planets during the 1960s and 1970s. At the same time, that knowledge had informed thinking about the Earth and its processes. The Earth is, at least from the standpoint of planetary scientists, just one of the several terrestrial planets in the solar system; its processes are not governed by different rules. Planetary research had also forced the scientific community to begin placing Earth in the context of its sister rocky planets Mars and Venus, and begin to think about why it had turned out so differently from what they believed was a similar beginning. This was one thread of the increasing interest in Earth's climate within the scientific community; Earth was a great deal closer than the other terrestrial planets, and due to the presence of water and life, it was also a good deal more complex.

The question of climatic evolution was, at one remove, a question of chemistry. On Mars and Venus, non-biologic chemical processes had produced very different outcomes; on Earth, however, the biosphere clearly played a substantial role in making the Earth chemically different from the other terrestrial planets. What the biosphere's role was could not be quantified, at least during the 1970s. But the mere claim that the Earth's climate was actively, if not consciously, regulated by life itself was highly influential. It caused great controversy in the scientific community, which was not prepared to accept that the thin green layer on the planet's surface could have such great impact.

It also caused great controversy in the public arena, but for a slightly different reason. Humans had achieved the power to fundamentally alter the conditions of life on Earth. The consequences of this ability were not yet known, and were not yet knowable, given the state of knowledge of the early 1970s. But recognition that humans might alter the atmosphere's chemistry enough to change the Earth's climate produced rapid demands by scientists and by political activists alike to figure out what those consequences might be.

CHAPTER FIVE

NASA Atmospheric Research in Transition

By studying and reaching a quantitative understanding of the evolution of planetary atmospheres we can hope to be able to predict the climatic consequences of the accelerated atmospheric evolution that man is producing on Earth.

—W. Wang et al., 1976

One major factor in the transition of NASA's meteorological satellite program from an applications program into a more scientific one was a change in the American political scene: environmentalism. The 1950s and 1960s had witnessed a few dramatic air pollution events that had resulted in mass deaths and many more that had generated clear, if apparently only temporary, hazards to citizens' health. Local and regional antipollution groups had formed in response, and then they "went national" because they had simply been unable to overcome resistance to regulation at lower levels of the nation's political system. Further, of course, they could not avoid the reality that the air was in constant motion. Controlling air pollution in one state meant nothing if its neighbors chose not to impose regulation.[1]

In the late 1960s, these local antipollution groups found common cause with the national wilderness conservation groups. This alliance proved very influential, achieving several major pieces of legislation. President Lyndon B. Johnson signed an Air Quality Act in 1966, and his successor, Richard M. Nixon, signed a series of stronger environmental laws: the Endangered Species Conservation Act in 1969, and in 1970 the National Environmental Policy Act and the Clean Air Act. Nixon also created the Environmental Protection Agency (EPA) and the National Oceanic and Atmospheric Administration (NOAA) the same year by executive reorganization.[2]

NASA played its own, if unwitting, role in promoting environmentalism. The photographs returned by its geosynchronous satellites and later by the Apollo astronauts were the first images ever taken of the whole Earth. NASA's imagery of the "Big Blue Marble"—the name given to a 1970s world news show aimed at children—presented the Earth as a spaceship floating in a black void. These images were politically powerful, depicting the Earth without the arbitrary political boundaries that were familiar to most citizens and revealing its finite, interconnected nature. Environmentalists incorporated the images into a worldview, using them to argue for a politics based upon internationalism and long-term sustainability. Widely deployed by Earth Day 1970, they became the root of a global environmental consciousness.[3]

In the scientific community, this larger public interest was paralleled by interest in disentangling the connections between the biosphere, atmosphere, and climate regulation. The planetary atmospheres research carried out during the first decade of the space age had made obvious that understanding these connections was important. And by the early 1970s, it was already clear to the leaders of the atmospheric science community that humans had the ability to alter the chemistry of the atmosphere. But it was not clear what the impacts of those changes were, or might be in the future. Driven by both highly public controversies and the scientific community's interest in new research questions, NASA began to construct new research programs around them. Hence, the confluence of environmental concern and planetary science produced a shift in NASA's Earthly atmospheric research program toward environmentally relevant research.

NEW DIRECTIONS IN ATMOSPHERIC RESEARCH

In July 1970, MIT had sponsored a workshop aimed at challenges the new emphasis on environmental quality posed for the scientific community. Organized as part of national planning for the 1972 United Nations Conference on the Human Environment, the Study of Critical Environmental Problems (SCEP) rapidly became a scientific classic. It represented an explicit acknowledgment by leading scientists that humans had achieved the capacity to alter the global environment.

The study's report, known as *Man's Impact on the Global Environment: Report of the Study of Critical Environmental Problems*, summarized what was known and, more important, what was not known about humanity's impact on natural systems.[4]

Meteorologist William Kellogg chaired a working group on climatic effects that a number of human activities might have on the Earth's climate.[5] Kellogg's group included Morris Tepper from NASA, Robert Fleagle from the University of Washington, Joseph Smagorinsky from the Geophysical Fluid Dynamics Laboratory (GFDL), Charles David Keeling from the Scripps Institution of Oceanography, Hans Panofsky from Penn State, and several others. Their tasking included reviewing the scientific knowledge about the potential impact of increasing carbon dioxide concentration in the atmosphere, the effect that a fleet of supersonic transports (SSTs) might have on climate and on the stratospheric ozone layer, and the possibility that aerosols—particulate matter in the atmosphere—might act to cool the Earth by reflecting sunlight away.

The SCEP's meeting was in part provoked by the early evidence that humans were changing the characteristics of the global atmosphere. During the International Geophysical Year (IGY), Wexler and Roger Revelle had arranged for measurements of carbon dioxide to be made by Keeling at the Mauna Loa Observatory in Hawaii. While Keeling had troubles sustaining support for this research over the ensuing decade, by 1970 the trend was already quite clear: carbon dioxide concentration as measured at the observatory was increasing. His measurements also compared very favorably with other sets of measurements made elsewhere. Due to the well-known fact that carbon dioxide is a greenhouse gas, the phenomenon of human-induced global warming became a matter of public debate.[6]

The numerical modeling community that had developed around weather modeling had begun experimenting with the use of models in climate studies during the 1960s. There were various types of models, used for various purposes. Atmospheric column models considered the impact of changes in various feedback processes on a single column of air, permitting detailed study of certain processes without high computing costs. Soviet researchers had used such a model in 1966 to examine the impact of removing the Arctic ice cap on Arctic surface temperatures, for example. Radiative-convective models permitted examination of radiative transfer, convection, and poleward transport within a two-dimensional sliver of atmosphere that stretched from equator to pole. These models achieved computational efficiency by assuming the average temperature at any point on a circle of latitude is essentially the same as any other point on that circle, which is roughly true of the real earth (topography, which the models did

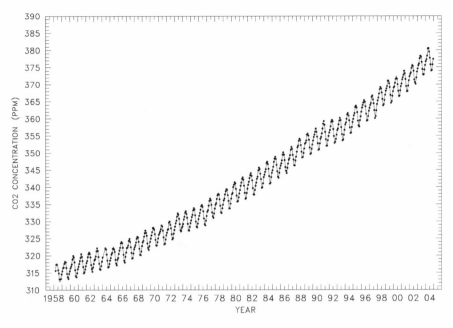

Keeling's record of average carbon dioxide concentration measurements at Mauna Loa. This version is updated to 2006. Courtesy: Ralph Keeling, SIO.

not contain, notwithstanding). At this level of abstraction, there was no need to waste computational resources simulating horizontal transport.

Finally, more complex, three-dimensional general circulation models derived from numerical weather prediction models had also been developed in several places during the 1960s to enable simulation experiments with the global climate. This was possible because the weather prediction models settled into representations of Earth's current climate if left to run long enough, regardless of the initial conditions they had been started with. Hence, while as weather models they were sensitive to initial conditions, as climate models they were not. They reflected both the short-term instability of the always-changing daily weather and the long-term stability of average weather patterns—climate.

To experiment with the Earth's climate, researchers could alter one of the inputs or processes within the model. At GFDL, Syukuro Manabe and Richard Wetherald had adapted Smagorinsky's weather prediction model to examine the sensitivity of the Earth's climate to several possible changes: in the solar constant, the amount of energy received from the Sun; in carbon dioxide concentration; in the stratosphere's water vapor content; or in the distribution of albedo. Their 1967 paper concluded, for example, that quintupling the concentration of water vapor in the stratosphere would raise surface temperatures by an average of 2 degrees C, while doubling the concentration of carbon dioxide would raise average surface temperatures between 1.3 and 2.3 degrees C, depending on various distributions of water vapor.

Their model contained many simplifications. It considered only steady-state conditions and did not deal with transitional periods. It did not have an ocean, which would act to slow the warming process through its vast heat capacity, perhaps by millennia. It used the average latitudinal values of albedo as they were then known from ground and satellite studies, and could not simulate albedo changes from, for example, changing patterns of cloudiness caused by warming. Hence the model was not the end product of a long-standing line of research that had replaced simplifying assumptions with observations and resolved all the potential ambiguities. Instead, it represented the beginning of a new research field.

These early results were controversial within the scientific community and on Kellogg's review panel, and many years later in an article that represents his own coming to terms with climate science, Kellogg explained one aspect of the controversy.[7] Climate warmers in the SCEP panel had argued that increases of radiatively significant trace gases, including carbon dioxide, chlorofluorocarbons (CFCs), and methane, would inevitably produce warming at the Earth's surface. A group of climate coolers, however, had focused on the reflective impact of aerosols—particulates including both anthropogenic pollutants and naturally occurring dust, soot, and the like. They had argued that increasing anthropogenic aerosols from industrial emissions would have a net cooling effect because they would reflect solar radiation back into space. As aerosol concentrations in the atmosphere increased, less and less solar radiation would reach the Earth's surface, and the Earth system would cool. The conflict in the group could not be resolved because while both effects would occur, no one knew which one would be dominant. Both mechanisms were physically plausible, but no one had made the necessary measurements to demonstrate which effect would be more powerful in the real atmosphere. Aerosols had also not been made part of the climate research models. In addition to lack of information on aerosol effects, researchers had insufficient computing power to simulate them in any case. Because of this divide, his working group could only produce recommendations calling for more research into greenhouse gases, aerosols, and climate models.

Kellogg's group also considered another question that in 1970 was of great controversy: whether SSTs, airliners that could fly faster than the speed of sound within the stratosphere, would alter the Earth's climate. There were a number of ways this might happen. All internal combustion engines produced water vapor as an exhaust product, and a fleet of SSTs would put millions of pounds of water vapor into the normally dry stratosphere. This water vapor might freeze into contrails, clouds of ice crystals that would reflect sunlight and thus induce cooling; conversely, increasing the water vapor content of the stratosphere could warm it, altering the thermal balance between the troposphere and the stratosphere and producing surface warming. Manabe and Wetherald had used their early climate model to simulate this second scenario as well, finding that a quintupling of stratospheric water vapor might raise average surface temperature about 2 degrees C. However, this amount of increase seemed unlikely to Kellogg's review committee, and they concluded that a fleet of five hundred SSTs would raise stratospheric water vapor content globally only slightly, from 3.0 parts per million to a projected 3.2 parts per million.[8]

The MIT study had been aimed at *global* environmental change, a subject that virtually required the space agency's participation, and Morris Tepper had been invited to take part in it as a member of Kellogg's climate panel. Just as space-based meteorological measurements were necessary to produce global meteorology, space technology was necessary to begin assessing the human impact on the environment. Yet the ability to measure atmospheric constituent gases and particles from space did not exist yet, and for these research areas the monitoring working group advocated the improvement of ground- and aircraft-based measurements. For gases, in situ measurements seemed best, while for particulates, a prominent new measurement technology was lidar. Like radar, lidar measured backscattered radiation, but lidar instruments transmitted energy at much smaller wavelengths. Lidar instruments, however, were far too large for space-borne use at this point in their development. A lidar system developed at the Langley Research Center between 1967 and 1969 utilized a 48-inch mirror to capture enough reflected radiation to enable measurement, much too large to squeeze onto any reasonable satellite. Hence the SCEP study only included a single recommendation for near-term space instruments, calling for a program to develop a follow-on to Vern Suomi's heat budget instrument that would be able to measure albedo changes on the order of 1 percent per ten years.[9]

The emphasis on the Earth's heat budget derived from research done by Thomas Vonder Haar, one of Vern Suomi's graduate students. Using instruments similar to Suomi's original experiment placed aboard the civilian Television-Infrared Observations Satellite (TIROS) satellites and aboard classified Defense Meteorological Satellite Program (DMSP) satellites, Vonder Haar had been able to construct seasonal maps showing the latitudinal variation of the Earth's energy budget.[10] In his data, the tropics received more energy from the Sun than the region radiated back into space (thus the tropics were the Earth's primary energy source), while the polar regions were energy sinks, radiating more energy to space than they received from the Sun. This was not at all surprising to meteorologists, or to oceanographers, for that matter. The Earth's weather patterns and ocean currents were universally believed to be energy conveyors that functioned to move energy from the tropics to the poles.

What was surprising about Vonder Haar's result was that his data showed the Earth was warmer and darker than the general scientific consensus held. In 1957 Julius London had published an albedo atlas of the Earth based on ground and airborne measurements, and he had developed an estimate that the Earth's over-all albedo was 35 percent.[11] In other words, the Earth reflected away 35 percent of the incoming solar radiation, absorbing the remaining 65 percent. Vonder Haar's data, however, gave a global average of 30 percent. While no one had expected London's estimate to be perfect, the size of this difference was very surprising. The 5 percent albedo difference translated into 40 percent more energy absorption by the Earth than expected.

The following year, MIT sponsored a smaller international conference to review the SCEP report. At this panel on "Inadvertent Climate Modification" thirty atmospheric scientists, climatologists, and oceanographers gathered at Wijk, outside Stockholm, for three weeks during June and July.[12] This group added detail to the research and monitoring agendas suggested by the earlier document. In addition to improved, and ongoing, measurement of the Earth's energy budget, they recommended monitoring global cloudiness patterns, the distribution of snow, ice, and atmospheric particulates, and sea surface temperatures, and determination of the solar constant (or, alternatively, determining its variation). They also recommended development of means to monitor the temperature distribution of the upper ocean layers, and of joint atmosphere-ocean climate models, recognizing that global atmospheric models alone could not adequately define all of the processes involved in regulating the Earth's climate.

Immediately after the Stockholm conference, NASA's Langley Research Center hosted a conference on the potential uses of remote sensing technologies in carrying out pollution- and climate-related research. Morris Tepper chaired the conference, while Will Kellogg, Vern Suomi, and oceanographer George Ewing headed working groups on detection of gaseous and particulate constituents of the atmosphere and water pollution. The resulting study detailed the techniques that might be used for measuring a wide variety of atmospheric substances, including oxides of nitrogen, hydrogen sulfide and sulfur dioxide, ozone, and fluorocarbons. It also specified the accuracies that would be necessary for effective monitoring. Published as *Remote Measurement of Pollution*, this conference report marked the beginning of NASA's transition toward environmental research.[13]

These three reports, taken together, represent the beginnings of sustained scientific concern that humans had obtained the power to change the fundamental conditions under which life on Earth existed. The global view offered by the space program had brought with it the price of recognition that modern, Western industrial civilization had the ability to permanently alter the global ecosystem—and quite possibly had already done so unwittingly. But it was not directly the climate controversy that got NASA involved in environmental research. Instead, the ozone conflict that broke out over SSTs, its own Space Shuttle, and CFC use was the trigger for NASA's involvement.

INSTITUTIONALIZATION OF ATMOSPHERIC CHEMISTRY

The pollution study done at Langley in 1971 had its roots in the controversy over whether SSTs would cause climate change. Kellogg's panel had concluded that they would not, but a new issue, ozone depletion, was raised shortly after the SCEP study was published. NASA leadership decided to apply the agency's skills to determining whether SSTs would destroy the stratospheric ozone layer, initiating a process of institutionalization of stratospheric research.[14] By the end of the decade, NASA was the lead agency for answering policymakers' questions about stratospheric ozone depletion.

The existence of the Earth's ozone layer was well-known within the scientific community, if not within the general public, by the time the SST controversy began. In 1881, Irish chemist W. N. Hartley had hypothesized that high concentrations of ozone existed at high altitudes and served to block ultraviolet radiation, and in 1913 French physicist Charles Fabry proved Hartley correct. In 1926, English physicist Gordon Dobson had designed an optical instrument to precisely measure these high-altitude ozone concentrations, and he established a small but worldwide network of observing stations during the ensuing decade. Finally, in 1931, geophysicist Sydney Chapman had established the mechanism by which ozone was created in the stratosphere.[15] He believed that oxygen molecules, which consist of two oxygen atoms bonded together, would be blasted apart by solar radiation. The resulting highly reactive oxygen atoms would then bond to other oxygen molecules, forming ozone. By 1940, therefore, the small number of scientists interested in the upper atmosphere believed they had a basic understanding of the ozone layer.

During the 1960s, stratospheric research received a boost from the Atomic Energy Commission (AEC). The AEC's above-ground nuclear weapons testing injected vast amounts of detritus into the stratosphere, including nitrogen oxides, and the agency sought understanding of how this affected the stratosphere. Further, the testing regime had produced an interesting conundrum. The weapons tests, AEC scientists thought, should have been producing a slight, but measurable, decrease in stratospheric ozone concentration, but data from the ground-based Dobson ozone monitoring network suggested that ozone concentration was actually increasing. The increase stopped in 1968 and concentrations stabilized for a few years, but this seemingly strange episode indicated that the stratosphere was more complex than scientists had previously thought.

Shortly after MIT's 1970 SCEP report was published, a scientist at Boeing Scientific Laboratories, Halstead Harrison, responded to the idea that water vapor from SST exhaust might change the Earth's climate with an article published in *Science*. Using new reaction constants derived by meteorologist Paul Crutzen and a computer model of the atmosphere he had adapted from one designed for Mars, Harrison argued that the water vapor produced by a fleet of 850 SSTs would deplete the ozone column by 2 to 3.8 percent. Most of this reduction would occur in the Northern Hemisphere due to its high concentration of air routes, producing a temperature rise on the Earth surface of about 0.04 degrees K.[16] This amount of change was trivial—indeed, it was unmeasurable, and hence it corroborated the SCEP report. One could not separate such a tiny change from all of the Earth's natural variations.

But Harrison's article unexpectedly opened a new and far more controversial issue. Atmospheric physicist James E. McDonald of the University of Arizona, a member of the National Academy of Sciences' Panel on Weather and Climate Modification, read Harrison's article and found his admission that ozone depletion *would* occur startling. McDonald had accepted his own panel's earlier conclusion that significant ozone depletion would not occur, and thus there were no biological effects to worry about. But Harrison's argument caused him to reconsider that conclusion. By 1970, medical scientists believed that ultraviolet radiation caused certain kinds of skin cancer.[17] If depletion of the protective ozone layer that Harrison predicted did occur, then skin cancer incidence would increase significantly. Indeed, McDonald believed there was a sixfold magnification factor. Each 1 percent reduction in ozone concentration would produce a 6 percent increase in skin cancer occurrence.

The Transportation Department's SST office had formed a climate impact study committee, and McDonald first approached this group with his newfound concern, but he was rebuffed. So McDonald made his analysis known to SST opponents—exactly how is unrecorded—and Representative Henry Reuss (D, Wis.) invited him to testify in House hearings on the SST held on 2 March 1971. The seriousness of McDonald's testimony was undermined, however, by his side-interest in UFOs and extraterrestrial visitation. McDonald had argued for what historian Steven Dick has called the extraterrestrial hypothesis, that the best explanation for UFOs was that they were from other planets. He had been quite outspoken during the late 1960s about this, earning the enmity of some of his colleagues and, of course, making himself an easy target of ridicule.[18]

The political expediency of attacking McDonald's credibility to undermine his scientific testimony in the public's eyes worked, but only briefly. The scientific community largely accepted that McDonald's assertion of skin cancer risk was correct *if* ozone depletion occurred. The controversy that exploded throughout the Western world in the years following McDonald's testimony was thus over whether the SST, or any other pollutant, would damage the ozone layer. The next round in what Lydia Dotto and Harold Schiff have called the "ozone war" was fired by Berkeley professor Harold Johnston, a specialist in ground-level ozone chemistry, after he attended a Transportation Department conference in Boulder, Colorado, in mid-March.[19]

Johnston had been invited by the conference's organizer, University of Wisconsin physicist Joe Hirschfelder. A member of the Transportation Department's stratospheric impact panel, Hirschfelder had come to believe that the Transportation Department was trying to bias the panel by limiting its data to that provided by Boeing. He demanded access to other sources of data, and eventually the department relented.[20] Hirschfelder was allowed to organize a conference and invite a small number of university scientists, including McDonald. The group convened on 18 March 1971, the same day the House was voting on whether to continue SST funding. Johnston rapidly became annoyed at the proceedings. In addition to a very tense atmosphere, the conferees seemed to accept the conclusions of the SCEP study that nitrogen oxides would not be a significant cause of ozone depletion. Johnston's knowledge of ozone chemistry suggested to him that this was wrong.

Johnston spent much of that night working out calculations showing that these gases would be far more potent ozone scavengers in the stratosphere than the group expected, and in the morning he handed out a handwritten paper that estimated NO_x–derived depletion of 10 to 90 percent. His effort did not help much, and at an impromptu "workshop" organized by Hirschfelder in the men's washroom and held in a small conference room that afternoon the discussion stayed on water vapor–induced depletion.[21] Participant Harold Schiff, a scientist from York University in Toronto, recalled that Johnston lashed out at the group for ignoring the NO_x reaction, and this finally prompted the other chemists to grapple with Johnston's idea — if only in self-defense. Schiff reported that the central question in that afternoon's argument was that no one knew what the stratosphere's natural concentration of nitrogen oxides was, as no one had ever measured it. Without that basic piece of information, one could not produce a credible estimate of NO_x–induced ozone depletion. If the stratosphere already had a high concentration of nitrogen oxides, then the amounts injected by a fleet of SSTs would not matter. If, on the other hand, the stratosphere had none at all, SSTs would be devastating. Johnston tended toward the "devastating" end of the spectrum, while his colleagues were not willing to make that leap. The conference thus ended with a recommendation that more research was necessary to determine whether or not the SST was a threat to the ozone layer.[22]

Johnston was not satisfied with that result. Back in Berkeley, he turned his calculations into a paper, which he sent out to several colleagues, including Hirschfelder, three of his colleagues at Berkeley, and David Elliot of the National Aeronautics and Space Council on 2 April 1971. On 14 April, Johnston sent a substantially revised version to *Science*. The journal's editors, in turn, sent it out for peer review in keeping with its policy, and the reviewers recommended that Johnston rewrite it. They deemed the paper unsatisfactory for two reasons. First, Johnston had not cited an article by Paul Crutzen, then working in Sweden, that indicated the stratosphere might have a high sensitivity to nitrogen oxides.[23] Second, Johnston's tone was unacceptable. Scientists were supposed to be coldly dispassionate in their writing, in order to appear unbiased and objective. Johnston was not, contending that an SST fleet could cut ozone concentration over the Atlantic corridor in half and allow enough radiation to reach the Earth's surface to cause widespread blindness. Publication of Johnston's paper was thus delayed while he rewrote it.

But reaction to it was not. Johnston's preliminary 2 April draft had been leaked to a small California newspaper, the *Newhall Signal,* causing the University of California's Public Relations Office to release it. Sensational summaries of Johnston's draft sped east on the wire services and on 17 May, two days before a Senate vote on an attempt to restart the SST program, the story made the *New York Times*. A lengthy follow-up article by the *Times'* famed science editor Walter Sullivan on 30 May put the issue solidly before the public.[24] According to Johnston, he reported, SSTs, no matter who built, sold, and operated them, could have devastating consequences for humanity.

Sullivan's article on Johnston's findings also drew senatorial attention. Senator Clinton Anderson (D, N.Mex.), who chaired the Senate Committee on Aeronautical and Space Sciences, wrote to NASA administrator James Fletcher about Sullivan's piece on 10 June.[25] After summarizing Sullivan's description of the NO_x reaction, Anderson commented that "we either need NO_x-free engines or a ban on stratospheric flight." Then he laid the ozone issue squarely on the space agency. Paraphrasing the NASA charter, he reminded Fletcher that it was NASA's job to ensure that the aeronautical activities of the United States were carried out so as to "materially contribute to the expansion of human knowledge of atmospheric phenomenon" and to "improve the usefulness, performance, safety, and efficiency of aeronautical vehicles." In short, figuring out the ozone mess was NASA's job, and Anderson encouraged Fletcher to establish a program to find out whether or not stratospheric flight was the kind of threat Johnston had made it out to be.

Science published Harold Johnston's paper in its 6 August issue, producing a public resurgence in the issue and substantial concern in the White House that the problem would not go away without some concrete action, especially after a National Research Council (NRC) analysis concluded that the hypothesis had merit.[26] And because the ozone controversy was framed in terms of human health effects, not mere environmental damage, it promised to have substantial political ramifications.[27] Certain Democrats with presidential ambitions would happily take ozone as a cause, and Democratic senators Birch Bayh of Indiana and Frank Church of Idaho quickly introduced the grandly named Stratospheric Protection Act of 1971.[28] Senator Henry Jackson also introduced a bill mandating a stratospheric research program, and his passed in late September.[29] The law placed responsibility for a four-year stratospheric research program in the Department of Transportation and provided a budget of $20 million to carry it out. This became known as the Climate Impact Assessment Program (CIAP).

CIAP

CIAP's central scientific question was whether stratospheric flight, by supersonic or subsonic aircraft, would cause ozone depletion. This depended on investigating a host of lesser questions regarding the chemistry and circulation of the stratosphere. CIAP's research program thus had several segments. The program office sponsored efforts to measure various trace gases in the stratosphere to determine their concentrations; it supported laboratory measurement of the reaction constants, or the speed, of key reactions; and it funded an effort to determine the emissions levels of aircraft engines at simulated flight altitudes. Finally, it supported efforts to produce theoretical models of ozone chemistry. The models would serve as syntheses of all the individual bits of knowledge about the stratosphere generated during the research program, and they would provide the prediction of future ozone damage that the panel was expected to provide.

CIAP was structured into a set of five subprograms, the largest of which (in terms of funding) was the atmospheric measurements program. The measurement program's strategy was to "focus ongoing work toward the specific objectives of CIAP," not to start new research programs or develop new measurement techniques. These would take longer to accomplish than available to the program. Instead, CIAP supported additional measurements by existing groups in order to accelerate research that would probably have been done anyway, but on a longer, less intensive schedule. At the University of Wyoming, for example, David Hofman, Ted Pepin, and James Rosen were experienced in measuring aerosols at high altitudes using balloon instruments, and in June 1972 they carried out launches to measure particulates in the stratosphere for CIAP. Barney Farmer of the Jet Propulsion Laboratory (JPL) used an infrared interferometer to measure nitrogen oxide in the stratosphere in March 1973; Harold Schiff of York University, one of the participants in the earlier SST conflict, made the first vertical profile measurements of nitrogen oxide for CIAP late in 1972 and reported the results in early 1973.[30]

These new measurements had the effect of substantially reducing the expected impact of an SST fleet, by demonstrating that the natural concentration of nitrogen oxides in the lower stratosphere was at the high end of the range of concentrations used in the projections made independently by Hal Johnston and Paul Crutzen. The high natural level of nitrogen oxides meant that the amount added by SSTs would be a relatively minor increase, leading to a lower SST impact. This also had the effect of explaining a minor scientific mystery remaining from the above-ground nuclear weapons testing program. These tests had injected very large amounts of nitrogen oxides into the stratosphere, particularly during a period of very intensive testing during 1961–62. Yet there had been no measurable decrease in ozone concentrations. The 1961–62 testing series had injected about the same quantities of nitrogen oxides into the stratosphere as several hundred SSTs would, providing the only significant real-world experiment available to test the chemical models. The authors of CIAP monograph #1 had assessed the utility of this data to predicting SST effects and found it wanting; the model predictions of depletion based on the amounts injected during the tests were within the measurement error of existing detection equipment, making it impossible to determine whether or not depletion had occurred.[31] Yet the high natural concentration of nitrogen oxides explained why the nuclear testing program had not produced measurable ozone reductions.

The nuclear weapons testing data did not lead to the rejection of SST impact, however. The weapons-derived injection was impulsive, occurring over a very short period of time and not continually, and hence its effects would differ significantly from SST injection. Continuous injection of nitrogen oxides by SSTs would produce a new steady-state level of nitrogen oxides in the stratosphere, which did not happen from the testing program. In addition, one of the key depletion reactions was quite slow, meaning that it did not produce depletion during the short-term weapons tests, but it would within the context of a new, higher steady-state level of nitrogen oxides. Hence, the CIAP final report found that, using the new measurements taken during the program, a post-1995 fleet composed of 40 Concorde-type SSTs and 820 of the Boeing-type SST could produce 13 to 17 percent depletion in the Northern Hemisphere.[32]

But that isn't what the CIAP executive summary said. Instead, it had been written to emphasize a "future technology" SST whose low-emission combustors would limit depletion to a "minimally detectable" level of 0.5 percent. This, in turn, had been based upon a new research program at the NASA Lewis Research Center, which had been established in 1971 to investigate the possibility of lower-emission jet engines. Its chief combustion researcher, Richard Niedzwicki, had believed that a sixfold reduction in nitrogen oxides emissions was possible in the next decade or so from new combustor technology, and this became the basis of the CIAP executive summary's "future SST." Because reporters only read (at most) the executive summary, and largely did not catch the subtle sleight-of-hand played within it, the news articles that came out of the press conference announcing the report's publication claimed that CIAP had exonerated the SST when it had done quite the opposite.[33] The Washington Post's headline for the story, for example, was "SST Cleared on Ozone."[34]

The CIAP scientists were understandably outraged at what they saw as a deliberate attempt to mislead the public about the SST's impact. The executive summary's whitewash of the findings led to vituperative attacks on their community, their competence, and their professionalism in newspaper editorials whose writers were mislead by the executive summary. A lively debate occurred in the "Letters" section of the premier scientific journal in the United States, *Science,* over the controversy as well. Tomas Donahue, a space scientist at University of Michigan, who had agreed on behalf of the American Geophysical Union to sanction a review of the study, disclaimed responsibility for the summary's published form, placing the blame squarely on the Transportation Department's Grobecker and Cannon.[35] Cannon, in turn, spent much of the year trying to repair the damage done to the department's, and his own, credibility within the scientific community by the executive summary's "technological optimism," as Harold Schiff put it.[36]

The Department of Transportation's apparent inability to keep its desired spin out of the program's findings was unacceptable within the atmospheric science community. It was their first taste of the politicization of science, a problem that had afflicted physics during the 1950s and 1960s. In physics, the conflict had been over nuclear weapons, nuclear policy, and, during the 1960s, the community's participation in weapons programs. In atmospheric science, the conflict shaping up was over damage to the Earth's ecosystems by modern technologies. By the time the CIAP report emerged from its committees in late 1974, there was already a claim that another common industrial product might destroy the ozone layer. This expanded the scope of the conflict considerably. It also led to NASA taking control of the ozone research program as the atmospheric science community sought a more neutral patron than the Department of Transportation had proved to be.

FOUNDING THE NASA RESEARCH PROGRAM

The CIAP report's conclusions were highly uncertain. There were a large number of still unmeasured rate constants and atmospheric quantities. The models used to synthesize the new data and make predictions were new and had not yet been rigorously checked against observations. Nor was there yet a statistically significant body of observations against which to check model performance. Finally, the ability to measure key trace species in the atmosphere was inadequate. New measurement techniques needed to be developed. Hence there was a great deal of research left to do to fully understand and characterize stratospheric processes. The CIAP effort was a beginning, in that sense, not an end.

Further, during CIAP's lifespan, a new ozone controversy had started that was unrelated to the SST but directly impacted NASA. Chlorine was implicated in stratospheric ozone depletion during 1973 by researchers working on a component of the Space Shuttle's Environmental Impact Statement. The Shuttle's solid rocket boosters would release chlorine compounds during their ascent into space, and according to this research these might also cause depletion like that suspected from SSTs. The combination of the remaining uncertainty from CIAP, the direct threat to the Shuttle program posed by the potential for chlorine-induced depletion, and both internal and external advocacy for a larger NASA role in environmental research led the agency to essentially take over CIAP's program.

NASA had been directly involved in CIAP. After Senator Anderson's initial request in 1971 that NASA begin a stratospheric ozone research program, NASA administrator James Fletcher, who was a chemist by training, had established an Ad Hoc Group on the Catalytic Destruction of Ozone at NASA headquarters to discuss how the agency might investigate the possibility of SST-induced ozone depletion. This group, which was co-chaired by Morris Tepper and George Cherry, included members drawn from the Lewis and Ames Research Centers, the Goddard Institute of Space Studies (GISS), Goddard Space Flight Center, and various offices at headquarters. It also included a representative from the Department of Transportation's CIAP, who briefed the group in October on what CIAP had planned. The agency leadership's goal for this group had been coordination of NASA's research with that undertaken by other organizations in order to avoid duplication and to ensure that NASA's unique capabilities were used to greatest effect. The Lewis Research Center in Ohio, for example, arranged to place sampling equipment on commercial airliners designed to help characterize the chemistry of the upper troposphere and lower stratosphere, while also arranging with NOAA officials to have that agency carry out the actual chemical analyses. Lewis also started efforts to examine combustion processes to identify ways to reduce nitrogen oxides production and to measure the actual constituents in jet engine exhaust, which was made difficult by the high-temperature, high-velocity flow conditions.[37]

NASA's other research centers initiated efforts as well. At the Langley Research Center, the ozone question had interested several members of the aerophysics division. Bill Grose, who had spent the 1960s calculating reentry trajectories for space capsules, recalls that he and others recognized that with the manned space program going largely out of business due to the gap between Apollo's end and the Space Shuttle's expected 1980 first flight, NASA would not need much reentry work. So Grose's group bootstrapped themselves into atmospheric science, initially working on the science necessary to produce sensors for trace gas detection and monitoring, including NO_x and ozone.[38] At JPL, NASA headquarters sponsored the development of new stratospheric remote sensing technologies, expanding a laboratory kinetics group under William B. DeMore, hiring Joe Waters, one of David Staelin's students, to develop a microwave-based instrument for detection of the active chlorine species chlorine monoxide, and supporting the development of a balloon-based infrared interferometer for measurement of a variety of stratospheric trace species.

A reorganization at headquarters in early 1972 reinforced this movement toward a more permanent effort. Administrator James Fletcher had split the Office of Space Science and Applications into an Office of Space Science and an Office of Applications in order to increase the stature of applications programs in the agency. Charles Matthews, a Langley veteran who had moved to Houston with the Space Task Group in 1962, was assigned to head the new Office of Applications in Washington.[39] He was not given a staff sufficient to run his programs, however, and devolved authority to lead centers. The Goddard Space Flight Center became the lead center for atmospheric sciences, Marshall Space Flight Center became lead for communications satellites, and after some struggle, Langley Research Center was designated lead center for environmental sciences. This was somewhat a misnomer, as technically environmental science was a subdiscipline of biology that investigated the relationship between organisms and their environments. But in practice, the research the Langley office supported concerned questions of human impact, investigating the relationship between human activities and the atmosphere.

NASA's movement into the study of human impact on the atmosphere was accelerated and made permanent by the agency's determination that its Space Shuttle and the solid rocket boosters used on certain expendable launch vehicles might also cause ozone depletion. The Shuttle program office at the Johnson Space Center was required to prepare an Environmental Impact Statement on the Shuttle by the 1970 National Environmental Policy Act, and for the atmospheric portion they had contracted with the University of Michigan. Ralph Cicerone and Richard Stolarski found that the exhaust from the Shuttle's solid rocket boosters would release chlorine, a highly reactive element known to destroy ozone, directly into the stratosphere.[40] Their June 1973 report was initially buried by the program office in Houston, Stolarski recollects, but NASA headquarters reversed that decision quickly and scheduled a workshop on the problem, held in January 1974.[41] In the meantime, at a conference in Kyoto, Japan, Cicerone and Stolarski presented a paper on volcanic chlorine as a potential ozone scavenger—omitting mention of the Shuttle and of NASA's support for their research at NASA's request. Paul Crutzen also presented a paper that discussed chlorine chemistry at this meeting. The Kyoto meeting thus put the chlorine problem in the air, so to speak.

By this time, a general consensus had formed within NASA that CIAP was not going to effectively answer important questions about the stratospheric ozone because it was too short-term. The program's time limitation had inhibited the development of sensors capable of sampling the many chemical species necessary to develop a complete understanding of ozone's complex chemistry because scientists could not design, build, test, and deploy sensors in CIAP's four years and therefore were not bothering to try. Bob Hudson, who briefed chief scientist Homer Newell, deputy administrator George Low, and administrator Jim Fletcher on the Shuttle problem on 13 February 1974, pointed out that of the ten chemical species that were important to the Shuttle's ozone problem, accurate sensors for only four existed.[42] Fletcher and Low left this meeting unhappy with the slow pace of sensor development and with the determination to implement a headquarters-directed program to continue stratospheric research once CIAP ended.[43]

The ozone problem became a statutory responsibility for the agency after the British journal *Nature* published a paper by F. Sherwood Rowland and Mario Molina, who argued in mid-1974 that photochemical decomposition of CFCs would release large quantities of chlorine monoxide into the stratosphere.[44] Lovelock had already documented the widespread presence of CFCs in the Earth's atmosphere, and he had calculated that based on the concentration of these gases in the atmosphere, virtually all of the billions of pounds that had been manufactured were still in the atmosphere. In other words, they were not being removed by chemical or biological processes. If, in fact, there were no chemical processes removing CFCs from the troposphere, eventually the atmosphere's circulation would move CFCs into the stratosphere. There, Rowland and Molina argued, they would decompose under the high radiation fluxes into fluorine and chlorine compounds, some of which were known from laboratory studies to be ozone scavengers.

Compared to the billions of pounds of CFCs produced every year for use in spray cans, air conditioners, and refrigerators, the Shuttle's exhaust was utterly trivial, and its impact was quickly dismissed. Revelation that mundane everyday items like CFC-propelled hair spray could be ozone destroyers — and thus cancer risks — quickly produced a media firestorm and caused Congress to move with unaccustomed speed. December 1974, the same month the controversial CIAP report was released, witnessed the first House hearings on the CFC issue and the following month, the Senate Committee on Aeronautical and Space Sciences convened hearings on the ozone problem.[45]

NASA leaders chose to seek an explicit leadership mandate over the nation's stratospheric ozone research from Congress after Hudson's February briefing. A Stratosphere Research Program that had started gelling in headquarters was initially funded by pulling small amounts of money out of several other programs, and it became the core of much larger research program once Congress took up the CFC problem. Deputy Associate Administrator John Naugle explained to Edward Todd at the National Science Foundation (NSF) in December that NASA was interested in both the leadership role in answering the short-term congressional concerns about CFCs and in constructing an ongoing stratospheric research program, and asked him to pass along the agency's interest to H. Guyford Stever, the president's science advisor.[46] NASA was rewarded with a congressional edict handing stratospheric ozone research to the agency, embodied in its fiscal year 1976 authorization bill and continued in 1977 amendments to the Clean Air Act.

Under the agency's new mandate, it became responsible both for carrying out an ongoing research program into stratospheric processes and for producing an assessment of the state of knowledge about the stratosphere every four years. NASA headquarters created the Upper Atmosphere Research Program (UARP) within the Office of Space Sciences that year, administered by Ron Greenwood, who moved from the Langley Research Center. Greenwood reorganized NASA's rather scattered stratospheric research activities, expanding the Langley Research Center's Environmental Quality Office, under James D. Lawrence.[47] At Goddard, Nelson Spencer perceived that the stratosphere was "the next big thing" and invited Robert Hudson and Richard Stolarski, then at the Shuttle Environmental Effects Project Office in Houston, to move north and establish a new stratospheric research branch.[48] The UARP's focus was on model, process, and laboratory studies, and in situ measurement technologies, while satellite instrument development remained in the Office of Applications. The UARP also became responsible for assembling a quadrennial assessment of the state of ozone science for Congress.

One of the first efforts the UARP undertook was expansion of NASA's laboratory capabilities in order to gain a better understanding of the rates at which many of the chemical reactions involved in the chlorofluorocarbon reaction sequence happened. It employed several laboratories at universities, at NOAA, and at JPL, to make the measurements. JPL's Bill DeMore hired a number of new laboratory kineticists to carry out the measurement effort, including Robert T. Watson, who

Plate 1. One of the first full-disk color images of the Earth from space, taken 18 November 1967 by ATS-3. Courtesy University of Wisconsin.

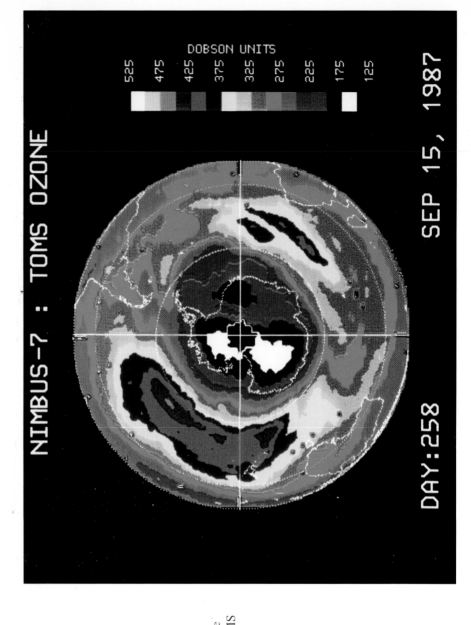

Plate 2. Antarctic ozone hole image, from TOMS instrument data, 1987. Courtesy NASA.

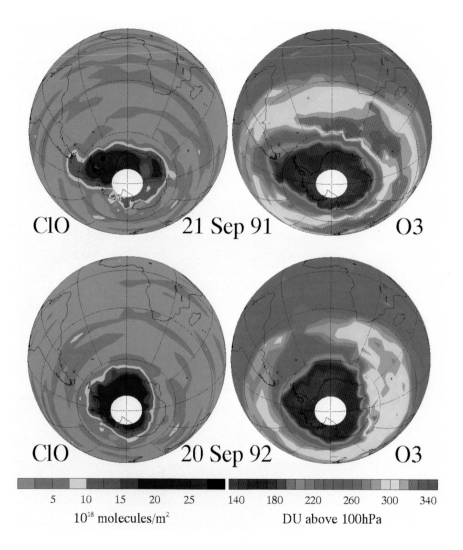

Plate 3. The 1991 Antarctic ozone hole, as seen by the JPL Microwave Limb Sounder. The white circle at the pole represents a data gap caused by UARS's inclined (non-polar) orbit. The images to the left represent concentration of the active ozone destroyer chlorine monoxide in the stratosphere, while the images to the right show ozone concentrations in the stratosphere. JPL Image P-45745, courtesy NASA/JPL/Caltech.

Plate 4. The 1997–98 El Niño as seen by TOPEX/Poseidon. The white area to the right (against the Peruvian and Central American coasts) is very warm water, while the purple area to far left is cold water. During an El Niño event, the large pool of warm water forms in the Western Pacific, then moves eastward along the equator to the west coast of North and South America. JPL image PIA01164, courtesy NASA/JPL/Caltech.

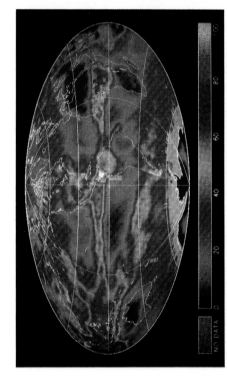

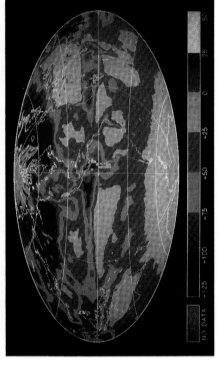

Cloud-radiative forcing (in W/m²). Monthly averages derived from ERBE for July 1985 are depicted for long-wave (top) and shortwave forcing (bottom). Long-wave cloud forcing is the reduction by clouds in the long-wave radiation that is emitted to space; hence it is the greenhouse effect of clouds. Clouds reduce emission to space because at their bases they absorb radiation emitted by the warmer surface and at their tops they emit to space at colder temperatures. Deep cirrus clouds such as the monsoon cloud systems over the Indian Ocean and Indonesia and the jet-stream cirrus clouds at midlatitudes give a large greenhouse effect. Because clouds reflect more shortwave solar radiation than the adjacent clear skies, the shortwave forcing is negative—a cooling effect. Surprisingly, the magnitude of the cooling is as large as the long-wave forcing over the tropical cirrus systems, and is even larger over the mid- and high-latitude oceans. The net cloud-radiative forcing (shown in the bottom image on the cover) is the sum of long-wave and shortwave cloud forcing. The averages range from −(100–140) W/m² (dark blue) to 10–40 W/m² (red). The net effect is largely negative; hence clouds have a cooling effect on the planet. The strongest cooling is caused by persistent stratus and storm-track clouds over mid- and high-latitude Atlantic and Pacific oceans. **Figure 7**

Plate 5. Top image, cloud long-wave forcing as calculated by the ERBE team from their data. In effect, this is the portion of the overall greenhouse effect attributable to heat absorption by clouds. *Bottom image,* cloud shortwave forcing, or reflectivity. ERBE's scientists found that cloud reflectivity outweighed cloud heat absorption, so clouds produce a net cooling of the Earth. Reprinted with permission from V. Ramanthan et al., "Climate and the Earth's Radiation Budget," *Physics Today* (May 1989), fig. 7. Copyright 1989, American Institute of Physics.

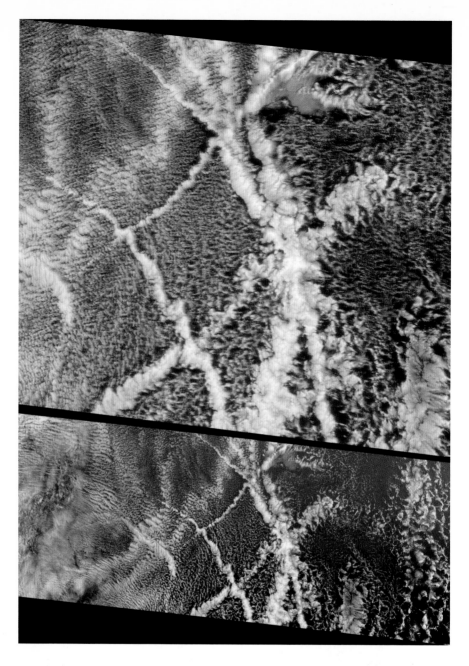

Plate 6. "Shiptracks," an example of the secondary aerosol effect. Sulfate emissions from ship engines cause the water droplets in clouds to be smaller, making the clouds more reflective. JPL PIA03433, courtesy NASA/JPL/Caltech.

Plate 7. CERES data visualization. This shows average outgoing long-wave radiation from Earth for a single day, 4 May 2003. White represents areas of low thermal emission; in the tropics, these are areas in which high-altitude clouds are the radiating surface. Yellow represents areas of the highest thermal emission, generally arid regions where lit-tle atmospheric water vapor exists to reduce the flow of heat from the Earth's surface to space. The image was produced jointly by Langley Research Center and Goddard Space Flight Center for *National Geographic* magazine. Color, courtesy NASA.

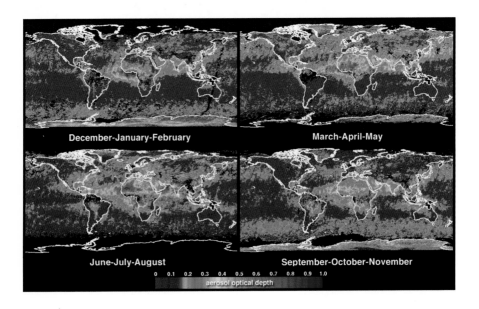

Plate 8. Seasonal aerosol distribution for 2001–2002, from the MISR instrument on EOS. PIA04333.tif, courtesy NASA/JPL/Caltech.

became manager of the UARP in 1980; Mario Molina, who joined the lab in 1980; Randall Friedl; and James Margitan.[49]

The UARP also funded instrument development. Moustafa Chahine, the chief scientist at JPL, had hired Joe Waters, a former student of David Staelin's, to develop a shuttle-borne microwave instrument. That project was quickly transformed into an aircraft- and balloon-based instrument to measure the critical species chlorine monoxide, which was most directly involved in ozone destruction. JPL already had a spectroscopy laboratory that specialized in molecules that were active in the microwave region, founded to enable planetary spacecraft to conduct remote sensing of other atmospheres. JPL also had an infrared spectroscopy laboratory established in the early 1960s; one of its founders, Barney Farmer, started developing an infrared-range Fourier transform spectrometer for balloon use. This measurement method promised the ability to measure most of the trace species involved in the suspected depletion reactions except chlorine monoxide, making it a valuable complement to the microwave instrument.

Not all of the instrument development was carried out in NASA facilities. One important example was another chlorine monoxide instrument, this one developed by James G. Anderson. Initially, this in situ instrument development started with funding from the Shuttle Effects Program Office, but moved to the UARP in 1978. Anderson, who had started developing rocket-borne instruments for measuring radical species under space scientist Charles Barth in Colorado, moved to the University of Michigan in 1975 and Harvard University in 1978. His series of instruments after 1976 were balloon- and aircraft-based.[50]

DEFINING POLLUTION SATELLITES

The 1972 conference on remote sensing of pollution had led NASA to define the final satellite in the Nimbus series as a "pollution patrol" satellite. While it retained its experimental meteorological functions too, it was the first NASA satellite to carry a number of instruments aimed at various pollutants. Launched in 1978, this mission was named Nimbus 7. Its instruments were primarily intended to examine questions related to stratospheric ozone depletion, albeit based on the chemistry of the nitrogen oxides depletion hypothesis of the early 1970s. Nimbus 7 had two successors, also devised in the late 1970s: a single-instrument mission called SAGE (Stratospheric Aerosol and Gas Experiment), and a very comprehensive mission, the Upper Atmosphere Research Satellite (UARS).

Don Lawrence, as head of Langley's Environmental Quality Office in the mid-1970s, was double-hatted as the head of the NASA applications program's effort to develop satellite sensors for environmental research. This was a product of the lead center concept, and it allowed Lawrence to demand that proposers of instruments be selected by peer review of instrument proposals. The Nimbus 7 payload was therefore selected by a review committee composed of NASA, NOAA, and university-based scientists out of a group of forty proposals. Of the four instruments chosen for the pollution aspect of the mission, Langley-based researchers had proposed three and Goddard one. This marked Langley's entrance into the Earth science business.

Langley's M. Patrick McCormick had proposed Stratospheric Aerosol Measurement (SAM) II, a solar extinction instrument, to measure aerosols in the upper atmosphere. This was based on a balloon instrument devised by James Rosen, David Hofman, and Ted Peppin at the University of Wyoming, whom McCormick had worked with during his PhD thesis work on lidar-based remote sensing. Because SAM II viewed the Sun directly through the atmosphere during a satellite's "sunrise" and "sunset," this instrument had an advantage of simplicity, high signal-to-noise ratio (the Sun being the most powerful signal in the solar system), and a correspondingly high vertical resolution. It was also self-calibrating. Its principal disadvantage was that it could measure only a narrow swath of atmosphere during each orbit, so it did not really produce global measurements. An earlier version, named Stratospheric Aerosol Monitor, flew in a single-channel, astronaut-tended form on the Apollo-Soyuz Test Program mission in 1975. As originally designed for Nimbus 7, SAM II had three channels, for detecting aerosols, ozone, and atomic species.[51]

The second ozone-related instrument selected for Nimbus 7 was jointly proposed by James Russell III of Langley and John C. Gille of the National Center for Atmospheric Research (NCAR). This was the Limb Irradiance Monitor of the Stratosphere (LIMS), and was based on an earlier instrument Gille had flown on Nimbus 6. The earlier instrument, Limb Radiance Inversion Radiometer, had been designed to measure emissions from carbon dioxide and ozone as a means of measuring stratospheric temperature and circulation. The LIMS instrument added channels to permit measuring vertical profiles of ozone, water vapor, nitric acid, and nitrogen dioxide concentrations in the stratosphere. Nitric acid and nitrogen dioxide were two of the chemicals proposed as ozone depletion mechanisms by Paul Crutzen and Hal Johnston, and Russell and Gille hoped to determine their concentrations in the atmosphere. Because the instrument measured emission of the thin crescent of atmosphere between the Earth's surface and space—the Earth's limb—getting a useful signal-to-noise ratio meant cryogenic cooling of the detectors. The cryogen would gradually evaporate, meaning the instrument had a short expected lifespan.

Russell and Gille also added an internal blackbody that the instrument's scan mirror would see once per scan, providing a known warm data point for use in calibration; a second known cold data point came from having the mirror look at space once per scan. These two data points would allow the project scientists to evaluate the instrument's degradation; since space's emissions never change, for example, a change in the instrument's measurement of space would actually reflect a change in the instrument. The addition of calibration capabilities to LIMS reflected the application program's increasing effort to obtain scientific credibility for space observations. Lawrence, McCormick, Russell, and their co-workers recognized that improving the standing of space science in the larger scientific community required renewed attention to the fundamentals of scientific practice. In operational terms, this meant engineering the instruments for credibility.

The third Langley instrument, MAPS, was proposed by Hank Reichle. Its goal was measurement of carbon monoxide, which could serve as a tracer. Carbon monoxide had a residence time in the troposphere of several weeks, and could therefore be used to follow air parcels as they moved through the atmosphere. Carbon monoxide was also an important part of the reaction chain for ground-level ozone, an important pollutant. This instrument, however, was deleted from the payload before launch. The high inflation of the 1970s affected NASA as it did everyone else, driving up costs and forcing the Nimbus program management to make cuts. An evaluation panel led by Vern Suomi reviewed all the instruments assigned to Nimbus 7, and MAPS, whose technology was deemed not ready, was one casualty. It later flew in 1981, on the second Space Shuttle engineering flight. This review also affected McCormick's SAM II, which had two of its three channels eliminated. The ozone and atomic spectra channels were deleted, leaving only the aerosols channel. Nimbus 7 already carried an ozone instrument, Donald Heath's TOMS/SBUV. McCormick had wanted the ozone channel so that he could use its data to separate ozone's signal from the aerosol signal, permitting him to evaluate particulate size. Since TOMS/SBUV would not see the same piece of atmosphere that SAM II would, its data could not be used for this. He did not win this argument, however, and had to wait for a future opportunity.

The non-LaRC instrument devoted to chemistry on Nimbus 7, Heath's TOMS/SBUV (Total Ozone Mapping Spectrometer/Solar Backscatter UltraViolet), was based on two earlier instruments Heath had flown on Nimbus 4, the

Monitor of Solar Ultraviolet Energy (MUSE) and the Backscatter Ultraviolet Spectrometer (BUV). The MUSE had been designed to measure solar ultraviolet output, which was an open scientific question during the 1960s. It could not be measured from the ground due to absorption by the ozone layer, and could only be estimated from balloon-based measurements. When this instrument had flown in 1970, there were no known sources that would permit onboard calibration checks, so its results, like many other NASA experiments during the decade, had been suggestive, not definitive. It had appeared to show extreme variability in solar ultraviolet output, which would be problematic for instruments using back-scattered ultraviolet radiation to measure atmospheric constituents. This had been the purpose of its companion instrument, the BUV. This had measured backscattered solar ultraviolet radiation at twelve wavelengths to produce vertical profiles of ozone concentration. But because of the apparent variability in solar UV output, removing the effect of solar irradiance changes from the backscattered data required simultaneous measurement of both incoming solar ultraviolet and the backscattered radiation as well as the ability to provide continuous calibration of both measurements. While the BUV instrument had a calibration source, the MUSE instrument did not, and hence the ozone profiles Heath's team produced were, again, suggestive, not definitive.

For TOMS/SBUV, Heath had added a solar-viewing capability to the BUV instrument and had devised a calibration method for it using a mercury-argon lamp. The Solar Backscatter Ultraviolet (SBUV) shared a diffuser plate (meaning only one subsystem could operate at a time) with the Total Ozone Mapping Spectrometer (TOMS) subsystem. The SBUV did not scan across the satellite track, however, so it did not produce a complete global map of ozone distribution. Instead, it provided vertical profiles of ozone concentration in the upper stratosphere. The TOMS instrument scanned, but did not produce vertical profiles. The TOMS produced total column measurements, reflecting the total amount of ozone in the column of air being viewed.

Nimbus G was launched 24 October 1978, becoming Nimbus 7 once in orbit. It was expected to carry out a five-year mission but remained in operation until 1994. In addition to representing NASA's adoption of environmental research as part of its fundamental mission, Nimbus 7 was the last of its series. NASA's primary project during the 1970s had been its Space Shuttle, which was supposed to provide routine, inexpensive, launch-on-demand service. All post-1980 payloads had to be compatible with the Shuttle's payload bay, and the Nimbus bus was not. New buses were being developed by several aerospace companies for deployment by the Shuttle using the Canadian Space Agency's robotic arm. This was also true

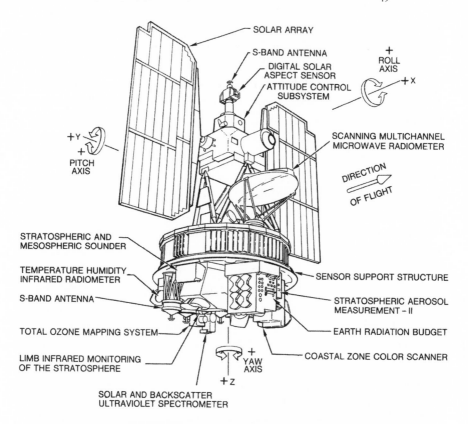

SOLAR ARRAY

S-BAND ANTENNA

DIGITAL SOLAR
ASPECT SENSOR

ATTITUDE CONTROL
SUBSYSTEM

+
ROLL
AXIS
+X

+Y
+
PITCH
AXIS

SCANNING MULTICHANNEL
MICROWAVE RADIOMETER

DIRECTION
OF FLIGHT

STRATOSPHERIC AND
MESOSPHERIC SOUNDER

TEMPERATURE HUMIDITY
INFRARED RADIOMETER

S-BAND ANTENNA

SENSOR SUPPORT STRUCTURE

STRATOSPHERIC AEROSOL
MEASUREMENT – II

TOTAL OZONE MAPPING SYSTEM

EARTH RADIATION BUDGET

LIMB INFRARED MONITORING
OF THE STRATOSPHERE

+
YAW
AXIS

COASTAL ZONE COLOR SCANNER

+Z

SOLAR AND BACKSCATTER
ULTRAVIOLET SPECTROMETER

NIMBUS – G OBSERVATORY

A wireframe drawing of the Nimbus 7 satellite. Courtesy NASA.

of robotic probes to other planets. Nimbus 7 was thus the end of an era, in one sense, representing the abandonment of expendable launch vehicles.

But it was only the beginning of NASA's stratospheric chemistry effort. Two other satellite projects were formulated during the final years of the decade. One was part of the short-lived Applications Explorer Program. A Goddard-initiated effort, this was aimed toward missions that required orbits that the Shuttle would not be able to reach. The satellites to be developed under it would be small, sized for launch on the Scout rocket. This was a very simple solid fuel rocket that could put a small satellite into high-inclination, near-polar orbit. The first satellite in this effort was the Heat Capacity Mapping Mission. The second was an improved version of Pat McCormick's SAM instrument, SAGE. Because of Nimbus 7's Sun-synchronous orbit and the SAM II's placement on the satellite, SAM II could obtain data only from the polar regions. But people affected by ozone depletion mostly lived in the mid-latitudes, and a different orbit was necessary to measure those parts of the atmosphere with an occultation-type instrument. In a high-inclination orbit, SAGE would be able to see into the mid-latitudes, and as the orbit precessed, it would sweep across most of the atmosphere. Between SAM II and SAGE, therefore, McCormick's aerosol data would cover most of the Earth.[52]

NASA also began developing a plan for a larger follow-on to Nimbus 7, this one aimed at the central questions of nitrogen oxides and CFC-induced ozone depletion. Eventually known as the UARS, this mission was comprehensive in scope. Shelby Tilford hoped that it would answer all the major scientific questions related to ozone depletion claims; it was therefore the largest Earth observation satellite flown to date. Late in 1977, Tilford assembled a working group at JPL to define the scientific requirements for the mission. His group had defined a two-satellite mission for launch in 1983 and 1984 into inclined orbits by the Shuttle. The satellites would carry instruments for monitoring solar ultraviolet output, which affected ozone production, and measuring stratospheric circulation and temperature as well as those for chemistry investigations. Simultaneous measurements of all these was necessary to improve understanding of stratospheric processes. To understand the relationship between stratospheric circulation and ozone concentrations, for example, researchers needed datasets that represented both at the same time. Similarly, one needed simultaneous measurements of the source molecules for the radical species suspected of causing ozone depletion and the sink molecules that served to remove them from an active state.

The question of how to accomplish some of the more difficult trace species measurements caused some controversy on the panel. The source species in particular were difficult to measure because they were weak emitters. Hence the panel report contained an explanation of the advantages and disadvantages of various measurement techniques for measurement of each chemical. Limb occultation-type instruments similar to the SAM and SAGE instruments were expected to be relatively easy to develop, and would have low data rates and potentially long lives, but because they only took measurements during satellite "sunrise" and "sunset," would require thirty to sixty days to produce a set of global measurements. Emission-sensing spectrometers could measure more species and produce a global dataset on nearly every orbit, but would be larger, would be more expensive to develop, and would require more power and a higher data transmission rate from the satellite. This would further increase the program's costs. They might also require cryogenic cooling, which would result in relatively short lifespans.

As finally approved in 1981, the UARS represented a compromise between the two approaches. Joe Waters's microwave limb-scanning emission radiometer was adopted. This was aimed at sensing of the critical radical chlorine monoxide as well as ozone and water vapor. Two different infrared spectrometers, one proposed by Lockheed and one by Oxford University, were accepted for sensing a variety of chlorine, nitrogen, and fluorine species. The Lockheed instrument was cooled by a solid cryogen, providing a limited lifespan but very high resolution; the Oxford instrument relied on a mechanical cooler that was less efficient but longer lived. One occultation instrument, the Halogen Occultation Experiment, by Langley researcher James Russell, overlapped some of the species sensed by these other two instruments, permitting evaluation of its utility as a monitoring instrument.[53] It did sense one unique trace gas, hydrogen fluoride, a critical measurement for silencing critics who insisted that CFCs were not the major source of chlorine in the stratosphere.

In addition, the UARS had chosen for it two instruments for measuring winds, one from the University of Michigan and one from York University in Canada. These were aimed at different levels of the atmosphere, overlapping between 70 and 120 kilometers altitude to permit comparison. Also chosen were four instruments to measure solar energy, in order to examine the relationship between changing energy input to the atmosphere and changes in the circulation and chemistry of the stratosphere and mesosphere. SOLSTICE, from Gary Rottman at the University of Colorado and SUSIM from the Naval Research Lab (NRL) were both designed to examine solar ultraviolet output, a key factor in ozone production. The Particle Environment Monitor from the Southwest Research Institute was aimed at high-energy radiation. Finally, an updated version of the Solar Max experiment ACRIM was accepted to continue monitoring total solar output.[54]

COOLING OF THE OZONE WAR

Between the launch of Nimbus 7 in 1978 and the year 1985, the political conflict over stratospheric ozone cooled considerably. This was due to several factors: the United States had banned the use of CFCs as propellants, although other nations had not; the laboratory and atmospheric measurements programs NASA sponsored produced a declining trend in the projected severity of CFC-induced depletion, and the supersonic future failed to win favor with the airline market, eliminating that point of conflict entirely. This did not result in a loss of support for the NASA research program, however. Because the CFC ban was neither total nor global, ozone depletion was still an important scientific, and policy, issue. Policymakers still needed a scientific answer to whether CFC production would result in substantial ozone depletion, and this helped assure funding.

One of the major questions left over from the CIAP program of the early 1970s was whether the reaction rate constants used by the ozone chemistry models were accurate. Many of the rate constants used in the models were known not to be constants but rather upper or lower limits, meaning that the actual rate could be significantly different. Hence during the late 1970s, NASA sponsored remeasurement of the major rate constants. In the short run, this remeasurement effort caused some problems for the research community by causing the ozone depletion predictions to vary widely over short periods of time. A new rate measurement from the NOAA Aeronomy Lab caused a sudden downward jump in depletion estimates for SST emissions, for example.[55] But the same new rate measurement caused a doubling of the depletion expected from CFCs. The rapid changes in estimates, as diplomat Richard Elliot Benedick somewhat dryly put it, caused some credibility problems for the atmospheric science community.[56] What should happen as a science matures is convergence—one normally expects that as better measurements are integrated into models, the range of model outputs would converge toward a correct answer. But instead, as more rates were measured, the model projections continued to vary widely. This produced a great deal of doubt in the policy community about the actual severity of the ozone problem, and whether steps to ameliorate it really needed to be taken.

The first of the quadrennial NASA assessments of the ozone question was due in 1981, and it approached the issue somewhat differently than its predecessors from the National Research Council (NRC). The NRC reports were contracted by EPA and contained policy recommendations. The person tasked with organizing the NASA assessment, Robert T. Watson, decided not to follow that precedent. Instead, he wanted a rigorous assessment of the state of knowledge without the policy aspect. He also wanted the report to speak for a considerably larger portion of the atmospheric research community. There had been competing reports on the ozone question. NRC had produced them for EPA, the United Kingdom's Meteorology Office had produced one for the British government, and the World Meteorological Organization (WMO) had produced them for the UN. Each report drew on different groups of specialists and arrived at different conclusions, producing still more controversy and uncertainty. Watson, who had been a postdoctoral fellow in Hal Johnston's laboratory before moving to JPL in 1976 and then to NASA headquarters in 1979, decided to invite all these different organizations to participate in a single assessment process and produce a single, comprehensive report.[57]

Watson intended the theme of the report to be the comparison of theory to measurements. The 1981 effort marked the first time that comprehensive satellite data was available for a number of trace species, and it offered the opportunity to compare satellite data to in situ and ground-based data as well as to theoretical predictions. It also marked the first availability of relatively comprehensive two-dimensional transport-kinetics models. The one-dimensional models used in the previous assessments contained chemistry but made little effort to account for the motion of the atmosphere, a limitation based largely on the lack of computing power available for chemical modeling. The two-dimensional models simplified circulation through the use of longitudinal averaging, taking advantage of the relatively small longitudinal variation of meteorological phenomena to reduce the computing burden imposed by fully three-dimensional circulation. They promised better assessment of the latitudinal variation of ozone production and loss, hopefully permitting more realistic comparison to the newly available global datasets.

These new global datasets contained significant uncertainties, and the authors of the assessment chose to explicitly explain them. The Nimbus 4 BUV instrument, for example, had shown a continuous and large decrease in stratospheric ozone, and the authors of chapter 3 reported that Donald Heath's assessment of this was that it was mostly an artifact of instrument degradation while in orbit. He had, they explained, tried two methods to compensate for the decay. In one attempt, he had normalized the data to one of the more reliable ground-based Dobson stations. In the other, he had compared the BUV data to the more recent Nimbus 7 TOMS/SBUV data, which should have been more reliable. Both attempts led to derived trends showing ozone decline in the lower stratosphere. But at higher altitudes, the two methods led to divergence. One trend became positive, while the other remained negative. Several explanations for this were possible, and the chapter authors went into them in some detail. They finally concluded that there appeared to be a downward trend in ozone concentrations that was in line with model predictions, but also stated that determining an actual ozone trend was "one of the most challenging research problems of this time."[58]

The explicit, detailed discussion of the uncertainties in the data was designed to help begin restoring credibility in the assessment process. This continued through their analysis of model results. The authors included discussion of ten models and their outputs, the scenarios used as inputs to them, and how recent changes in rate constants had altered the output. One model that Watson recalled many years later as having been particularly good had been developed by a DuPont researcher; this one showed steady-state decline in ozone concentrations (which might occur sometime in the mid-twenty-first century) of between 7 and 24 percent, depending on the choice of rate constants.[59] For 1980, it had projected a decline of 0.5 to 0.7 percent, generally in line with the other models and with what the observational data suggested had occurred.

The predictions of ozone decline made in this assessment were considerably lower than those of the 1979 assessment, helping to cool the ozone controversy. In one sense, it deprived activists of a propaganda tool. But the deliberate effort to explain the sources of uncertainty in both data and models, and the decision to avoid making policy recommendations, helped begin restoring scientists' credibility among policymakers. The lack of an executive summary removed a potential flashpoint of controversy, for even done honestly, such summaries inevitably simplified discussion in ways that could be construed as enhancing or reducing uncertainty. And there were considerable uncertainties remaining.

The process of constructing the assessment also provided some less obvious benefits to the research community. NASA's effort to draw university, corporate, and foreign researchers into the process brought disparate voices and viewpoints together, permitting them to work out disputes and to assess each other's competence and knowledge of the issues at hand. Resolving differing interpretations at the workshop also helped keep controversies out of the scientific journals, where they could be exploited by political and economic interests. This was important because while the United States had enacted a partial ban on CFCs, no other nation had. European governments were highly resistant to the idea of a ban, while certain American manufacturers were beginning to perceive that it might be in their interests to promote a ban, which would position them to gain a large share of a new market for substitutes to CFCs. They had a head start because of the U.S. ban, and, of course, European manufacturers were well aware of that fact. Bringing European scientists into the assessment process helped reduce suspicions that the motivation behind the American scientific effort was really an economic one; while it did not guarantee that the European governments would accept their scientists' results, it helped assure they would receive essentially the same information.

The process also served to further NASA resolve to improve the quality of its observational data. The problem of instrument decay in space was a challenging one. It was not clear what all the sources of degradation were. Decay in its optics was one possibility that was widely suspected to have occurred with the BUV instrument but was very difficult to prove. The steps taken with the Nimbus 7 TOMS/SBUV to improve its calibration abilities would not necessarily detect degradation in the instrument's optical path, for example. Optical decay was also difficult to compensate for, since it would not necessarily be linear, or affect all wavelengths that traveled through the optical material equally. To fully examine this potential problem essentially required using an independent measurement of the same radiances being received by TOMS/SBUV, which could only be done by another instrument viewing the same scene at the same time. This meant a second satellite instrument, or, better yet, an instrument that could be returned from space so that it could be checked after its flight.

The Space Shuttle promised such a capability; indeed, the ability to retrieve satellites from space had been one of its major selling points. Tilford and Watson had begun discussing its use for such comparative measurements the previous year, and Heath at Goddard had started investigating the possibility of a Shuttle-based version of the SBUV instrument. Because it was relatively small, it could be flown in conjunction with other payloads and therefore be flown relatively frequently. For its first few years, the Shuttle would only be capable of equatorial and inclined orbits, not polar ones, but careful choice of launch windows could still provide opportunities for overlapping measurements with Nimbus 7 during one- to two-week Shuttle missions. This instrument became known as SSBUV.[60]

Tilford also funded the development of a Shuttle-borne infrared interferometer that had been proposed by Crofton B. "Barney" Farmer. Farmer had been hired by JPL in 1966. He had done his PhD research at Kings College, London, on the infrared spectrum of the Sun by dragging an infrared interferometer up Mount Chacaltaya in Bolivia to get above as much of the atmosphere's water vapor as possible. His first task at JPL had been building an infrared spectroscopy lab, which JPL's management had thought necessary to support their planetary exploration goals. Later he built balloon-based interferometers for measuring the solar spectrum, for nitrogen and chlorine trace species in the stratosphere, and an instrument for the Viking project to measure the global distribution of water vapor in the Martian atmosphere. Farmer had proposed his Shuttle instrument after not being selected for the Voyager mission to the outer planets. Known as ATMOS (for Atmospheric Trace Molecule Spectroscopy), this was a very high spectral resolution Michelson interferometer that promised the most comprehensive measurement of nitrogen and chlorine trace species in the stratosphere yet carried out. Tilford thought ATMOS would be a revolutionary instrument, and supported it over the next decade even as its development overran significantly.[61]

Tilford and Watson also supported a set of Balloon Intercomparison Campaigns in 1982 and 1983 that were designed to demonstrate new instrument capabilities and to examine instrument errors. Because many previous balloon-based measurements had been made by different investigators using different instruments at different times, it was impossible to know whether differences in measured values represented natural variation in the atmosphere's chemistry or merely measurement error. Watson's purpose in holding the balloon intercomparisons was to get as many instruments on the same balloon gondola as possible, so that they would all measure the same piece of atmosphere at the same time. That way, the measurement errors could be quantified.[62]

The Balloon Intercomparison Campaigns were flown out of NCAR's balloon facility in Palestine, Texas. They used some of the largest stratospheric balloons to date in order to hoist the multi-instrument gondolas. Not all of these flights were successful. On two occasions, balloons disintegrated right after launch, causing Watson to turn to JPL's machine shops to make new gondolas so the instruments could be reflown. Another balloon, this one with JPL spectroscopist Barney Farmer's high-resolution infrared interferometer onboard, performed properly until the ground station commanded the gondola's release. The balloon was too high for the parachute to deploy properly, and when it finally did the metal joint connecting the parachute to the gondola shattered. The gondola and its instruments were crushed into the ground. Farmer, whose interferometer was destroyed, remembers that Watson got him the $2 million he needed to build a better one the day after the accident.[63] Known as the Mark IV, the new instrument would get its first field trial in a cargo container in Antarctica in 1986.

JPL's Joe Waters, whose microwave spectrometer flew on a different balloon in that series, commented many years later that they had been pushing the balloons' limits to operate at the 40 kilometers altitude the chemistry required. The accidents weren't all that surprising. But the campaigns accomplished their primary purpose, with the results narrowing the measurement uncertainty range for several key trace species. They also demonstrated the capabilities of several new instruments, some of which were intended to check the performance of the UARS after its launch by balloon- and aircraft-based underflights.

The 1970s witnessed two new major scientific questions emerge, climate change and ozone depletion. Ozone depletion became highly controversial during the decade, and NASA leaders chose to seek leadership of the issue. At least during the decade, the agency did not suffer for this decision. Instead, it used the opportunity to improve the quality of its research and the respectability of its program within scientific circles. It embarked upon a series of efforts to develop new instrumentation and expand laboratory facilities, improving its research capabilities through the next decade.

The agency also attempted to fix a science assessment process that was not working, and in fact, was harming the science community by producing a multiplicity of divergent knowledge claims. The public image of science was that it produced an objective certainty—reliable knowledge. This was an illusion. Working scientists all know that scientific knowledge is uncertain, particularly in fields undergoing rapid change, as stratospheric chemistry clearly was. But science that was to be a basis for costly economic decisions needed to achieve some degree of certainty, and a very significant degree of credibility. It had not done that during the 1970s, and would not until late in the 1980s. But it was a crucial beginning toward building a consensus view.

Finally, the beginning of the ozone wars led NASA to start studying ways to improve the reliability and credibility of its instruments. Its leaders sought ways to provide calibration records for their instruments, either from more effective onboard calibration mechanisms or by supporting ground-, airplane-, or balloon-based instruments that could be used to check space-borne measurements' quality. This raised their costs, of course, but it also meant the agency was better prepared when the ozone wars reerupted.

CHAPTER SIX

Atmospheric Chemistry

The Martian surface is fried by ultraviolet light because there's no ozone layer there. So the nearby planets provide important cautionary tales on what dumb things we should not do here on Earth.

—Carl Sagan, 1992

During the 1980s, NASA's planetary program essentially ended. The first planetary launch of the decade was the 1989 Galileo mission, which did not arrive at its target planet, Jupiter, until 1995. Due to the Challenger explosion, the Mars Observer mission approved in 1985 did not launch until 1993. It disappeared right before arriving at Mars, an apparent victim of an explosion.[1] The Magellan high-resolution radar-mapping mission to Venus did not launch until 1989. There were two causes of this cessation of a vibrant and successful planetary science program. The first was the budgetary environment of the first several years of the decade. President Ronald Reagan had not initially been a supporter of space exploration, and he and his budget director had imposed budget rescissions that ended a number of programs and blocked all new mission starts.[2] While this changed eventually, the interruption ensured no data until the 1990s.

The second reason for the hiatus in space science was the failure of the Space Shuttle to deliver on its promise of inexpensive, reliable space access. Instead, the Shuttles were relatively quickly recognized as unreliable and extraordinarily expensive. Late in 1984, the Reagan administration began to look for a new launch vehicle, and after the 1986 Challenger accident began to force payloads off the Shuttle and onto expendable rockets again.[3] The Upper Atmosphere Research Satellite (UARS) had to be redesigned after the Challenger explosion, producing a cost explosion that put its ultimate price tag over the $1 billion mark. This money came out of other programs, as did the cost of making fixes to the Shuttle and of delaying and modifying other missions like Galileo.[4] The increasing costs of approved missions meant fewer new approvals.

With no new missions, the only planetary data scientists had to work with for the decade was from the Voyager outer planet flyby missions launched in 1977, which had encounters with Jupiter in 1980, Saturn in 1981, Uranus in 1986, and Neptune in 1989. Many planetary scientists therefore turned to more Earthly questions, in search of intellectual stimulation and funding. There were plenty of available scientific questions, but two stood out that were of global interest and had national support: atmospheric chemistry and climate change. NASA began to devise a global climate observing system in the late 1970s, but this did not begin to garner political support, and therefore funding, until 1989. Instead, NASA's major atmospheric science effort during the 1980s was its atmospheric chemistry program. There were two primary efforts: Robert Watson's Upper Atmosphere Research Program (UARP) and Robert "Joe" McNeal's Tropospheric Chemistry Program.

These programs did not rely on space hardware, however. The UARS that was supposed to be the backbone of the stratospheric research program during the 1980s did not get a ride into orbit until 1991. A single flight of the Jet Propulsion Laboratory (JPL) Atmospheric Trace Molecule Spectroscopy (ATMOS) instrument in 1985 and the aging TOMS/SBUV onboard Nimbus 7 were the only sources of space-based ozone chemistry data during the decade, and they were insufficient to answer the major scientific questions. Aircraft-, balloon-, and ground-based research were the basis of NASA's atmospheric chemistry program in the 1980s. When UARS finally launched, it provided corroboration and demonstrated the global extent of stratospheric conditions measured locally by these other means. But it was not the source of fundamental advances in knowledge. The head of NASA's Earth Observations Program, Shelby Tilford, believed that a proper research program needed to be comprehensive, involving laboratory-, aircraft-, and model-based studies as well as providing for spacecraft instrument development; these proved the salvation of the agency's scientific reputation as UARS sat in storage.

TROPOSPHERIC CHEMISTRY

In 1978, Jack Fishman, a researcher at the National Center for Atmospheric Research (NCAR), and Paul Crutzen published an analysis of tropospheric ozone that launched an extensive series of investigations during the 1980s.[5] Ground-level ozone, while a pollutant that caused health problems for humans, was largely considered to be chemically inert in the troposphere. In the troposphere, ozone was produced by photolysis of nitrogen oxides, which are industrial emissions and are also generated within internal combustion engines. It is destroyed by plant life. The scientific community believed until the late 1970s that the primary natural source of ozone in the troposphere was the stratosphere. Ozone-rich stratospheric air descended into the troposphere somewhere in the high mid-latitudes (although the "where" was rather speculative) and was eventually removed at the ground.

Fishman and Crutzen argued that this could not be true. There was much more land in the Northern Hemisphere than in the Southern Hemisphere, and thus greater ozone destruction. For them to have roughly the same ozone concentrations, stratospheric descent into the Northern Hemisphere had to be much larger than into the Southern Hemisphere. But there was no evidence that this was true, and, they contended, no theoretical basis for believing it either. There had to be a significant ozone production in the troposphere to make up the difference. This had to be particularly true for the Southern Hemisphere, where much less industrial activity occurred but where tropospheric ozone levels were nonetheless similar to Northern Hemisphere levels. They speculated that photolysis of carbon monoxide could provide some of the additional ozone, but there were other possibilities as well, including as yet unidentified sources of nitrogen oxides.[6]

Combined with interest in chemical cycling triggered by James Lovelock's speculations about climate regulation, this argument set in motion a tropospheric chemistry program at NASA that effectively paralleled the UARP. Called originally the Air Quality Program, and later the Tropospheric Chemistry Program, like the UARP, it was comprehensive in nature, involving laboratory studies, model development, and field experiments. The field component was the Global Tropospheric Experiment (GTE), a name that failed to reflect its true nature as a continuing series of instrument development and field studies lasting from 1982 through 2003.

The manager for the Tropospheric Chemistry Program from its founding through 1999 was Robert "Joe" McNeal. McNeal had started the National Science Foundation's (NSF) tropospheric chemistry program in 1978, after working as an atmospheric chemist at the Aerospace Corporation in Santa Monica for more than a decade. NSF's program managers rotated every two years, and in 1980 Shelby Tilford asked McNeal to come to NASA to create a tropospheric chemistry program there. Atmospheric chemistry was just beginning to become institutionalized within the government, and McNeal recalls that his initial focus was on figuring out what NASA's contribution to the field could be. Whereas NASA had been made lead agency for stratospheric chemistry, no one had been for tropospheric chemistry. So his first task was to assess what NASA's capabilities were, what unique abilities it possessed vis-à-vis other research agencies, and what it could contribute to the new research area.[7]

Atmospheric chemistry was, in McNeal's view, a "measurement limited" field. The trace species that were important were present in the atmosphere in minute quantities, at one part per billion and in some cases one part per trillion levels. The ability to measure at these tiny levels was new, and just as was true for stratospheric chemistry, the number of chemical species that could be measured was small. Further, no one had attempted to systematically evaluate which techniques gave the best results—the intercomparison problem. Finally, the laboratory instruments that could measure to these levels could not be taken into the field, which was where researchers wanted to make the measurements, and so development of instruments that could be put into an airplane or on a balloon was an obvious priority. The Upper Atmosphere Research Office was performing this role for stratospheric chemistry, and the development and intercomparison of instruments for field research became one part of McNeal's area.[8]

McNeal also believed that one of NASA's primary strengths as a research organization was management resources. The field experiments that he anticipated carrying out during the 1980s would involve the efforts of hundreds of people from many different universities and government agencies, and hence coordination was a significant challenge. Further, a global-scale effort would involve collaboration with other governments, which was something that NASA, which maintained its own foreign affairs staff, could also handle. Finally, it also had research aircraft that were relatively underutilized and thus available in the earlier years of the program, stationed at the Ames Research Center in California and at Wallops Island in Virginia.

McNeal chose Langley Research Center to manage the day-to-day operation of the Tropospheric Chemistry Program. He had met Don Lawrence and his chief scientist at the time, Robert Harriss, during a "get acquainted" visit, and had been impressed with their knowledge and interest in this general research area.

Harriss had been hired away from Florida State University for an ocean science program at Langley Research Center, but this was terminated when the oceans program manager at NASA headquarters had decided to centralize physical oceanography at JPL. Harriss had then become Lawrence's chief scientist. Harriss's own specialty was biogeochemistry, and he focused on the exchange of gases between the ocean surface and the tropospheric boundary layer, the perfect skill set for the program McNeal intended to forge. So McNeal asked Harriss to be the project scientist for the effort, and assigned the project management function to Langley, with the caveat that Lawrence had to keep the project management functions separate from the science functions. That way, Langley's scientists would have to compete alongside researchers from other NASA centers, other agencies, and universities.[9]

Early the following year, a small group of atmospheric chemists and meteorologists met at NCAR in Boulder to discuss the scientific questions pertinent to the new field. This group sent a letter report to NSF calling for a coordinated study of tropospheric chemistry; NSF then asked the National Research Council (NRC) to form a committee to draft this plan. NRC formed a Panel on Global Tropospheric Chemistry, chaired by University of Rhode Island biogeochemist Robert Duce, to carry out this task. Harriss and McNeal formulated their scientific program from this committee's deliberations and report.[10]

One of the central themes of the committee's discussions was that while knowledge of tropospheric chemistry was growing explosively, the research being done was crisis-driven. It was formulated in response to short-term policy needs. It lacked the comprehensiveness that was necessary to produce a well-integrated understanding of the full range of the atmosphere's chemical processes and fluxes. Anthropogenic sources of ozone-generating trace gases were well-inventoried in North America and Europe due to the decades of pollution research carried out there, but the policy focus of the research had resulted in relative neglect of source gases of natural origin as well as of origins outside the developed world. Hence, one of the committee's recommendations was that the proposed program be long-term in nature, not focused on the immediate problems of the early 1980s.[11]

In the NRC committee's view, progress in atmospheric chemistry was also being inhibited by the tendency of individual scientists to focus on a small piece of the overall challenge. In one sense, of course, this was vital. As was the case in stratospheric chemistry, for example, one needed individual reaction rate measurements in order to lay the foundations for chemical models. But building a better understanding of biogeochemical cycles also required that studies be carried out at larger scales. Investigation of sources and sinks within the biosphere, of transport from one region to another, and of the chemical transformation and removal processes all needed to be performed to fully understand the complex chemical cycles. Such studies would require the participation of many scientists in organized field experiments, drawn from a variety of specialties.

The biogeochemical cycles that were of most interest were those of nitrogen, sulfur, and carbon. The committee recommended investigating potential natural sources of these trace gases in places relatively remote from industrial society: in Arctic tundra, tropical rainforest, the open ocean, and African savanna. Further, they recommended studying the chemical impact of biomass burning on the atmosphere. From basic chemistry, biomass burning had to be a source for these trace gases, but the magnitude of its impact on the global atmosphere was unknown.[12]

Finally, the committee argued that new instruments with faster response times needed to be developed and validated. Existing instrumentation largely did not respond quickly enough to changing levels. This made it difficult to link chemical concentrations to specific air masses so that the chemicals could be traced to their sources. Since one focus of the proposed program was on transport, being able to credibly trace gases to their sources was an important factor. Fast-response instruments seemed to be possible for a number of important species, and the committee sought support for their development.[13]

Thus, the first part of the program McNeal and Harriss assembled was instrument development and comparison. This became known as the Chemical Instrumentation Test and Evaluation (CITE) series of missions. The first of these, carried out in July 1983, will serve to illustrate the CITE series. Harriss recalled later that the initial focus was on getting an instrument to measure the hydroxyl radical. Hydroxyl was suspected of being the atmosphere's cleanser, able to bond with and convert other chemically active trace species and remove them from the atmosphere. But it was extraordinarily difficult to measure because of its chemical reactivity. It was also present in minute quantities, in the parts per quadrillion range. A number of scientists believed they had effective hydroxyl instruments, yet there was a great deal of doubt in the chemistry community that they really produced good results. Hence, the first instrument comparison done by the new GTE for hydroxyl and two trace species related to it, carbon monoxide and nitric oxide.[14]

The first CITE experiment was held at Wallops Island, Virginia, and took place in two phases, ground-based and airborne. GTE's project office had arranged for a cluster of trailers to be set up at the northern end of the island, open to the expected sea breeze, with the trailers equipped with necessary test and air handling equipment to support the experimental equipment. McNeal had accepted proposals from researchers at ten different institutions for this experiment, including the Ames, Langley, Wallops, and Goddard centers, the University of Maryland, National Oceanic and Atmospheric Administration (NOAA), Georgia Institute of Technology, Washington State University, and Ford Motor Company. Three different measurement techniques for each chemical were chosen, with all but one being in situ sampling instruments. The one remote sensing instrument was a lidar proposed by Ford Company researcher Charles C. Wang; this was intended to measure hydroxyl via laser-induced fluorescence.[15]

The experimental procedure that the GTE project team established was to test the nitric oxide and carbon monoxide instruments against samples of known test gas concentrations as well as against ambient air samples fed to them through common manifolds. Hence, the three carbon monoxide instruments, for example, would sample essentially the same atmosphere at the same time via a single air duct. This could not be done for the hydroxyl instruments, however, because one was a remote sensing lidar. Further, no laboratory standard gas mixtures for hydroxyl existed because of the molecule's extremely short lifespan to test these instruments against. For hydroxyl, the strategy was simply to compare the ambient measurements in the hope that they would at least be within the same order of magnitude.

The ground test results for carbon monoxide and nitric oxide were definitive in the eyes of the GTE experiment team. The two groups of instruments agreed to within 10 percent, with no detectable biases in any of the instruments. Given that atmospheric variability of these trace species was much higher than the instruments' demonstrated error levels, these were excellent results. The hydroxyl measurements, however, were disappointing. The three instruments all had operational difficulties, and there were few periods of overlapping data by the end of the experiment. It was therefore impossible to draw any conclusions about their levels of agreement that would have any statistical relevance. The conclusion the GTE team drew from this was that none of the hydroxyl instruments they had tested was capable of producing reliable measurements.[16] In fact, hydroxyl proved so difficult to measure that GTE did not get a reliable hydroxyl instrument until 1999.

After the disappointing hydroxyl results, but with good results from many other instruments, Harriss convinced McNeal that it was still worth conducting field experiments to begin characterizing fluxes of carbon, sulfur, and nitrogen in and out of the biosphere. This led to measurement campaigns in Barbados, Brazil, and the Alaskan and Canadian Arctics. These missions went by the acronym ABLE (for Atlantic [Arctic] Boundary Layer Experiment), and involved use of the NASA Electra aircraft to measure trace species at very low altitudes in the atmospheric boundary layer as well as the establishment of ground measurement stations for comparison and to establish the existence of specific sources.[17]

BOUNDARY LAYER EXPERIMENTS

In July 1981, while the new tropospheric chemistry program was being organized, Harriss had arranged for a new instrument from the Langley Research Center, the Differential Absorption Lidar (DIAL), to be flown on the NASA Electra from Wallops Island, Virginia, to Bermuda. The DIAL had been developed by a group led by Edward V. Browell to detect aerosols via a backscatter technique. It could also measure ozone. The DIAL system produced a continuous profile of the aerosol and ozone content of the air either above or below the aircraft, and still more useful, it produced its output in real time.[18] It could therefore be used to guide the aircraft toward interesting phenomena. And because it produced a continuous readout, it could also be used to examine the continuity of atmospheric structure.

On the July 1981 flight, and a second set of flights to Bermuda in August 1982, Harriss, Browell, and some of their colleagues used the DIAL and some companion instruments designed to sense ozone and carbon monoxide to trace the movement of haze layers from the continental United States eastward into the Atlantic. These haze layers extended more than 300 kilometers into the Atlantic from the U.S. East Coast, clearly demonstrating the existence of long-range transport of pollutants. More interesting to the team, the lidar returns clearly showed that the layers of aerosols maintained a consistent vertical structure over great distances. They did not blend together as the air mass moved. This fact offered the ability to link individual layers to specific sources. Hence chemical transport could be studied in detail. Based on the vertical distribution of ozone and aerosols, their initial data supported the Fishman-Crutzen hypothesis that ozone was produced within the boundary layer due to surface emissions.[19]

In 1984, Harriss's group mounted a similar expedition to Barbados also named ABLE. Their target was study of the Saharan dust clouds that blew westward over the island. These had been known since at least the 1840s, when naturalist Charles Darwin had witnessed them, but Browell's DIAL system allowed investigation of the transport mechanism. The lidar revealed that the dust actually formed many very thin layers that remained distinct over very long distances. This was useful knowledge, as the lack of mixing meant that specific air parcels could conceivably be traced to their origins.

The next field mission, ABLE 2, was considerably more complex an undertaking than its predecessors. In November 1981, Hank Reichle's Measurements of Air Pollution instrument had flown aboard Space Shuttle Columbia, and had returned a surprising result: there appeared to be high concentrations of carbon monoxide over the Amazon basin. This was unexpected, as carbon monoxide is a combustion product that at the time was typically associated with industrial emissions.[20] Instead, this appeared to be from biomass burning. In 1977, for example, the National Academy of Science had evaluated biomass burning's contribution to global carbon monoxide concentrations as about 3 percent, a number that could not possibly be true given Reichle's new data. Figuring out where this anomalous concentration of carbon monoxide was coming from was the scientific basis for ABLE 2. To accomplish this, McNeal worked with Luis Molione of the Brazilian space agency, Instituto Nacionãl de Pesquisas Espaciais, to arrange logistics, identify potential Brazilian collaborators, and gain all the necessary permissions. This was not an easy thing to do, as Brazil's government at the time was a military dictatorship and not particularly interested in science or in having its territory overflown by the aircraft of other governments. But Molione was able to get the government to grant permission for the experiment, and was also able establish a parallel ground-based program that continued after the GTE portion was over.[21]

The ABLE 2 field experiment was carried out during two phases, in order to capture data from the two dominant tropical seasons, wet and dry. Chemical conditions would obviously be different, affecting biogenic emissions. Hence, the first field phase of the experiment was carried out during 1985's dry season, July and August, with the NASA Electra operating out of Manaus, Brazil. Surface measurement stations were established at Reserva Ducke, a biological preserve about 20 kilometers northeast of Manaus, on the research vessel R/V Amanai, and on an anchored floating laboratory on Lago Calado. A tethered balloon station, radiosonde and ozonesonde launches, and a micrometeorological tower completed the experimental apparatus. The surface measurements included enclosures designed to identify specific emission sites; by establishing the location of specific emissions, the science teams could link surface emissions to the airborne measurements. They also provided some of the most interesting results of the experiment.

The science team encountered astonishing chemical conditions as the dry season evolved. In the nineteen research flights, the GTE group measured carbon monoxide levels above the forest canopy that slowly increased through the experiment period. These eventually averaged 3 to 6 times that of the "clean" tropical ocean atmosphere. The primary source for these high levels were agricultural fires that had been set to clear fields; the haze layers produced by burning had carbon monoxide concentrations more than 8 times that of the clean atmosphere. By early August, the haze layers were clearly visible in imagery from the Geosynchronous Operational Environmental Satellite (GOES) satellites, and covered several million square kilometers.[22] This finding was the most significant of the expedition, strongly suggesting that biomass burning was capable of influencing tropospheric chemistry on a global scale.

There were other interesting results. Steven Wofsy of Harvard found that the rainforest was a net source of carbon dioxide at night and a net sink during the day, with the forest soil appearing to be the dominant emission source. Rivers were net sources of carbon dioxide regardless of diurnal affects, and wetlands showed a weaker diurnal cycle than the forest soils. Another discovery by Wofsy's group was that the forest soil was a large producer of nitric oxide and isoprene. This was a surprise because the forest soils in the mid-latitudes were not significant producers of these chemicals, and thus were not implicated in ozone production. But the levels being emitted by the Amazonian soils were high enough to initiate substantial ozone production, leading the researchers to conclude that the natural emissions of the rainforests influenced the photochemistry of the global troposphere.

The wet season expedition, carried out in April and May 1987, was considerably less dramatic. The science teams deployed a similar arrangement of ground-, tower-, balloon-, and aircraft-based measurements to examine the chemistry of the boundary layer in the rainy season. This permitted them to measure the "respiration" of the rainforest, via monitoring carbon dioxide levels within and above the canopy, as well as the fluxes of other chemicals. One significant finding was that nitric oxide emissions from the soil were much higher from pasturelands than from the rainforest soil itself. Hence continued conversion of rainforest to crop or pastureland could itself affect the chemistry of the atmosphere.

GTE's next set of field expeditions were to the Arctic. Harriss and his colleagues believed that the lightly inhabited Arctic region was an area that was likely to be very sensitive to the effects of anthropogenic changes in the atmosphere. The soils in the region were high in carbon content that might be released as the Earth warmed, providing a potential positive feedback effect, and ground data suggested that even the most remote areas of the Arctic were being affected by air pollution from mid-latitude sources. Because of the complex linkages among atmospheric trace gases and the biosphere, and because in winter Arctic air masses moved southeastward across North America, the changing chemistry of the Arctic could also impact air pollution in the mid-latitudes. The Arctic, like the Amazon, had largely been neglected, however, justifying field research in the region. These expeditions became known as ABLE 3A, carried out during July and August 1988, and ABLE 3B, carried out during July and August 1990.[23]

The first phase of the Arctic boundary layer expedition took place primarily in Alaska, with the Wallops Electra operating out of Barrow and Bethel for most of the experiment. As in the Amazonian expeditions, the GTE project office had erected a micrometeorological tower as well as placing enclosure measurements in selected areas to sample soil emissions. The mission scientists had been particularly interested in the methane emissions of the lowland tundra, which was dominated by peatland and shallow lakes, and placed their ground instrumentation in the Yukon-Kuskowkwim Delta region for this study. Methane emissions were known to be widely variable, dependent on the wetness or dryness of the soil and on soil temperature. Other recent examinations of tundra emissions had indicated that methane emissions increased as temperatures rose; since methane is a greenhouse gas, this would provide a positive climate feedback. In the eyes of Harriss, it would also provide an early warning system for global environmental change, as one could monitor the methane emissions as a proxy measurement for soil warming. The ABLE 3A results from the enclosure and aircraft measurements of methane emissions confirmed this earlier work; the Arctic tundra was very sensitive to temperature, exhibiting a 120 percent increase in methane emissions for a 2 degree C increase in temperature.[24]

The mission scientists also measured nitrogen species. The Arctic region was widely believed to be a net sink for tropospheric ozone, with destruction processes outweighing the combined effects of tropospheric production and intrusions of high-ozone air from the stratosphere. The ability of the region to continue destroying ozone depended upon concentrations of nitrogen oxides remaining low so that the Arctic troposphere itself did not become a source of ozone; given their growing awareness of the ability of long-range transport to move pollutants across thousands of miles, it was not clear this would be the case. McNeal had chosen two groups of scientists to make the nitrogen oxides measurements, one led by Harvard University's Stephen Wofsy, the other by John Bradshaw from Georgia Institute of Technology. Their results indicated that the region was, as expected, a net sink for nitrogen species and ozone during summer. However, doubling the nitrogen oxides levels would transform the Arctic into a net source of ozone, and growing levels of industrial pollution being transported in could achieve that.[25]

ABLE 3B expanded on these results two years later with a deployment to the Canadian subarctic region around Hudson Bay. This area was the second largest wetland in the world and thus a major source of natural methane emissions; studying it, the study's scientists believed, would contribute to overall understanding of what they called the "chemical climatology" of North America. Their results here, however, were remarkably different from the ones in the Alaska experiment. Methane flux from the wetlands region was much lower than expected, while stratospheric intrusions of ozone were higher than they had been in the previous expedition. Ed Browell's DIAL laser system showed that aerosol plumes from forest fires affected a significant portion of the troposphere in the region. Steve Wofsy concluded that these fires were the primary source of hydrocarbons and of carbon monoxide in the high latitudes, with industrial emissions contributing less than a third of these contaminants. The DIAL laser's ability to distinguish air masses by their aerosol and ozone composition also led the mission scientists to conclude that they had found several examples of air parcels transported in from the tropical Pacific, however, raising questions about very long-range transport. The chemical age of these air masses was about fifteen days, and they contained much lower amounts of most of the trace species the mission scientists were looking for than the Arctic background air.[26] This led Harriss and McNeal to propose shifting future missions to the Pacific, in order to examine the Pacific basin's chemical climate. These expeditions took place later in the 1990s, after some other important investigations had expanded knowledge of atmospheric chemistry.

The final major field experiment prior to GTE's Pacific shift took place in 1992. This was TRACE-A, the Tropospheric Aerosols and Chemistry Expedition-Atlantic. Planned in conjunction with a major international field experiment to characterize the atmospheric impact of biomass burning in Africa, the Southern African Fire-Atmosphere Research Initiative (SAFARI), the TRACE-A mission was conducted out of Brazil to enable study of transatlantic transport properties. The overall initiative had grown out of a meeting in November 1988 at Dookie College, Victoria, Australia, that had been held to develop the scientific goals for a new International Global Atmospheric Chemistry Program. GTE had already shown that biomass burning and the associated land-use changes had global-scale impacts, but its two Brazilian field expeditions had been too limited in scope to establish longer-range processes. With TRACE-A in Brazil, equipped for long overwater flights, and SAFARI's scientists to the east in Africa, the scientific community could begin to quantify more fully the chemistry and transport of fire emissions.[27]

For TRACE-A, GTE switched from the low-altitude, short-range Electra aircraft to the higher-altitude, longer-range DC-8 from Ames Research Center. The expedition's goal was to trace the movement of burning-derived aerosol plumes as they moved eastward across the Atlantic, and this happened at higher altitudes than the Electra was efficient at. The instrumentation on the DC-8, however, was essentially the same. It was equipped to measure tracer species and aerosols as well as the reactive gases that produced ozone. The greater range enabled it to make flights across the entire tropical Atlantic to Africa, permitting it to trace air masses through the whole region. This ability enabled the science teams to determine that the high levels of ozone that Ritchle's MAPS instrument and that Donald Heath's Total Ozone Mapping Spectrometer (TOMS) instrument both saw over the tropical Atlantic came from reactive nitrogen species, some of which flowed in from Africa, some from South America, and some, in the upper troposphere, that seemed to be generated in situ. This led the scientists to speculate that lighting was a significant source of active nitrogen in the upper troposphere. But the fact that polluted air flowed into the Atlantic basin from both directions (at different altitudes) was surprising; the meteorology of the tropics was more complex than they had expected.[28]

The 1980s, then, witnessed a dramatic shift in scientific understanding of tropospheric chemistry. While NASA was hardly the only organization studying the question, it had the ability to mount large, multi-investigator experiments to examine the full range of scales in the atmosphere. McNeal, Harriss, and the many other GTE experimenters used that capability to develop new knowledge about the biosphere's interaction with the atmosphere, demonstrating in the process that nonindustrial, but still anthropogenic, emissions were substantial suppliers of ozone precursors to the atmosphere. While there is no doubt that the local residents of South American and African regions already knew that their activities produced polluted air, these expeditions demonstrated the global reach of these fire emissions and caused Western scientists to begin accounting for them in global studies. After GTE's missions of the 1980s and early 1990s, pollution was no longer merely a problem of modern industrial states. Instead, the increasing scale of biomass burning, tied directly to increasing population pressures in the underdeveloped world, contributed at least as much to global ozone pollution as the industrial regions.

In the process of carrying out its own research agenda, the Global Tropospheric Chemistry Program also contributed to the scientific capacities of the host nations. Many of the investigators who participated in the field expeditions were funded by the Brazilian and Canadian governments. McNeal's strategy had been to use the visibility of a NASA visit to raise the visibility of local scientists to their own governments, improving their status and funding prospects.

VOLCANIC AEROSOLS

Another area of research NASA entered during the late 1970s was aerosols, particles suspended in the atmosphere. In trying to understand Venus's atmosphere, James Pollack's group at Ames Research Center had undertaken detailed studies of sulfate aerosol chemistry and radiative impacts. In a similar vein, Brian Toon had written his doctoral thesis on a model comparison between Martian climate shifts and potential shifts in Earth climate caused by volcanic explosions. He had employed data from the Mariner 9 mission and data from historic volcanoes on Earth, to conduct his study. He had moved to Ames in 1973 from Cornell, to work with Pollack, on the Pioneer Venus project.

Sulfate aerosols also tended to be injected into the Earth's atmosphere by both human sources (coal-fired power plants, for example) and volcanoes, and as the Mars and Venus efforts wound down, Pollack's group turned their efforts toward understanding how volcanic eruptions on Earth might affect climate. There was some historical evidence that large volcanic explosions did cause global cooling. In 1815, Mount Tambora in the Dutch East Indies had exploded, turning 1816 into a "year without a summer." Half a world away, in New England, frosts continued throughout the summer of 1816, causing widespread crop losses. Speculation at the time connected the sudden cold to dust from the explosion; the much better recorded (although smaller) Krakatoa eruption of 1883 had been identified as the probable cause of a weaker cooling the following year.[29]

The Krakatoa explosion was the first from which useable information about the distribution of volcanic debris was available, primarily from a scattered handful of astronomical observatories; a better set was available from the 1963 eruption of Mount Agung in the Philippines. Pollack's group used their model to examine the response of the model climate system to elevated levels of volcanic debris in the stratosphere, finding that their model results were consistent with the observed cooling from the Agung explosion. This they traced largely to the radiative effects of the sulfate aerosols, not to the ash component of the ejecta plume. They concluded that a single, large explosion could produce globally average cooling of up to 1 degree K. If a series of such explosions over a period of years occurred, this cooling effect could be deepened and prolonged. In their published paper, they also discussed a number of sources of potential errors, including the limited available data and the inability of their model to simulate cloudiness changes (a flaw common to all such models).[30]

The 18 May 1980 eruption of Mount St. Helens in Washington State happened to be well timed. NASA had two satellites with aerosol instruments aboard in orbit, Stratospheric Aerosol Measurement (SAM II) on Nimbus 7 and Stratospheric Aerosol and Gas Experiment (SAGE) on AEM 2, with which to study the evolution of the volcanic plume as well as aircraft-based instruments and ground-based lidar instruments. Beginning late in 1978, Pat McCormick, Jim Russell, and colleagues from Wallops Flight Center, Georgia Institute of Technology, NCAR, NOAA, and the University of Wyoming had carried out ground truth experiments to evaluate how well the SAM II and SAGE instruments characterized the stratosphere's aerosol layers. To accomplish these intercomparisons, they had employed a variety of balloon and aircraft sensors. NASA's P-3, a large, four-engine aircraft originally developed for ocean surveillance, carried Langley's lidar instrument to examine aerosol size and density by laser backscatter measurements. NCAR's Sabreliner, which could fly in the lower stratosphere, was equipped with direct sampling instruments. The University of Wyoming's balloon group, finally, deployed dustsondes, which provided measurements from ground level through about 28 kilometers.[31]

The ground truth experiments were international in scope, starting with flights from Sondrestrom, Greenland, and then moving to White Sands, New Mexico; Natal, Brazil; Poker Flat, Alaska; and finally Wallops Island, Virginia. International partners in Britain, France, Italy, and West Germany had also developed ground-based lidars to use in evaluating the SAGE data on their own. The researchers had chosen the sites to permit examining performance at both high and low latitudes; as luck would have it, while the teams were in Brazil, the volcano Soufriere on St. Vincent erupted, permitting the P-3 and SAGE satellite to examine its plume.[32] The Soufirere measurements suggested that the mass of the material lofted in the stratosphere by the eruptions represented less than half of 1 percent of the global stratospheric aerosol loading, leading McCormick to conclude in a 1982 paper that the eruption was unlikely to have had any significant climate effect.[33]

Analysis of the Soufriere data had been delayed by the eruption of St. Helens, whose location made it a prime opportunity to document the eruption of a large volcano. St. Helen's plume would move west-east across North America, where the existing meteorological and astronomical observing network would be able to track and record it in great detail. NASA had also been in the process of finalizing an agreement with NOAA and the U.S. Geological Survey to initiate a program called RAVE (Research on Atmospheric Volcanic Emissions), another of Pollack's ideas, when the eruption began. McCormick's team at Langley, using the Wallops P-3, and groups from Ames Research Center, using a U-2, and the Johnson Space Center, using an RB-57, all flew missions to characterize the volcano's emissions as the plume moved east, the P-3 from below via airborne lidar and the other two aircraft from inside the cloud. The SAGE satellite's orbit carried it over the plume between 20 and 28 May, adding its larger-scale data to the aircraft and ground measurements.[34]

In November, Langley hosted a symposium to discuss the findings of the St. Helen's effort.[35] The St. Helens explosion was the first Plinian-type eruption to be subjected to modern measurement technology, and it contained some surprises. The plume from the initial explosion had reached to 23 kilometers, well into the stratosphere. Most of the silica ash had fallen out of the stratosphere quickly, with researchers from Ames, Lewis Research Center in Cleveland, and the University of Wyoming's balloon group all finding that the ash was no longer present after three months. Sulfate aerosols from the eruption, as expected, remained in the stratosphere six months later, but St. Helens had turned out to be unusually low in sulfur emissions. Based on this, Brian Toon's group at Ames had therefore predicted the eruption would not have any significant climate impact. Another surprise to the conferees was that chlorine species in the stratosphere had not increased as they had anticipated. Throughout the several eruption events, chlorine remained at essentially normal background levels. Many types of rock contained chlorine, and they had assumed that some of this would be transported into the stratosphere by the ejecta plume. That had not happened. Instead, the missing chlorine became a bit of a mystery.

If St. Helens's eruption had not been expected to have a measurable impact on the Earth's climate because it had not propelled large enough amounts of sulfates into the stratosphere, the late March 1982 eruption of El Chichón in Mexico was another question entirely. It was not a particularly large eruption, but whereas St. Helens had been unusually low in sulfur, El Chichón proved unusually high.[36] Located in the province of Chiapas, this explosion produced the largest aerosol cloud of at least the previous seventy years. The cloud impacted the operation of satellite instruments that had not been designed to study aerosols—it played havoc with the Advanced Very High Resolution Radiometer (AVHRR) instrument on the operational weather satellites, for example, invalidating much of its surface temperature data, while suggesting that AVHRR might be able to provide information on global aerosol density once appropriate algorithms were developed—but the eruption offered the opportunity to study the climate effects of these aerosols in a way that St. Helens had not.

Writing for *Geofísica Internacional*, Pat McCormick explained that the eruption's potential value for testing models of stratospheric transport, aerosol chemistry and radiative effects, and remote sensors had caused NASA to organize a series of airborne lidar campaigns to examine the movement of the volcano's aerosols from 56S to 90N latitudes, and supporting ground and airborne measurements. (The SAGE satellite had failed late in 1981 due to contamination of its batteries and was not available to help follow the movements of the aerosols.) The first airborne mission, in July 1982, flown from Wallops Island to the Caribbean, was exploratory, with the researchers trying to determine whether the edge of the aerosols cloud was detectable. For a later effort in October and November, they orchestrated a series of ground and other in situ measurements timed to coincide with their aircraft's flight path to provide data for comparison purposes. These flights extended from the central United States to southern Chile, to assess the spread of the aerosols into the Northern Hemisphere mid-latitudes. Two more series of flights in 1983 examined the aerosol dispersion into northern high latitudes, into the Pacific region, and a final series of flights into the Arctic in 1984 reached the North Pole.[37]

Even without SAGE, however, satellites proved able to provide unexpected information about the volcanic plume. Reflecting the very high density of the cloud, the geosynchronous imaging satellites had been able to track it throughout its first trip around the world. During this first circumnavigation, it spread to cover a 25 degree latitude belt. After it could no longer be followed on the visible-light imagery, other satellites could still detect it. The Solar Mesosphere Explorer, which had a stratospheric infrared water vapor sensor, was able to follow the mass via its infrared signature. It showed that most of the aerosol mass remained south of 30N for more than six months, the result of an unexpected stratospheric blocking pattern. Similarly, the TOMS instrument on Nimbus 7, which utilized ultraviolet backscatter for detection, proved able to trace the sulfur dioxide from the eruption.

And unlike the St. Helens eruption, the El Chichón eruption was followed by substantial ozone losses extending into the mid-latitudes. It was not clear what the cause of this was. Using an infrared spectrometer, two researchers at NCAR had measured greatly increased levels of hydrogen chloride in the leading edge of the cloud as it passed over North America six months later. This was one of the reservoir species resulting from the complex ozone depletion reaction, and its presence was evidence that chlorine was responsible for the missing ozone.[38] It was not clear how it had gotten there, however. While they assumed that it was derived from chlorine species released from the volcano, that was not necessarily the case. No one had actually measured this chlorine species close to the source, and, of course, the previous investigation of Mount St. Helens had shown that chlorine levels had not been significantly affected. In the El Chichón case, the chemistry mystery was further deepened by a clear correlation between maximum aerosol density and maximum ozone loss in time and space. This suggested that aerosols were somehow involved, but in ways that were not at all obvious or easy to parse out. The data from El Chichón itself could not resolve the mystery.

Finally, the predicted climate impact of the eruption was not apparent. In his review of El Chichón's impact on scientific knowledge, David Hofmann noted that whatever climate impact El Chichón might have had was masked by an unusually strong El Niño the following year. El Niño, a phenomenon that causes dramatic, short-term meteorological changes, begins with formation of a large, unusually warm body of water in the Pacific. Eventually, that warm pool forces a temporary reversal of the Pacific equatorial current, bringing the warm water to the west coast of North and South America. There it produces torrential rains, which effectively transfer the excess heat of the upper ocean into the troposphere. Hence, a strong El Niño translates into general tropospheric warming over a period of about six months. By removing El Niño's effect mathematically, James K. Angell of NOAA had argued that the El Chichón cooling *had* occurred. In effect, it had slightly weakened the El Niño that had been forming. The cooling effect he found had been slightly weaker than predicted by the climate models, but was the same order of magnitude, providing a measure of confirmation.[39] This result met with some skepticism, however. While mathematical adjustments to data were normal practice in science, they could lead one astray. The only way to prove Angell's point definitively would have been to do the eruption over again while preventing an El Niño from forming, thus eliminating one variable from contention. That, of course, was not possible. Confirmation would have to wait for another volcanic explosion to occur.[40]

HOLES IN THE OZONE LAYER

In May 1985, as the first international assessment of the state of ozone science was being prepared under the World Meteorological Organization's (WMO) auspices, researchers at the British Antarctic Survey revealed that there were huge seasonal losses of ozone occurring over one of the Dobson instruments located in the Antarctic.[41] The 30 percent losses they were seeing were far more than expected under existing theory; at this point in time, the consensus was that there had been about 6 percent depletion globally. Nor had the TOMS/SBUV science team reported the existence of deeply depleted regions. If the Dobson instruments were to be believed, both theory and satellites were in error somehow.

The Antarctic Survey's paper came out when many of the leading stratospheric scientists were meeting in Les Diablerets, Switzerland, to review the draft of the next major international ozone assessment, which was due that year. This document updated the status of the laboratory-based efforts to refine the rate constants of the reactions involved in the nitrogen and chlorine catalytic cycles involved in ozone chemistry and their incorporation into chemical models as well as examining the recent history of stratospheric measurements. The Balloon Intercomparison Campaigns (BIC) that Bob Watson's office had funded in the early 1980s and the fact that the most recent several years of laboratory measurements had not produced major changes to reaction constants had resulted in a belief that the gas-phase chemistry of ozone was finally understood. In reviewing and summarizing all this, however, the 1985 assessment, which was also the first multinational assessment sponsored by WMO, had grown to three volumes, totaling just over a thousand pages.

The Antarctic Survey's announcement thus came as a bit of a shock. Some of the conferees at Les Diablerets were already aware of the paper, as Nature's editor had circulated the paper to referees during December 1984.[42] The paper's authors had raised the question of a link to chlorine and nitrogen oxides, in keeping with the prior hypotheses regarding depletion mechanisms, but did not provide a convincing chemical mechanism. The gas-phase chemistry of ozone did not appear to enable such large ozone losses. The paper was too late to incorporate into the 1985 assessment, however, and the conference did not formally examine it. It was much discussed, however, in the informal hallway and dinner conversations that accompany conferences. Adrian Tuck, then of the U.K. Meteorological Office, recalls that many of the attendees were inclined to ignore the paper, as it was based on only the measurements of one station.[43] Several earlier attempts to find an overall trend in the Dobson data had not succeeded, defeated by the combination of quite significant natural variation in ozone levels and the inability to resolve differing calibrations. The network's data had gained some disrepute in the community because of this. But Tuck had found Farman's paper well enough done to be taken seriously. He knew Joseph Farman, the paper's lead author, to be a very careful researcher. There was clearly something wrong with either the Dobson instrument or the stratosphere.

Farman's paper caused Goddard Space Flight Center's Richard Stolarski to look again at the TOMS/SBUV data. Farman had written to Donald Heath, the TOMS/SBUV principal investigator, well before publishing his paper but had not gotten a response. But the TOMS/SBUV should have detected the depleted region described in the paper if it were real and not an artifact of the Dobson instrument, and Stolarski found that it had. The TOMS/SBUV's inversion software contained quality-control code designed to flag ozone concentrations below 180 Dobson units as "probably bad" data.[44] Concentrations that low had never been seen in Dobson network data, and could not be generated by any existing model. It was impossible as far as anyone knew, and it was a reasonable quality-control setting based on that knowledge. But the Antarctic ozone retrievals had come in well below the 180 unit setting. Their map of error flags for TOMS had showed the errors concentrated over the Antarctic in October. They had ignored it, however, assuming the instrument itself was faulty.

Stolarski had reexamined their data and found that the depleted region encompassed all of Antarctica—the "ozone hole" that rapidly became famous—by the end of June that year. In August, at a meeting in Austria, Heath showed images generated from the data for 1979–83 depicting a continent-sized region in which ozone levels dropped to 150 Dobson units.[45] Plate 2 shows the phenomenon two years later. It was after these images began circulating that a great many atmospheric scientists (and policymakers) began to take the ozone question seriously again. On one level, the images offered confirmation that Farman's data reflected a real phenomenon and not an instrument artifact, and thus it merited scientific investigation. They also demonstrated that it was not a localized phenomenon. The Dobson measurements were point measurements, taken directly overhead of the station, while the TOMS/SBUV data covered the entire Earth.

The TOMS data images placed the depleted region into perspective, in a sense, showing the geographic magnitude of the phenomenon. On another level, like the older images of the Earth from space, these images were viscerally powerful. They evoked an emotional response, suggesting a creeping ugliness beginning to consume the Southern Hemisphere. JPL's Joe Waters, for one, saw it as a cancer on the planet.[46] While almost no one lived within the boundaries of the depleted region, if it grew very much in spatial extent, it would reach populated landmasses. And since no one knew the mechanism that produced the hole, no one could be certain that it would not grow.

Susan Solomon at the NOAA Aeronomy Lab had also been struck by the Farman paper, but also recalled that a group in New Zealand had measured unusually low nitrogen dioxide levels.[47] In the stratosphere, nitrogen dioxide molecules react with chlorine molecules to form sink species, thereby removing the chlorine from an active role in destroying ozone. She speculated that the relative lack of nitrogen dioxide could mean increased chlorine levels through some unknown mechanism. She also drew on two other bits of recent work to forge a model of what that mechanism might be. In 1982, Pat McCormick had published a paper based on SAM II data showing the presence of what he called Polar Stratospheric Clouds (PSCs). While these had been known since the late nineteenth century, they had been thought to be rare—there had been few written accounts of them. But they were actually quite extensive, according to his data.

The other bit of information she drew on was work done during 1983 and early 1984 by Donald Wuebbles at Lawrence Livermore and Sherry Rowland. They had begun investigating the possibility that chemical reactions in the stratosphere might happen differently on the surface of aerosols than they did in a purely gaseous phase. At a meeting in mid-1984, they had shown data from laboratory experiments that suggested that the presence of aerosols did serve to alter the chemical reaction pathways. Their model, which was very preliminary and based on a number of hunches, had suggested this heterogeneous chemistry could be responsible for depletion rates of up to 30 percent.[48]

The ozone hole extended back to 1979, and therefore could not be related to either El Chichón or volcanic activity more generally. There had been no significant eruptions between 1963 and 1979, eliminating the possibility that volcanic aerosols were to blame. But McCormick's PSCs were an obvious alternate suspect. While the composition of the PSCs could not be determined from the SAM II data, the temperatures they seemed to be forming at were consistent with water ice and with nitric acid trihydride—also mostly water. The mechanism Solomon and her co-workers proposed depended upon the presence of water, which would react with chlorine monoxide to release chlorine. This could only happen if nitrogen oxide species concentrations were extremely low, however, because they would normally remove the chlorine monoxide into a reservoir species more quickly. To get the very low concentrations of nitrogen oxides, a reaction on the surface of PSC crystals involving chlorine nitrate was necessary. This reaction created nitric acid and chlorine monoxide; this would decompose under the Antarctic sunrise to release chlorine.[49]

As the TOMS satellite images of the stratospheric ozone spread through the atmospheric research community during late 1985, the number of potential mechanisms expanded. A group led by Michael McElroy at Harvard University proposed a chemical mechanism that also included reactions involving PSC surfaces, but was focused on the release of bromine. Various meteorologists proposed several dynamical mechanisms, generally postulating ways that low-ozone tropospheric air might have ascended into the stratosphere, causing the hole simply by displacing the normally ozone-rich stratospheric air. Finally, Linwood Callis at Langley Research Center proposed a solar mechanism. In his hypothesis, odd nitrogen produced by high-energy particles hitting the upper atmosphere might be descending into the stratosphere, where it would destroy ozone.[50] The solar cycle had reached an unusually high maximum in 1979, coinciding with the formation of the ozone hole. This mechanism had the happy consequence of eliminating the hole naturally. As solar activity returned to normal in the late 1980s, the hole would disappear on its own.

There were, then, three general classes of mechanisms proposed to explain the ozone hole by mid-1986: anthropogenic chemical (i.e., variations on chlorofluorocarbon [CFC] depletion chemistry), natural chemical (odd nitrogen), and dynamical. These hypotheses were testable, in principle, by measurements. During the preceding years, NASA and NOAA had fostered instruments for stratospheric chemistry that could look for chemical species required by the chemical hypotheses and had also developed the capacity to examine stratospheric dynamics. The instruments aboard the UARS had been intended to measure the key species as well as stratospheric temperature and circulation, and had it been available would have made selection of the most appropriate hypothesis far simpler than the process that actually played out. But while finished, it could not get into space. Instead, NASA resorted to field expeditions to resolve the controversy.

In February 1986, shortly before a workshop scheduled to discuss the proliferation of depletion hypotheses, the supporters of a chemical explanation for the ozone hole gained a significant boost from JPL. Barney Farmer's Atmospheric Trace Molecules Spectroscopy (ATMOS) instrument had flown on Space Shuttle Challenger (STS-51B) during the first two weeks of May 1985. Capable of measuring all of the chlorine and nitrogen species involved in the photochemical depletion hypothesis except the key active species chlorine monoxide, the instrument had permitted the science team to produce the first complete inventory for them in the May stratosphere. This included several first detections of some of the trace species, and included the entire active nitrogen family and all the chlorine source species, including the manmade CFC-11, CFC-12, HCFC-22, and the primary natural chlorine source methyl chloride. It also measured the primary sink species, hydrogen chloride, and the data showed the expected diurnal variation in concentrations.[51]

The ATMOS measurements were also important in that they were the first simultaneous measurement of all the trace species. Previous measurements of the various chemicals had been made at different times and places, by different investigators, using different techniques. This had made it very difficult to claim that differences between measurements were chemical in causation and not a product of experiment errors or natural variations. ATMOS effectively eliminated those sources of error. It provided confirmation that some of the species predicted to exist in the stratosphere actually were there, that they existed in approximately the expected ratios, and that they varied in the course of a day the way theory said they should. While it could not see the poles from the Shuttle's orbit, and it flew at the wrong time of year for the Antarctic phenomenon in any case, by demonstrating that all of the chemical species required by the CFC thesis existed in the stratosphere, it provided a substantial credibility boost.

At a meeting at the Aeronomy Lab in March 1986, the proponents of each of the ozone hole theories had their chance to explain it to their peers. While the meeting had not been called to plan a research program to demonstrate which, if any, of these hypotheses happened to be true, the collected scientists came up with one anyway. Adrian Tuck recalls that Arthur Schmeltekopf, one of the laboratory's senior researchers, pointed out that the instrumentation necessary to select between the hypotheses already existed in one form or another.[52] For either of the chemical mechanisms to be correct, certain molecules had to be found in specific ratios relative to other molecules. Barney Farmer's infrared spectrometer could measure most of the chlorine and nitrogen species in question. It could not fly on any of the Shuttles, which were grounded due to the destruction of Challenger that January during a launch accident, and balloons large enough to carry it could not be launched from the American Antarctic station at McMurdo. But his balloon version, known only as the Mark IV interferometer, would work perfectly well as a ground-based instrument. David Hofmann's ozonesonde and dustsonde balloons could provide measurements at various altitudes within the polar vortex, permitting evaluation of the dynamical hypothesis, and he was already a veteran at making these measurements in the Antarctic. Robert de Zafra at the State University of New York, Stony Brook, had developed a microwave spectrometer that could remotely measure chlorine monoxide, the key species Farmer's instrument could not sense. Finding high levels of chlorine monoxide in the Antarctic stratosphere was crucial to verifying the anthropogenic chemical hypotheses. This too was a ground instrument. Finally, the Aeronomy Laboratory's Schmeltekopf had developed another remote sensing instrument to measure nitrogen dioxide and chlorine dioxide. This was a visible-light spectrometer that employed moonlight.

The meeting participants did not expect that a single, primarily ground-based expedition would be conclusive. It would not, for example, provide the kinds of data necessary to disprove the dynamical thesis. Demonstrating that the upwelling proposed by the dynamics supporters was not happening would require simultaneous measurements from a network of sites within the vortex, very similar to the meteorological reporting networks used for weather forecasting in the industrialized nations. NASA's Watson wanted to launch this first expedition in August 1986, much too short a time to arrange for additional ground stations. Further, expedition scientists at McMurdo would only be making their measurements from the edge of the polar vortex, not deep within it. This limited the utility of the results. Finally, the remote sensing instruments being used would not necessarily be seen as credible. Much of the larger scientific community still resisted remote sensing, preferring in situ measurements. At the very least, in situ measurements provided corroboration of potentially controversial results.

To better address these potential criticisms, Watson, Tuck, Schmeltekopf, and others also sketched out a plan for a second expedition using aircraft. This was based upon the payload designed for a joint experiment planned for early 1987 that was designed to examine how tropospheric air was transported into the stratosphere. Known as STEP, for Stratosphere-Troposphere Exchange Project, this had been the idea of Edwin Danielsen at Ames Research Center. Danielsen had conceived of ways to use tracer molecules to investigate vertical air motion, including ozone and nitrogen oxides, and had assembled an instrument payload for the NASA ER-2 (a modified U-2 spyplane). A key unknown in the transport process was how moist tropospheric air dried as it moved into the stratosphere, and this question was STEP's principal target. Two new instruments, an in situ ozone sampler devised by Michael Proffitt at the Aeronomy Lab, and an in situ NO_y instrument built by David Fahey, also of the Aeronomy Lab, had been chosen for this tracer study, supplemented by a water vapor instrument built by Ken Kelly at the Aeronomy Lab, aerosol instruments from NCAR and the University of Denver, and nitrous oxide instruments from Ames Research Center and NCAR.[53] Finally, a new version of James Anderson's chlorine monoxide instrument rounded out the payload.

The first ground-based expedition to figure out the ozone hole was carried out between August and October 1986. The NSF, which ran McMurdo Station, handled the logistics of moving the thirteen members and their equipment down to the Antarctic. Susan Solomon had volunteered to be the expedition leader after Art Schmeltekopf had not been able to go. The four experiment teams were all able to make measurements successfully, with Hofmann's dustsondes showing that aerosols were descending, not ascending, tending to refute the dynamical hypotheses, while the NOAA spectrometer showed high levels of chlorine dioxide and very low levels of nitrogen dioxide, in keeping with the anthropogenic chemistry hypothesis and in opposition to the solar cycle thesis. The SUNY Stony Brook instrument recorded high levels of chlorine monoxide, again as expected under the anthropogenic hypothesis. Only the JPL team could not reduce their instrument's readings to chemical measurements in the field. They needed access to computers back in California. But the information from the other instruments was sufficient to convince the researchers that the anthropogenic hypothesis was probably correct. The data they had was clearly consistent with the chlorofluoro-carbon theory, and clearly inconsistent with the others.

Before the team left Antarctica, they participated in a prearranged press confer-ence to explain their results, and here they raised what was probably an inevitable controversy. Solomon, who was too young to have been a participant in the ozone wars of the 1970s, made the mistake of giving an honest answer: that the evidence they had supported the chlorofluorocarbon depletion hypothesis, and not the others. Widely quoted in the mainstream press, her statement outraged propo-nents of the other hypotheses, and they were quite vocal in complaining about it to reporters. The most aggrieved parties were the meteorologists who had pro-posed the dynamical theses. The evidence against the dynamical thesis, Hof-mann's aerosol measurements, had not been circulated in the community (it was still in Antarctica), so no one could check or absorb it. The team thus returned to what appeared to be a vicious little interdisciplinary conflict, carried out via the mainstream press.[54]

Yet this conflict in the popular press did not really have an impact on the re-search effort, suggesting that it had far less reality than press accounts at the time suggested. British meteorologist Adrian Tuck, for example, who had been involved in the planning for the expedition, did not doubt that both dynamics and chem-istry had roles in the hole's formation. The relevant questions involved the details of the processes and relative contributions of them, considerations that were not well described in the press.

Hence, the media controversy did not much affect planning for the airborne experiment. The expedition's goals had included closer attention to dynamical concerns in any case. It was obvious that even if dynamics were not solely respon-sible for the depleted region, they certainly played a role in establishing the con-ditions that formed and maintained it.[55] The use of aircraft, particularly the fragile and difficult to fly ER-2, required a greater meteorological infrastructure for the mission whose data would also be available to test the dynamical thesis. Watson had been able to convince the Ames Research Center's leadership to provide the

ER-2 as well as the DC-8, equipped with many of the instruments deployed by GTE, as well as Barney Farmer's Mark IV interferometer. Estelle Condon was assigned to be the project manager for the expedition. Watson had also chosen Adrian Tuck, who had joined the Aeronomy Lab in 1986, as the mission scientist. Tuck had gained considerable experience at carrying out airborne sampling missions in the U.K. Meteorological Office, where he had been originally been hired to help figure out the Concorde's impact on the stratosphere.[56] Brian Toon from Ames was his second, due to his long-standing aerosol research interests.

The airborne mission was flown from Punta Arenas, Chile, beginning in August 1986. Condon had arranged for the conversion of one of the hangars at the local airfield into a laboratory and office complex. Art Schmeltekopf had convinced Watson that the principal investigators should be made to convert their instrument's data into geophysical variables (i.e., into temperature, pressure, concentrations of a particular molecule) within six hours after a flight, and post them for the other scientists to see. This was intended to solve a perennial problem. In many other NASA and NOAA field campaigns, investigators had sent their graduate students and not come themselves, meaning data did not get reduced until long after the expedition was over. Sometimes, it disappeared entirely and no one ever saw results. This expedition was too important to permit that. Further, the expedition's leaders needed to know what the results of one flight were before planning the next. In this effort, some of the instruments needed daylight, and some of the observations needed to be done at night. Hence sound planning demanded a quick data turnaround. One consequence was that the hangar at Punta Arenas had to be converted into fairly sophisticated laboratory complex, complete with computers to do the data reduction. Ames's Steve Hipskind shipped four standard cargo containers' worth of gear there, and rented an air force C-141 cargo aircraft to carry the scientists' equipment down in August.[57]

The expedition's leadership had also established a satellite ground station at Punta Arenas just as had been done for GARP Atlantic Tropical Experiment (GATE) in 1974. Up-to-date meteorological information was necessary to plan the aircraft missions. The ER-2 was very limited in the range of winds it could take off and land in, and both it and the DC-8 were at risk of fuel tank freezing if temperatures at their cruise altitudes were too cold. Further, the scientists wanted the aircraft to fly into specific phenomena, requiring an ability to predict where they would be when the aircraft reached the polar vortex.

Tuck arranged to borrow a pair of meteorologists who had done forecasting for the Royal Navy during the Falklands Islands war from the U.K. Met Office. He also chose to use forecast analyses from the Met Office, and a meteorological

AIRBORNE ANTARCTIC OZONE EXPERIMENT
DC-8 PAYLOAD

WHOLE AIR SAMPLER
VEDDER-NASA Ames
HEIDT-NCAR

OZONE CHEMILUM. DET.
GREGORY-NASA Langley

LYMAN ALPHA HYGROMETER
KELLY-NOAA

OZONE/AEROSOL LIDAR
BROWELL-NASA Langley

F.T. INTERFEROMETER
FARMER-JPL

F.T. SPECTROMETER
MANKIN-NCAR
COFFEY-NCAR

UV/BLUE SPECTROGRAPH
WAHNER-NOAA

Fig. 1. The DC-8 NASA 717, with principal investigators, their affiliations, and their instruments. Left hand list consists of in situ experiments; right-hand list shows remote sounders.

AIRBORNE ANTARCTIC OZONE EXPERIMENT
ER-2 PAYLOAD

MULTIFILTER SAMPLER, HNO$_3$, HCl
GANDRUD-NCAR
ClO, BrO
ANDERSON-HARVARD
CLOUD PARTICLE SPECTROMETERS
FERRY-NASA Ames
MICROWAVE RADIOMETER
GARY-JPL
AEROSOL SIZE SPECTROMETER
FERRY-NASA Ames
CN COUNTER
WILSON-U. DENVER
LASER SPECTROMETER, N$_2$O
LOEWENSTEIN-NASA Ames

AIR MOTIONS, P & T
CHAN-NASA Ames

WHOLE AIR SAMPLER
VEDDER-NASA Ames
HEIDT-NCAR

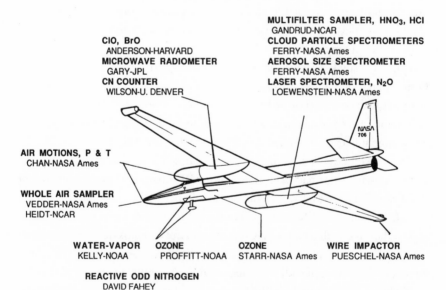

| **WATER-VAPOR** | **OZONE** | **OZONE** | **WIRE IMPACTOR** |
| KELLY-NOAA | PROFFITT-NOAA | STARR-NASA Ames | PUESCHEL-NASA Ames |

REACTIVE ODD NITROGEN
DAVID FAHEY

Fig. 2. The ER-2 NASA 706, with principal investigators, their affiliations and their instruments. All experiments are in situ, except the microwave radiometer, which sounds the temperature profile above and below the aircraft.

facsimile link was established to permit near-real-time transmission from Brack-nell. Further, to ensure the flight plans carried the two aircraft into the ozone hole, and through and under PSCs, the expedition leaders and pilots wanted access to real-time TOMS and SAM II data, and to data from the second-gener-ation SAGE instrument, launched in 1984. Because the TOMS, SAM, and SAGE instruments all required sunlight to operate and the mission leaders wanted the DC-8 to make nighttime flights, the Met Office also borrowed an algorithm from the Centre Nationale des Recherches Météorologiques that computed total ozone maps from the High Resolution Infrared Spectrometer (HIRS) 2 instru-ment on the NOAA operational weather satellites.[58] Because HIRS sensed emis-sion, and not absorption or backscatter, it was independent of the Sun.

About 150 scientists took up residence in Punta Arenas for the expedition that August. During the two-month expedition, the research aircraft flew twenty-five missions, twelve for the ER-2 and thirteen for the DC-8, surprising many of the participants. The difficult weather conditions in the early spring Straits of Magel-lan, the thirty-year plus ages of the two aircraft, and the great distance from spare parts had caused the expedition's leaders to hope they would get half that many before the winter vortex broke up. Meteorologically, they were lucky, and the aircraft ground crews provided sterling service in keeping the aircraft ready. There were incidents that colored future expeditions, however. The predicted winds for the DC-8's cruise altitude were off by half (too low) on two occasions, forcing emergency aborts due to insufficient fuel. And the temperature at the ER-2's cruise altitude also tended to be overestimated, leading to the wing-tip fuel tanks freezing. The chief ER-2 pilot, Ron Williams, had expected that, however, and had calculated the rate it which it would thaw and become available for the return trip. These incidents served as reminders that this research was also dangerous business.

The principal technological challenge during the mission turned out to be Anderson's chlorine monoxide instrument. His group at Harvard had had about six months to prepare it, and had assembled and flight-tested it for the first time in June. In the expedition's first ER-2 flight, however, it had failed just as the air-craft reached the polar vortex. But it worked again when the aircraft landed, lead-

(Opposite) The experiments carried during AAOE. AAOE Equipage: A. F. Tuck, et al., "The Planning and Execution of ER-2 and DC-8 Aircraft Flights over Antarc-tica, August and September 1987," *Journal of Geophysical Research* 94:D9 (30 August 1989), 11,183. Reproduced with the permission of the American Geophysical Union.

ing the team to suspect that the intense cold was triggering the failure. One of his assistants wrote new software prior to the second flight that logged all of the instrument's activities in hope of determining the fault; this led them to a space-qualified connector between the instrument and its control computer that was opening under the intense cold.[59]

Hence the third flight, 23 August, produced the first useful data from the instrument. This flight showed chlorine monoxide approaching levels nearly 500 times normal concentrations within the polar vortex, while ozone, as measured by Proffitt's instrument, appeared to be about 15 percent below normal. As flights through September continued, the ozone losses deepened, and the two instruments demonstrated a clear anti-correlation between chlorine monoxide and ozone. The most striking correlation occurred on 16 September. The ER-2's flight path took it through a mixed area in which ozone and chlorine and ozone moved repeatedly in opposition as if locked together, leaving little doubt among the experimenters that chlorine was responsible for the ozone destruction. By the end of the third week of September, the ER-2 was encountering parts of the polar vortex in which nearly 75 percent of the ozone at its flight altitude had been destroyed.[60] While correlation did not in and of itself prove causation, none of the other hypotheses could explain this piece of evidence.

There was quite a bit of additional data gathered during the expedition that was also relevant to theory selection. David Fahey's experiment produced data that strongly suggested that the PSCs were composed of nitric acid ice. It had found highly elevated levels of nitric acid while flying through them; cloud edges were clearly visible in his data. JPL's Mark IV spectrometer's data showed vapor-phase nitric acid increasing toward the end of September, as the stratosphere warmed, also suggesting that it existed in a condensed form prior to that. Its measurements of the active nitrogen family also clearly showed that these trace species were substantially reduced, corroborating the in situ measurements. Measurements made by Max Lowenstein from Ames Research Center, by Michael Coffey and William Mankin from NCAR, and by Barney Farmer's Mark IV provided clear evidence that stratospheric air within the vortex was descending throughout the period, not ascending as required by the dynamics theories.[61]

The evidence gathered during the Airborne Antarctic Ozone Experiment (AAOE), then, was clearly consistent with one of the three hypotheses the investigators had carried into the Antarctic, and equally clearly inconsistent with the other two. This left the researchers with little choice but to accept the hypothesis that anthropogenic chlorine was the proximate cause of the ozone hole, with the major caveat that particular meteorological conditions also had to exist to enable

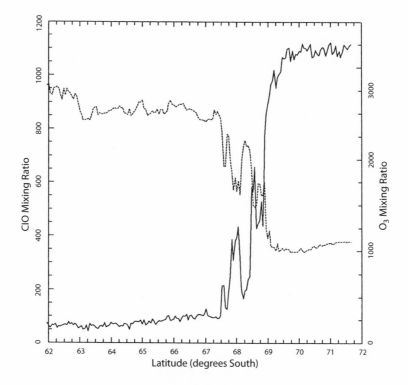

The anti-correlation between chlorine monoxide and ozone from 16 September 1987. This chart is often referred to as the mission's "smoking gun" result. From: J. G. Anderson and W. H. Brune, "Ozone Destruction by Chlorine Radicals Within the Antarctic Vortex: The Spatial and Temporal Evolution of ClO-O₃ Anticorrelation based on in Situ ER-2 Data," *Journal of Geophysical Research* 94:D9 (30 August 1989), 11,475. (figure 14). Reproduced with the permission of the American Geophysical Union.

it. This did not necessarily mean that the hypothesis was *true* in an absolute sense. It remained possible that a fourth hypothesis might emerge to explain the observations even better. But in the absence of a better hypothesis, the expedition's leaders had little choice but to accept the anthropogenic thesis as the correct one. The participants in the AAOE thus drafted an end of mission statement that was released 30 September, two weeks after the end of negotiations over an international protocol to ban CFCs. The statement concluded that the "weight of observational evidence strongly suggests that both chemical and meteorological mechanisms perturbed the ozone. Additionally, it is clear that meteorology sets up the special conditions required for the perturbed chemistry."[62]

A series of scientific meetings after the expedition served as forums to discuss its results and those of related efforts that had gone on during it. One other significant result had to be considered, and its implications for the AAOE's results thought through. Mario Molina and his wife, Luisa Tan, working at JPL, had carried out a series of elegant experiments that demonstrated a third chlorine-based mechanism could be the primary cause of the ozone hole. This thesis proposed that if the PSCs were mostly nitric acid ice instead of water ice, they would scavenge hydrogen chloride and hold the molecules on their surfaces, where they became more available for reaction with another chlorine species (chlorine nitrate). This second reaction released a chlorine monoxide dimer while trapping nitrogen dioxide in the ice. Because nitrogen dioxide was necessary to convert active chlorine into inert reservoir species, its removal would permit very high concentrations of chlorine to occur. At virtually the same time, researchers in Britain isolated the difficult-to-measure dimer and quantified the rates of its catalytic cycle.[63]

At a meeting in Dahlem, Germany, that November, the dynamical and odd nitrogen hypotheses were again discarded in favor of a new synthesis centered on chlorine chemistry, with meteorology providing some necessary preconditions— relative confinement of the vortex and very cold temperatures. The participants also attempted to assess the relative contributions of the chemical mechanisms suggested by the laboratory efforts. The AAOE data had shown low levels of bromine oxide, limiting the potential impact of the bromine catalytic cycle that McElroy had proposed. Solomon's could not be evaluated. The telltale species for her mechanism was hydrogen dioxide, which the expedition had not measured. Finally, the Molinas's mechanism was able to explain the observed results very closely, leaving it the dominant thesis at the end of the meeting.[64]

This meeting's participants also sketched out where the uncertainties still lay in their understanding of the ozone hole. The PSCs were one locus of uncertainty. While their composition seemed to be known, what particles served as the nuclei for their formation, what temperature they formed at and over what range they were stable at, and how large the PSC crystals could grow before they sedimented out of the stratosphere were all unknown. The details of the various catalytic cycles were unclear, of course, as the inability to test Solomon's idea suggests. Further, none of the chemical models, even those including the heterogeneous chemistry, could replicate the observed depletion pattern. The depleted region extended downward in altitude more deeply than models predicted, and the overall ozone loss predicted by the models remained less than observed. This suggested that feedback processes were at work that the models did not capture.

Finally, it was not clear whether the Antarctic ozone hole was relevant to the mid-latitudes where most humans lived. Models tended to treat the hole as if it were completely contained in a leak-proof vessel, but to many of the empirical scientists, this was absurd. Most of the Earth's stratospheric ozone was produced in the tropics (where the requisite solar radiation was strongest) and was transported to the poles. That also meant the Antarctic's ozone-poor air would be transported out after the vortex breakup in October. The difficulty, as it had been throughout the ozone conflict, lay in proving quantitatively what was qualitatively obvious.

While the AAOE and National Ozone Expedition (NOZE) II expeditions had been in progress, diplomats had been in Montreal negotiating a treaty that would cut CFC production by 50 percent. It had been deliberately isolated from the expedition's findings to prevent biasing it with undigested data; scientific briefings to the conferees had been provided by Daniel Albritton, director of the Aeronomy Laboratory, who did not go on the expedition.[65] The resulting Montreal Protocol, of course, had no force until ratified by national governments, and this was where the science results could have an impact.[66] Further, the protocol contained a clause requiring the signatory nations to occasionally revisit and revise its terms in the light new scientific evidence — an escape clause, if the science of depletion fell apart. Instead, this became a point of conflict as additional research suggested that high depletion rates might be possible in the northern mid-latitudes, producing pressure for rapid elimination of the chemicals.

In the absence of the Antarctic data, the scientific basis for the Montreal Protocol had been the 1985 WMO-NASA assessment. While this had been limited to the gas-phase only chemistry that had been the basis of the prior decade's research, it had clearly documented that CFCs and their breakdown products were rising rapidly in the stratosphere. By this time, the laboratory kinetics work organized by NASA had also resulted in stabilization of the rate constant measurements. These no longer showed the large swings of the late 1970s, giving confidence that the gas-phase chemistry for the nitrogen and chlorine catalytic cycles was reasonably well understood.[67] Finally, the assessment had also established a clear scientific position that ozone depletion in the mid-latitudes *should* be happening. From a policy standpoint, however, where the assessment was weak was in its demonstration that the depletion expected by the chemists was *actually* happening in the atmosphere.[68] There were solid economic reasons to keep the CFCs flowing, and powerful economic interests relied upon the lack of evidence for real-world depletion to insist that there was no basis for regulation of CFCs.

But during the busy year of 1986, Donald Heath had circulated a paper prior to publication claiming to find very substantial mid-latitude loss based upon the

Nimbus 7 TOMS/SBUV instrument. Separately Neil Harris, one of Sherry Rowland's graduate students, had deployed a new method of analyzing the Dobson data. In previous analyses, the annual data from all of the stations had been lumped together, which had the effect of masking potential seasonal trends. Because the stations were not all of equivalent quality, it also tended to contaminate the dataset, making data from the good stations less reliable. Harris decided to reexamine each of the twenty-two stations northward of 30N individually, using monthly averages instead of annual averages. By comparing pre-1970 and post-1970 monthly averages, he found a clear wintertime downward trend.[69] Their trend was half that indicated by Heath's data, but they were at least pointing in the same direction.

Hence Albritton and Watson established a new group, called the Ozone Trends Panel, to resolve these conflicting bits of evidence to see if there really was a trend revealed in the data. The twenty-one panel members reanalyzed the data from each of the thirty-one Dobson stations with the longest records by cross-checking them against ozone readings from satellite overflights and using the result as a diagnosis with which to correct the ground station data. This revised data revealed a clear trend of post-1970 ozone erosion that was strongest in winter, and that increased with latitude. They also performed their trend analysis with the unrevised data, which showed the same general trend, but less clearly.[70] They were not able to confirm the Solar Backscatter Ultraviolet (SBUV) findings, but it remained possible that the instrument was correct. The group found clear evidence that ozone in the troposphere was increasing, which would partially mask stratospheric ozone loss from the Dobson instruments. This would result in an underestimate of ozone erosion from the Dobson network data. Tropospheric ozone would not be detectable to the SBUV, however. It might be the more accurate, in this case, but the assembled scientists did not have a basis on which to evaluate the SBUV's own degradation.

The panel's findings were formally released 15 May 1988, one day after the American Senate voted to ratify the Montreal Protocol requiring a 50 percent reduction in CFC production. Its general conclusions were already well known, as Watson and Albritton had briefed policymakers and politicians on them, and inevitably, they had been leaked. Yet their formal press conference still drew great attention. At it, Watson, Sherry Rowland, and John Gille stated unequivocally that human activity was causing rapid increases in CFCs, and halons in the stratosphere, and that these gases controlled ozone. They reported that the downward trend found in the reanalysis was twice that predicted by the gas-phase models. However, they also felt the need to specifically reject the TOMS/SBUV data as an independent source of knowledge. The large downward trend in the TOMS/SBUV data was, in the panel's opinion, primarily an artifact of instrument degradation.[71]

Watson also took the occasion to call for more stringent regulation than that contained in the protocol, and this caused some controversy. As it stood in 1988, the protocol would reduce CFC production by half, and it would be capped at that level. But this was not enough to him. The inability of models to provide a credible prediction of ozone destruction meant that it was impossible to forecast a safe level of CFC production that was not zero. Watson believed a ban on the chemicals was necessary to restore the stratosphere to its original state.

The research effort did not stop with ratification of the Montreal Protocol by the Senate. There were considerable remaining uncertainties over the precise chemical mechanisms behind the unexpectedly high levels of chlorine in the Antarctic, and over the meteorological conditions that were necessary anteced-ents to the hole phenomenon. Further, the protocol contained a clause requiring that its terms be reexamined regularly so that it could be adjusted in the light of new scientific evidence. Since the protocol had been negotiated on the basis of the gas-phase chemistry of 1985, it might well need to be altered to reflect the heterogeneous chemistry of 1987. When added to the revised Dobson data show-ing clear indications of greater ozone loss in the wintertime Arctic than expected, these left NASA's Watson in need of more data. Conditions on the periphery of the Antarctic hole were similar to those in the core of the Arctic polar vortex in terms of temperature, for example, raising the possibility that as chlorine contin-ued to increase an Arctic version of the hole might form. Far more people (includ-ing all the people paying for this research) lived in the Northern Hemisphere than did in the Southern Hemisphere, and both the human and political implications were obvious.

Two indications that the chemistry of the Antarctic phenomenon might also exist in the Arctic had already been found. On 13 February 1988, the Ames ER-2 had carried its Antarctic payload on a flight from its home at Moffet Field north to Great Slave Lake, Canada. This was still south of the Arctic polar vortex, but the data from the flight clearly showed highly elevated levels of chlorine monox-ide appearing suddenly north of 54N.[72] The second indication that Arctic chem-istry might also be perturbed came from Aeronomy Lab scientists, who had car-ried the spectrometer used during the NOZE and NOZE II expeditions to Thule, Greenland, the last week of January 1988. This was inside the Arctic vortex during the time they were present. They had found elevated levels of chlorine dioxide and very depressed levels of nitrogen dioxide, suggesting that the wintertime chemical preconditioning that happened in the Antarctic was also happening in the Arctic. This did not mean that a similar ozone hole would form, however. As Susan Solomon pointed out in the resulting paper, the Arctic stratosphere was much warmer, and in the years since records started in 1956, average monthly temperatures in the Arctic stratosphere had never fallen to the minus 80 degrees C that seemed to be the point at which PSCs formed.[73] While PSCs did form during shorter periods of extreme cold in the Arctic, the lack of prolonged periods of extreme cold suggested that depletion would not reach the extreme levels found in the Antarctic.

These findings caused Watson to seek the opinions of Adrian Tuck, Art Schmeltekopf, Jim Anderson, and some of the other mission scientists on whether to try to do an Arctic expedition modeled on AAOE in the winter of that year or whether to wait until the winter of 1989. The principal investigators associated with the instruments had been involved in either expeditions or post-expedition conferences of one sort or another since STEP in January 1987, and it was asking a lot to send them back into the field in the winter of 1988. They decided—or, rather, Watson, Tuck, and Anderson convinced the rest—to mount the Arctic expedition sooner rather than later, in January 1989. This mission became the Airborne Arctic Stratospheric Expedition (AASE), flown out of Stavanger, Norway.[74]

The AASE made thirty-one flights into the northern polar vortex in January and February 1989. This particular winter proved to be unusually warm and windy at the surface, correlating to an unusually cold, stable stratosphere. The expedition's scientists were therefore able to collect a great deal of information about the expedition's primary targets, the PSCs. The resulting special issue of *Geophysical Research Letters* contained twenty-three papers on PSCs. The observations confirmed that both nitric acid and water ice clouds formed, with the nitric acid clouds dominating as expected from thermodynamic considerations, and several instruments provided characterizations of the nuclei around which the ice crystals formed. The expedition did not settle all of the questions surrounding PSCs, unsurprisingly; it was still not clear, for example, how they facilitated the process of denitrification. Richard Turco, in his summary of the expedition results, remarked that a "consistent and comprehensive theory of denitrification remains elusive."[75]

As expected, the expedition found that the chemistry of the Arctic polar vortex was highly disturbed. The low levels of nitrogen species and high levels of chlorine species mirrored those in the Antarctic, and the final ER-2 flight in February actually found higher chlorine monoxide levels than had been measured in the Antarctic. This confirmed to the science teams that the same chemical pre-processing that happened in the Antarctic had happened in the Arctic. Based on measurements by Ed Browell's lidar instrument on the DC-8, this resulted in ozone destruction just above the ER-2's cruise altitude. This finding was corroborated by Mike Proffitt's ozone instrument and the Ames Research Center nitrous oxide tracer measurements. It did not result in a hole like that in the Arctic, however, because extensive downwelling was simultaneously bringing ozone-rich air in from higher altitudes, and because the polar vortex broke down before the Sun was fully up.[76]

Hence the message that the expedition's scientists took out of the AASE was that all of the chemical conditions necessary to reproduce the Antarctic ozone hole existed in the Arctic, but that unusual meteorological conditions would have to occur in order for one to actually happen. Very cold temperatures would have to prevail into March, and the atmospheric waves that normally roiled the Arctic stratosphere would have to be quiescent. Such conditions were not impossible.

The combined results of the Ozone Trends Panel and the four field expeditions caused the Montreal Protocol to be renegotiated. In his history of the protocol, Edward Parson explains that the results finally produced industry acceptance that actual harm had been done by CFCs. CFCs would therefore be regulated based on what had already happened, not on what might happen in the future. And because the chemicals had lifetimes measured in decades, there was no longer any doubt that damage would continue to happen. And further, of course, there was still no way to determine what level of chlorine emissions might be harmless. It continued to be the case that the models did not reflect reality—reality was worse. Hence in a series of meetings culminating in London in June 1990, the protocol was revised to include a complete ban on the manufacture of CFCs, as well as other anthropogenic chemicals that introduced chlorine into the stratosphere. CFC production was scheduled to cease in 2000; the other chemicals had deadlines ranging from 2005 to 2040.[77]

THE GLOBAL VIEW, FINALLY: CORROBORATION

On 15 September 1991, the crew of Space Shuttle Discovery finally deployed the long-awaited UARS. As it had with the SAGE satellite in 1979, NASA arranged a field mission to provide data against which to compare its results. This was planned as a second deployment by the ER-2 and DC-8 to the Arctic during January and February 1992. As fortune would have it, however, the catastrophic eruption of Mount Pinatubo in the Philippines in June 1991 provided another natural laboratory in which to study the thorny question of mid-latitude ozone loss. The eruption was followed by substantial mid-latitude ozone depletion, and several lines of evidence, from the second Arctic expedition, from UARS, from a 1992 flight of the ATMOS instrument on Shuttle Atlantis, and from balloon-based measurements, all pointed to sulfate aerosols as additional actors in ozone chemistry. But the eruption of Pinatubo also launched a final outbreak of the ozone wars, leading political critics of the anthropogenic chlorine hypothesis to return to long-discredited claims that volcanic chlorine was primarily responsible for stratospheric depletion.

During 1991, more disturbing evidence of Northern Hemisphere depletion came from Goddard Space Flight Center. In 1989, the TOMS science team had found a way to correct the TOMS/SBUV data for instrument degradation by exploiting differences in the way the individual channels degraded. This permitted them to produce a calibration that was independent of the Dobson network, and they had then spent two years revising the data archive in accordance with the new calibration. This revised data brought the TOMS globally averaged depletion measurement to within the instrument error of the Dobson stations. The data showed a clear poleward gradient to the ozone loss, with no significant reduction in the tropics but loss in the mid-latitudes, increasing toward the poles. It also displayed the expected seasonality, with considerably greater loss in winter than in summer.[78]

The vertical distribution of the mid-latitude loss, however, was confusing to the research community. The gas-phase depletion chemical models indicated that most of the ozone loss should be in the upper stratosphere, but it clearly was not. Instead, the ozone loss was concentrated in the 17 to 24 kilometers region, the lower stratosphere, and the same altitude that the polar ozone destruction happened. This suggested that the primary mechanism for ozone destruction in the mid-latitudes was heterogeneous chemistry, which required the presence of particulates. But certainly the PSCs could not be at fault in the mid-latitudes, where the stratosphere was much too warm to permit their formation. Some of the mid-latitude loss could be attributed to movement of ozone-poor air from the collapsing polar vortex into the mid-latitudes, but this did not seem a complete explanation for the phenomenon. So some researchers turned back to the sulfate aerosols that were prevalent throughout the stratosphere at this altitude.[79]

The second Arctic expedition, AASE II, was designed somewhat differently from the preceding two airborne missions. It was designed to examine the evolution of the stratosphere from fall until the end of winter, and at a latitude range covering the tropics to the Arctic. The mission's research goals were primarily directed at understanding the mid-latitude depletion phenomenon and whether

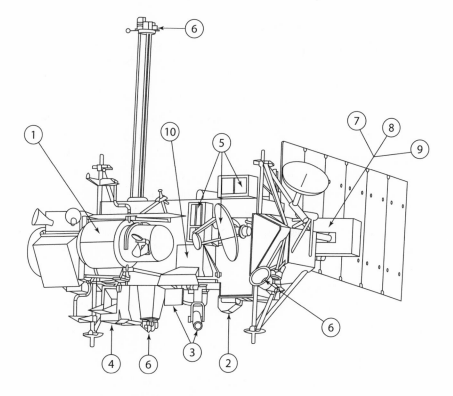

1. Cryogenic Limb Array Etalon Spectrometer
2. Halogen Occultation Experiment
3. High Resolution Doppler Imager
4. Improved Stratospheric and Mesospheric Sounder
5. Microwave Limb Sounder
6. Particle Environment Monitor
7. Solar Ultraviolet Spectral Irradiance Monitor
8. Solar/Stellar Irradiance Comparison Experiment
9. Active Cavity Radiometer Irradiance Monitor
10. Wind Imaging Interferometer

Wireframe of the Upper Atmosphere Research Satellite. Because this satellite was deployed by the Space Shuttle, it could not reach a polar orbit. Its instruments leave a data gap at the highest latitudes. Courtesy NASA.

the severely perturbed Arctic chemistry had anything to do with it. The ER-2 was flown first out of Ames Research Center in October 1991 for six missions, and then shifted to Bangor, Maine, at the beginning of October for the Arctic flights. These continued through the end of March, in order to examine the breakdown of the vortex and the distribution of its air to lower latitudes. While Bangor was too far from the vortex edge to permit deep penetration, this site had been chosen for its lower risk.[80] The wind and icing conditions at Stavanger had been more danger-ous than warranted by a repeat mission, particularly since the major goals of this mission did not require reaching far inside the polar vortex.

The AASE II expedition provided a great deal of new information about the conditions necessary to produce large-scale ozone destruction in the Arctic. It confirmed the suspicion from the first Arctic expedition that extensive down-welling of higher altitude air continually introduced fresh ozone, limiting the total column destruction. It also demonstrated that the Arctic stratosphere fos-tered more rapid destruction of chlorine monoxide than did the Antarctic, limit-ing the chemical's southward movement as the vortex broke up. This was a prod-uct of nitric acid photolysis once the Sun had risen far enough to light the stratosphere. Because the photolysis rate increased with temperature, the warmer Arctic stratosphere destroyed the chlorine monoxide more rapidly. This had the effect of further reducing the rate of ozone destruction. In their expedition sum-mary, Jim Anderson and Brian Toon pointed out that large-scale ozone losses in the Arctic would require substantial removal of nitrates via extensive PSC forma-tion, and that had not happened during the expedition. The Arctic stratosphere had been too warm that year.[81]

However, there had been enough PSC formation that the expedition was able to determine that they were not primarily ice after all. Using a lidar instrument aboard the DC-8, a team led by Edward Browell of Langley Research Center demonstrated that the PSCs were largely composed of droplets in a supercooled liquid state. Brian Toon had suggested in 1990 that this might be the case based on a thermodynamic study; other measurements made during the AASE had also suggested the possibility. The difficulty in proving the case simply lay in the fact that both states existed. There were PSCs composed of ice and others composed of liquid, and many PSCs contained both. Disentangling the true state of the clouds was difficult due to the limited number of measurements available and the complexity of the cloud phenomenon. This sent the chemical kineticists back to the laboratory, to investigate how the chlorine and nitrate reactions differed between the liquid surface and the ice-phase surfaces they had been working with previously.[82] It also made modeling the cloud phenomenon far more difficult, since the relative abundance of the different types of clouds, and the ratio of liq-uid to solid particles in the clouds, now mattered, but there were not enough observations available to determine the ratios empirically.

The Mount Pinatubo eruption of 1991 further complicated the chemical pic-ture. The Pinatubo eruption was the largest volcanic explosion of the century, and its position in the tropics, where tropospheric air ascends into the stratosphere, had ensured that its ejecta cloud was rapidly transported throughout the stratosphere. Large increases in stratospheric sulfate aerosols were measured in the volcano's wake, and widespread ozone depletion had followed. A tentative mechanism to explain sulfate-aerosol mediated ozone destruction had been proposed in 1988 based on laboratory studies. This postulated that hydrolysis of certain nitrogen species on sulfate aerosol droplets could alter the balance between the chlorine and nitrogen catalytic cycles, accelerating the release of active chlorine. This might occur outside the polar regions anywhere the stratosphere happened to cool enough, explaining some of the observed mid-latitude loss. By increasing the sulfate aerosol loading of the stratosphere (and therefore the total surface area available for the reaction), a volcanic eruption would accelerate the conversion of chlorine from inactive reservoir species to active species, thereby increasing ozone loss.[83]

While the Pinatubo eruption provided new observations that tended to confirm a role for sulfate aerosols in ozone loss, it also produced a reeruption of the ozone war. There was a line of argument in conservative circles that the anthropogenic chlorine depletion hypothesis was being hyped by the scientific community in the interests of increased funding, and this was revived after the eruption.[84] The eruption, of course, had been followed by extensive ozone loss and thus could easily be blamed for it directly. One simply had to contend that the volcano, and not humans, was the original source of the chlorine.

Second, the mission scientists for the AASE II had made an unwise decision: they held a press conference midway through the expedition. On 3 February 1992, James Anderson, Brian Toon, and Richard Stolarski had announced that the chemical conditions in the polar vortex were primed for an ozone hole, and one might extend as far south as Bangor if conditions within the polar vortex remained cold enough through March. This statement was taken up in the press, often without the important caveat *if* conditions remained cold enough, and broadcast throughout the nation. It was picked up by Senator Albert Gore, Jr. (D, Tenn.) the next day, who used it to attack then-President George H. W. Bush for resisting a resolution before Congress to accelerate the phase-out of CFC production by five years. Senator Gore proclaimed that Bush would be responsible for an ozone hole forming over his home of Kennebunkport, Maine.[85]

At virtually the same time, but much more quietly, the first chemical maps of the Arctic stratosphere produced by Joe Water's Microwave Limb Sounder team at JPL made their way to Washington (reproduced in Plate 3). These put in a visual form the relationship between temperature, active chlorine species, and ozone throughout the Arctic stratosphere. Combined with the announced intention of a major producer, Du Pont, to cease producing most CFCs by 1996 regardless of additional regulation, Bush rescinded his instructions to Senate Republican leaders to block the resolution, and it passed without dissent 6 February.

The volcano and the press conference created what *Science* writer Gary Taubes labeled the "ozone backlash." The Artic hole did not happen; instead, the Arctic stratosphere warmed within days of the press conference, eliminating any chance of a substantial hole forming. The nonappearance of Senator Gore's Kennebunkport hole triggered a series of attacks by conservative writers. But these were not directed primarily at Gore for his exploitation of ozone depletion for political gain—of which he was certainly guilty. The attacks were directed at the expedition scientists and at the larger thesis of CFC-induced ozone depletion. The arguments these writers made was that volcanoes, not CFCs, were the primary source of stratospheric chlorine, and included the false claim that no one had ever measured CFCs in the stratosphere.[86]

This line of attack, broadcast to millions of people on Rush Limbaugh's radio show, was disturbing enough that Sherry Rowland felt compelled to make it the subject of his 1993 presidential address to the American Association for the Advancement of Science. Framing the controversy as a failure of the scientific community to properly educate the public, Rowland deconstructed the root of this claim. A 1980 paper in *Science* had argued for a volcanic origin for most chlorine in the stratosphere. This argument was based upon measurements of chlorine gas trapped in bubbles within the ash fall of an Alaskan volcano that had erupted in 1976, not measurements of chlorine species in the stratosphere. Rowland then sketched the ways in which chlorine might be removed from the volcanic plume on its way to the stratosphere that the author had not considered, and he pointed to the 1982 measurements made by Mankin and Coffey of NCAR in the El Chichón plume. They had documented an increase in hydrogen chloride of less than 10 percent after this eruption, and that had not appeared until six months after the eruption. He then pointed to all the measurements of CFCs in the stratosphere that had been made since 1975 to discredit the claim that they had not been measured.[87]

In April 1993, Joe Waters's Microwave Limb Sounder group at JPL published the results of their first eighteen months of operations in *Nature*. Their data, displayed as colored maps of temperature, chlorine monoxide, and ozone, clearly showed the expected anti-correlation of ozone and chlorine monoxide in a way that made clear the spatial extent of the stratosphere's altered chemistry. The simultaneity of the instrument's data also provided important corroboration of the temperature dependence of the PSC-accelerated reactions and followed the complete cycle of evolution and collapse of the vortex.[88] Eventually, they began to make movies from the data showing the complete lifecycle of the phenomenon.

The following year, the ATMOS science team published an inventory of carbonyl fluoride based on the measurements taken on the 1985 Spacelab 3 flight and the late March 1992 Atlas 1 flight. Carbonyl fluoride had no natural source in the stratosphere, and was produced solely by breakdown of CFCs. It could therefore be used to trace the buildup of anthropogenic gases in the stratosphere. The team found that between 1985 and 1992, the amount of carbonyl fluoride increased by 67 percent, in keeping with estimates of the amount of CFCs released.[89] Similarly, the ATMOS team argued in a separate paper that their measured 37 percent increase in hydrogen chloride and 62 percent increase in hydrogen fluoride between 1985 and 1992 were in agreement with other ground measurements and with the estimates of CFC release. Further, after pointing out that there had been no measured increase in the stratospheric burden of hydrogen chloride immediately after the Pinatubo eruption, they argued that the sizeable increase in hydrogen fluoride during this seven-year period gave the game away in any case.[90] The primary origin of these gases was human. The HALOE instrument on the UARS, which also measured fluorine species worldwide, finally, corroborated their results.

In 1995, Paul Crutzen, Mario Molina, and Sherry Rowland received the Nobel Prize in chemistry for their 1970s work in ozone chemistry. While the gas-phase chemistry that had been the center of their effort during those years had not turned out to be the primary mechanism for ozone destruction, they, Hal Johnston, Richard Stolarski, and Ralph Cicerone had set in motion a complex series of research efforts that led eventually to the correct mechanism. These efforts were comprehensive in time, space, and methodologies, encompassing laboratory kinetic studies, numerical models, balloon and aircraft measurements employing direct sampling and remote sensing, and space-borne observations. The twin spines of this effort were NASA's UARP and the NOAA Aeronomy Laboratory; without their support, the difficult question of ozone depletion would not have been resolved as quickly.

There were weaknesses, however. As the *space* agency, NASA could have been expected to make better use of the global view than it did, and if one merely counts up the dollars spent, space viewing was its emphasis. But the long delays in getting the UARS into orbit left the agency without the space hardware it had intended to use to solve the ozone question during critical years. It deployed a fallback science program using aircraft to carry out its statutory responsibilities; when finally deployed, the expensive space assets wound up confirming what the scientific community already believed. The vibrant aircraft research program NASA had maintained primarily as a means of testing and validating new space instrumentation wound up being its primary science program. Tying its science programs to the Shuttle program had clearly been a mistake. What the science program had needed was reliable space access, and the Shuttle had not delivered on that promise.

Yet the global view did produce important corroboration, for while the active research community seemed comfortable with the results of in situ measurements, the larger scientific and policy communities did not find them convincing. Satellites provided confirmation that point measurements made inside the atmosphere adequately reflected the global atmosphere. Their principal weaknesses during the 1980s (besides the lack of reliable space access) had been in the reliability of their calibration. While Pat McCormick's occultation instruments had not had this problem, they also either did not measure the all-important ozone (SAM II and SAGE) or did not yet have a long enough record to make a convincing case (SAGE II). TOMS/SBUV, which fortunately operated for fifteen years, took many years to get a reliable calibration technique. Hence despite the determined efforts of the UARP to get better satellite data in the 1970s, they did not really succeed until the early 1990s.

Modeling, too, proved to be a weakness, although one that was not limited to NASA. The theoretical models deployed to study the ozone question, if evaluated as means of *prediction*, were failures. No one generated a model of ozone chemistry that accurately predicted ozone depletion. During the 1970s, the real atmosphere showed much less depletion than the models did, while during the 1980s it showed much more depletion than did the models. The atmosphere proved much more complicated than the chemistry community had believed when they began this research effort, and contained far more phenomena (and variations of phenomena) than the collective intelligence of the research community could program into a model. Instead, the models were numerical thought experiments, serving as guides to the observational research that needed to be done and helping to clarify the relative importance of various chemical and dynamical processes.[91] In the field expeditions, they also helped guide decisions about where to send the aircraft to obtain desired measurements. They were essential to the ozone research program for these reasons. But like the weather forecast models developed during GARP, their ability to predict had limits.

Beginning with the GTE's early missions, NASA developed the ability to carry out large-scale, multi-instrument atmospheric studies that combined in situ and remote sensing measurements. New in the late 1970s, by the end of the 1980s they were well-established means of gaining new knowledge of the atmosphere. They were methodologically complex, requiring not only large-scale participation from scientists during the experiment but significant post-expedition coordination as well. Program managers began to draw scientists into workshops to compare and study data, and in the most contentious areas, they began to organize formal assessments of the state of relevant knowledge. These served multiple purposes. They served to accelerate the normal process of scientific argumentation that takes place in journals, providing forums where interpretations could be argued over before being committed to print. They also served to identify areas that were not well understood, helping direct future efforts. Finally, in addition to these scientific purposes, they increasingly served policy purposes. The assessments in particular were written to provide scientific bases for policy action; by the end of the decade, they began to include "Summary for Policymakers" sections, making this purpose clear.

The scientific community's knowledge of the Earth's atmosphere changed dramatically during the 1980s. In the 1970s, tropospheric ozone was believed to be largely derived from stratospheric injections into the troposphere, with an additional human industrial component to the ozone budget (smog); by the end of the 1980s, atmospheric chemists had demonstrated that photochemical production of ozone in the troposphere had a significant biogenic component as well. They had also shown that agricultural burning had global-scale chemical impacts on the atmosphere. Combined with the dramatic imagery of the ozone hole, they had conclusively demonstrated that humans had achieved the ability to fundamentally alter the chemistry of the atmosphere.

CHAPTER SEVEN

The Quest for a Climate Observing System

Human beings are now carrying out a large scale geophysical experiment of a kind that could not have happened in the past nor be reproduced in the future. Within a few centuries we are returning to the atmosphere and oceans the concentrated organic carbon stored in sedimentary rocks over hundreds of millions of years.

—*Roger Revelle and Hans Suess, 1957*

NASA's atmospheric chemistry programs were only one manifestation of the agency's effort to improve scientists' understanding of the Earth's atmosphere, albeit a very important one. Another was its effort to establish a climate observing system. In 1977, after a National Academy of Sciences study chaired by oceanographer Roger Revelle argued that increasing greenhouse gas emissions would raise global temperatures several degrees by the middle of the twenty-first century, the U.S. Congress had passed a National Climate Program Act directing NASA, the National Oceanic and Atmospheric Administration (NOAA), the National Science Foundation (NSF), the Department of Energy, and other agencies to formulate a program of research into climate.[1] NASA's part of this effort was the formulation of a space-based research infrastructure for studying global climate processes, drawing on its expertise in remote sensing and planetary studies. It commissioned its first study of what a comprehensive climate observing system should include in 1979; it launched the first piece of hardware for its Earth Observing System (EOS) in 1999, twenty years later.

Interest in climate research was not new to the late 1970s, of course. What had changed was the subject's immediacy. Scientists had tended to see Earth processes as slow and relatively insensitive to change before the 1970s. The Earth system, in this view, was basically stable and could be perturbed only with great difficulty. But a few pioneering scientists during the decade had argued that this view was false. Based on analysis of the Camp Century ice core drawn from the Greenland ice sheet, a team of Danish scientists concluded that Earth's climate could change radically in the space of a century. In 1973, Nicholas Shakleton had concluded from a million-year sediment core record that there had been many glaciations in the Earth's recent past, not the four previously held to have existed, and these were linked to small variations in the Earth's orbit—the now-famous Milankovich cycles. These small variations in the orbit caused equally subtle changes in received solar energy, driving vastly larger changes in climate through feedback processes that were not well understood. The following year, drawing on a range of evidence, including ice core, tree ring, and pollen research, Reid A. Bryson of the University of Wisconsin had argued that only small changes in either solar output or in atmospheric transmittance of solar radiation would have large climatic effects. He had also argued that those effects would appear relatively quickly, finding that the major changes leading to the end of the last glacial period had occurred within a century or two.[2] If these interpretations of Earth's climatic past were true, dramatic climate change could happen on a timescale that humans could witness—and that politicians might have to address.

NASA itself funded some of the earliest studies of the potential severity of human-induced warming, at its Goddard Institute for Space Studies (GISS) in New York. These made GISS one of the leaders of the reinvigorated field of climate science. The GISS studies, in company with others made at its parent, the Goddard Space Flight Center in Maryland, and still others at Ames Research center by James Pollack's modeling group, all led NASA to propose a Mission to Planet Earth. By 1989, when NASA finally got approval to begin building it, this had become the most expensive science program in American history, expected to cost $17 billion through fiscal year 2000.

CLIMATE MODELING AT GISS

James E. Hansen, a graduate student from the University of Iowa, had gone to work at GISS on various problems related to the Venusian atmosphere. Hansen had gotten interested in planetary atmospheres at Iowa, where James Van Allen had been an informal mentor, and had developed a mathematical model of aerosol scattering in Venus's cloud cover. This got him involved with both the Pioneer Venus mission that was in slow progress during the mid-1970s, and with the development of the GISS nine-level weather forecast model, for which he developed the solar radiative heating scheme. This task also, he reflected many years later, allowed him to hire Andrew Lacis, also from Iowa State, who became the group's principal radiative transfer specialist.[3]

At first, like many modelers, Hansen and Lacis had developed a one-dimensional radiative-convective model in order to provide important pieces of the physics coding while also being able to study the impact of various greenhouse gas concentrations on the model's equilibrium climate. Working with W. C. Wang, they published their first such study in 1976. But the one-dimensional model had prescribed feedbacks, which meant that they would not change as the model climate did. It wasn't realistic enough. Hansen wanted to transform the GISS nine-level general circulation model that had been used for the Global Atmospheric Research Program (GARP) Observing System Simulation Experiments into a three-dimensional climate model to overcome this limitation. He proposed a task to do this to NASA in 1975, but it was not initially funded. His philosophy had been to develop a coarse-resolution version of the model to permit long integrations within the center's computing capacity. Unlike a weather model, which need only run the equivalent of a few days, a climate model had to be run over years and preferably decades. But it was not clear that one could achieve stable integration of a general circulation model over decades of time at a coarse resolution. One of Hansen's Venusian co-workers decided to try it anyway. It worked.[4]

The demonstration convinced NASA to support the effort, which started in earnest in 1977. Others joined the effort. When Milt Halem moved the weather modeling effort down to the Goddard Space Flight Center, Gary Russell, their mathematician and principal programmer, stayed at GISS. David Rind came over from the Lamont-Dougherty Geological Observatory, bringing over stratospheric dynamics expertise as well as strong interest in paleoclimate studies, prefiguring one of GISS's major contributions to climate modeling as a science: the comparison of model results to observations, something Jule Charney had also demanded of weather modelers. This had been the whole point of GARP.

More than five years elapsed before Hansen's team published their first paper based on the three-dimensional model, which they eventually named the GISS Model II. Years later, they explained that there had been several reasons for the long gestation. First, Jastrow had wanted them to focus on the "farmer's forecast," the thirty-day forecast, so they experimented on the influence of things like sea surface temperature on monthly forecasts. Second, they had faced resistance from reviewers who would not accept that a low-resolution model could be valid. Hansen had used the lowest spatial resolution that would still define the large-scale features of the general circulation, so that the limited computing power that he had could be applied toward better model physics. There was little point in having sufficient spatial resolution for precise storm tracking when that was not the purpose of the model.[5]

Third, they carried out a great many experimental runs before publishing, which they ascribed later to their innate conservativeness. In 1979, well before their first publication, Charney had been asked to head another National Academy of Science summer study of the possibility of global warming, and knowing about the GISS effort, he had asked for some of these results. Charney had begun the study using the only other extant three-dimensional model, Suki Manabe's at the Geophysical Fluid Dynamics Laboratory (GFDL) at Princeton, and had wanted Hansen's for comparison. The two models differed substantially in their sensitivity to changes in carbon dioxide levels, with the GISS model more sensitive than Manabe's. This had led to what Hansen considered the most useful scientific outcome of Charney's effort: a detailed comparison of the two models' component processes aimed at understanding the different outcomes.[6] This analysis, too, had also taken time to carry out.

The differences between the models did not undermine Charney's confidence that carbon dioxide–induced warming was inevitable, however. The underlying theory was unimpeachable: one could not change the radiative characteristics of Earth's atmosphere without a climate response. There was already a century's worth of radiative transfer physics research backing this conclusion. And this is what his report ultimately announced, casting the potential warming in a range of temperatures based on the two models' sensitivities. Charney subtracted a half degree from the lower result (Manabe's), and added a half degree to the higher result (Hansen's), to get what has since become the canonical range: 1.5 to 4.5 degrees C. Writing his preface as chairman of the Atmospheric Sciences Committee, Vern Suomi stated that the group had found "no reason to doubt that climate changes will result and no reason to believe that these changes will be negligible."[7]

The final reason their first three-dimensional model–based publication took so long was that they continued experimenting with one-dimensional models to better define some of the components of the global model, and in parallel they began to develop sets of observational data against which to compare the model outputs.

One of these led to the first paper to get Hansen into political troubles, "Climate Impact of Increasing Atmospheric Carbon Dioxide" (1981).[8] In this study, Hansen's group had studied the model climate's response to doubled carbon dioxide while changing a series of individual feedback processes. By doing this they hoped to characterize each feedback process's relative importance to the model climate's response. They included various humidity, cloud, sea ice, and vegetation feedbacks in their testing.

They had also examined the model climate's sensitivity to its ocean. The Earth's oceans, by storing heat away from the atmosphere, delayed surface warming by a rate that was difficult to determine. The ocean's surface layer, generally called the mixed layer because it was affected by surface winds, would warm in a few decades. Because there was thought to be little exchange between the mixed layer and the deep ocean, however, the vast heat storage of the deeps would not necessarily further slow the rate of surface warming. Then again, there would be some mixing of the mixed layer and the deeps over the decades it would take the Earth to equilibrate to its new carbon dioxide–enriched climate, so Hansen tried to bracket the oceans' impact on warming by running the model using a mixed-layer only case and a case in which the mixed layer was coupled to a thermocline layer.

Finally, they had developed a set of global temperatures to compare against their model. They started in the 1880s, when there were sufficient instrumented stations to define global temperatures, although crudely; after 1900 they felt on firmer ground. One of their biggest revelations in this study was from this temperature data, in fact. Their dataset showed that global temperatures were "nearly as high today as [they were] in 1940. The common misconception that the world is cooling is based on Northern Hemisphere experience to 1970." In other words, while the Northern Hemisphere had been cooling, the Southern Hemisphere had been warming strongly, offsetting nearly all the Northern Hemisphere cooling. This also meant that the world had warmed about 0.4 degrees C in the past century, which was "roughly consistent" with the model-calculated result. The temporal pattern of the model warming was not the same, however, until they included additional forcings from the known major volcanic eruptions and solar variations.

Stating that "the global warming projected for the next century is of almost unprecedented magnitude," they concluded their article with a statement regarding the potential consequences of global warming, positive and negative.[9] They proposed that due to polar amplification of the warming, there was some danger of rapid disintegration of the West Antarctic ice sheet, which unlike the primary

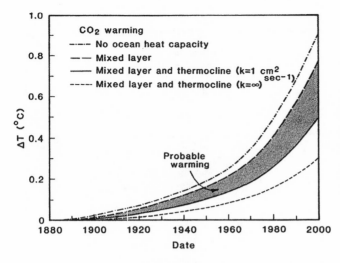

Fig. 1. Dependence of CO_2 warming on ocean heat capacity. Heat is rapidly mixed in the upper 100 m of the ocean and diffused to 1000 m with diffusion coefficient k. The CO_2 abundance, from (25), is 293 ppm in 1880, 335 ppm in 1980, and 373 ppm in 2000. Climate model equilibrium sensitivity is 2.8°C for doubled CO_2.

Hansen's description of the probable delay in near-term surface warming produced by ocean heat storage. Source: Hansen, 1981, fig. 1. Reprinted from *Science*.

Antarctic ice sheet was partly grounded below sea level. The 2 degrees C global warming that the model indicated was very likely would place the air temperature of this area above freezing, leading to rapid breakup. This ice sheet contained 5 to 6 meters' worth of higher sea level, meaning that breakup in the twenty-first century would inundate large portions of Florida, Louisiana, New Jersey, and other states as well. Rapid declines in Arctic sea ice, however, would finally open the "Northwest Passage" that mariners had sought for centuries, speeding ship-based commerce.

Finally, they pointed out that all this would become clear relatively soon. Even with the lowest possible climate sensitivity, Hansen calculated that the carbon dioxide warming signal would emerge from the noise of natural variability by the end of the twentieth century. Indeed, these calculations showed that there was little difference between the high- and low-sensitivity cases in terms of the next two decades. One would only be able to judge the sensitivity of the Earth's real climate from direct observations in the first decades of the twenty-first century. This realization caused Hansen's team to look for evidence of the climate's sensitivity to carbon dioxide–induced warming in the Earth's record of its past climates.

First they had to endure political dislike of their results, however. The *New York Times* science writer, Walter Sullivan, had gotten a copy of the study and was able to get his story onto the newspaper's front page. Hansen had won a grant from

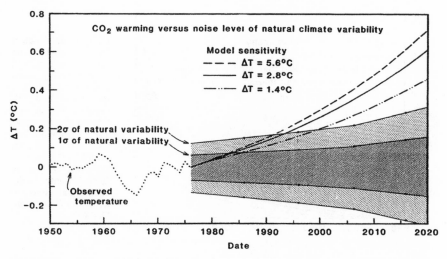

Fig. 7. Comparison of projected CO_2 warming to standard deviation (σ) of observed global temperature and to 2σ. The standard deviation was computed for the observed global temperatures in Fig. 3. Carbon dioxide change is from the slow-growth scenario. The effect of other trace gases is not included.

Hansen's demonstration of the point at which he thought human-induced surface warming would become larger than natural variability. Note that this study included only the effect of carbon dioxide forcing, not those supplied by other greenhouse gases such as methane and chlorofluorocarbons. From Hansen 1981, fig. 7. Reprinted from *Science*.

the Department of Energy during the final year of the Carter administration to study the carbon dioxide problem; these results caused the new Reagan administration's Energy Department officials to terminate it, costing GISS five researchers. Further, at the meeting at which Hansen was informed of this, he was also warned not to share his work with collaborators at Pennsylvania State University, lest they also lose their funding.[10] It had been unwise, perhaps, to speak of the potential human consequences of warming during the first of America's anti-environmental administrations.

Despite the loss of Department of Energy funding, Hansen and his team continued to work on the three-dimensional model and study its results in the context of paleoclimate data under GISS's NASA funding. In April 1983, they described the details of their three-dimensional Model II in an extensive *Monthly Weather Review* paper, and in 1984 they reported on their first efforts to evaluate the Model II against Earth's climatic past.[11] Using the data from the Climate: Long range Investigation, Mapping, and Prediction (CLIMAP) study of climate at the peak of the most recent ice age, 18,000 years before present, they established the climatic boundary conditions (trace gas concentrations, sunlight, etc.) needed to drive their model. These boundary conditions produced an ice age climate in the model to compare to the CLIMAP atlas of ice age temperatures. They then varied each of the model feedback processes to see how each affected the ice age Earth's climate, a familiar form of parameter variation.

This produced an interesting insight into the CLIMAP data that team members David Rind and Dorothy Peteet published independently; CLIMAP, according to the model, seemed to have tropical sea surface temperatures too high.[12] These temperatures had been inferred from plankton embedded in sediment cores, leading to further investigations of the CLIMAP methodology and quite a bit of controversy over the magnitude of tropical response during warming and cooling phases. By the mid-1990s, the GISS model's analysis had largely been accepted as the more accurate picture on the basis of ice cores drawn from tropical mountain glaciers and other forms of observations. In the 1980s, however, Rind and Peteet's finding was not the most controversial outcome of the GISS model studies.

The most controversial outcome came from a different part of the analysis. Hansen and his colleagues used their temperature record of the industrial era to produce a climate sensitivity for Earth to compare to the model and to that suggested by ice age temperatures. This analysis led them to conclude that all of the Earth's feedbacks combined were positive, on all timescales. In other words, whatever errors might exist in Model II, the record of the recent past, and the record of the ice age past, all agreed that the increasing amounts of carbon dioxide in the atmosphere *would* cause Earth to warm significantly. They put it this way: "We infer that, because of recent increases in atmospheric CO_2 and trace gases, there is a large, rapidly growing gap between current climate and the equilibrium climate for current atmospheric composition. Based on the climate sensitivity we have estimated, the amount of greenhouse gases *presently in the atmosphere* will cause an eventual global mean warming of about 1 C, making the global temperature at least comparable to that of the Altithermal, the warmest period in the past 100,000 years."[13]

They also argued that the most recent National Academy of Sciences study of climate, *Carbon Dioxide and Climate* (1983), had significantly underestimated the potential warming. This study, chaired by the director of the Scripps Institution of Oceanography, had not accounted for the response time of the ocean adequately in its analysis. The researchers had chosen a fixed fifteen-year response time for the ocean, which was much too short based on the GISS analysis of climate sensitivity. A fast ocean response meant that much of the carbon dioxide–

induced warming would already be apparent in surface temperature records, and the study's authors had therefore contended that not much more warming would result. The GISS analysis showed a slower ocean response, which meant that much of the induced warming was still in the pipeline, and would not become apparent in surface temperature records for another decade or two.[14] Their analysis suggested that there already was an additional 1 degree C global warming built into the climate system by preceding decades' anthropogenic emissions that would slowly manifest in the surface temperature record over the next several decades.

This time lag between emissions and the resulting warming meant that waiting for global warming to be revealed in the surface temperature records before committing to policy actions also meant committing to even more warming. Here, then, was the point around which scientific controversy would rage for another decade, and political controversy for at least two more (the political controversy was still going on as I write this in 2008): is the Earth getting warmer? Hansen's group already thought it was, based on their analyses. Yet this was a difficult question to answer, as temperature is difficult to measure accurately, particularly over Earth's vast oceans. Hansen's peers would not agree with him until 1995. A great many things had changed by then.

NUCLEAR WINTER, OR, THE COLD WAR POLITICS OF CLIMATE

The aerosols research NASA had carried out during the early 1980s became highly controversial in 1983, after publication of a paper by Richard Turco of R&D Associates; Brian Toon, Tom Ackerman, and Jim Pollack at Ames Research Center; and Carl Sagan at Cornell University. Based on research they had done on the Martian and Venusian atmospheres, and on the relationship between volcanism and climate on Earth, they proposed the nuclear winter thesis. This postulated that the smoke and soot from large-scale nuclear exchange would initiate a sudden, dramatic surface cooling by blocking incoming visible light from reaching the surface while permitting infrared energy to radiate from the surface into space—exactly what happened during the Martian dust storms. While no one much cared what happened on Mars, the same thesis applied to 1983 Earth was politically salient. Reagan administration officials were trying to convince the public that nuclear war was "winnable"; this group of scientists was arguing that even a "smallish" nuclear conflict was probably suicidal.

The nuclear winter hypothesis had evolved directly out of research the Ames group had done regarding the then-new idea that a large asteroid impact had caused the extinction of the Cretaceous dinosaurs 65 million years ago. In 1980, physicist Luis Alvarez and his son Walter had proposed that a 10-kilometer-diameter asteroid had collided with the Earth, throwing enough debris into the stratosphere to cause a sudden, catastrophic cooling and darkening of the Earth for a period of about three years.[15] This would have effectively collapsed the Earth's ecosystems, depriving the dinosaurs of their food supplies. No large animals of any type had survived this extinction—the largest surviving land animal appeared to be a mammal the size of a modern field mouse.

The Alvarezes' thesis was controversial, however. While they had some geologic evidence to support it, their cataclysmic explanation violated an important ideological precept in geology: things do not happen quickly. Geologists worked in very long timescales, and the Earth was rarely considerate enough to record very short-term phenomena. Further, modern geology (i.e., since the mid-nineteenth century) had been built upon the rejection of catastrophism, which was tainted by its association with religious explanations of the Earth's origin. Volcanism was the only geologically acceptable catastrophic process—otherwise, reputable geologists studied gradual processes.[16] The few impact craters on the Earth's surface were curiosities, not then subject of serious study, and in any case none were large enough to be the remains of the Alvarezes' killer.

The asteroid impact thesis, however, was backed by enough evidence to deserve serious attention, and the Ames group had the ability to address the climatological claims made in their paper. Toon and his colleagues challenged the hypothesis on the grounds that the silica ejecta would, as it did in volcanic explosions, fall out of the stratosphere relatively quickly, meaning that it would not achieve a global distribution. The nuclear weapons testing program had shown that material injected into the stratosphere took about a year to become distributed globally, and the ejecta from this impact would fall out over about three months. What caused longer-term cooling from a large volcanic explosion was sulfate aerosols, and an asteroid impact would not inject these into the stratosphere. Hence the global cooling scenario the Alvarezes had proposed originally could not be right. As Alvarez recounts in his memoir, however, a group from Los Alamos demonstrated in 1981 that a very large impact could create a new transport mechanism that could spread the ejecta cloud worldwide in at most a few days, salvaging the thesis.[17] As reformulated, the impact would generate a substantial cooling for a shorter period, less than a year, which had the effect of bringing it into better alignment with the geological evidence that existed.

While the impact thesis remained controversial through the next decade, it sparked a great deal of thought about other, more obvious and more likely catastrophes. In the late 1970s, leaders of the American anticommunist movement had set out to destroy the policy of détente with the Soviet Union, preferring to advocate a more aggressive policy of rollback. They achieved political power in 1980 with the election of Ronald Reagan, who ran for the presidency on a platform of rearmament. After his election, Reagan initiated new nuclear and non-nuclear weapons development, including the development of mobile intercontinental ballistic missiles that would "hide" (in their initial form) in American communities. Because few in Congress really believed that the American nuclear deterrent was inferior to the Soviet Union's despite the best efforts of conservative propagandists, the administration's officials constructed a new strategic policy to justify the new weapons. Deterrence of the Soviet Union was no longer sufficient. Instead, the new policy was "nuclear warfighting."[18] Reagan officials sought to emphasize the usability of nuclear weapons in future wars, with some publicly commenting on the "winnability" of nuclear war and the "acceptability" of millions of dead (Americans).[19]

While the rhetorical aggressiveness of these Reagan officials was seen in conservative policy circles as merely a way of demonstrating American resolve to the Soviet leadership, it had the effect of sparking an almost instantaneous revolt inside the United States. A nuclear disarmament movement crystallized around the concept of a nuclear freeze in 1981. Both the United States and the Soviet Union were to agree to cease building and modifying nuclear weapons and delivery systems for them, which, because the weapon cores decayed over time, would lead relatively quickly to mutual disarmament. By the end of that year, there were twenty thousand freeze activists in forty-three states. The proposal was endorsed by major religious denominations as well as local and state governments, and by early 1982 was being openly debated in the U.S. Congress. The suddenness of the movement was stunning, and its scale directly threatened the administration's foreign and military policies as well as its reelection hopes for 1984.[20]

The American government's promotion of the acceptability of mass death caused a great deal of questioning in other governments, particularly those neutral states that would not participate in the war but would still suffer. Paul Crutzen, then at the Max Planck Institute for Chemistry, was asked to prepare an article for the Royal Swedish Academy of Sciences journal *Ambio* on the consequences of global nuclear war to the stratospheric ozone layer. Nuclear war would, of course, inject large amounts of ozone-depleting nitrogen oxides into the stratosphere, and unlike the single test detonations of the 1950s would inject it from thousands of explosions throughout much of the Northern Hemisphere. In the process of preparing the paper, however, Crutzen found that no one had ever considered the impact of soot, smoke, and aerosols from the post-strike burning of cities, crop-land, and forests on the hemisphere's climate. In the context of the Alvarezes' claims about dinosaur extinction, it was an entirely reasonable question to ask if the climate effects of nuclear weapons use would be so severe as to cause extinc-tion of the human species. Working with John Birks at the University of Colorado, Crutzen prepared an estimate of the amount of smoke and soot that might be result from a nuclear exchange. This, in turn, caused the Ames group to experi-ment with the nuclear war thesis using their model.[21]

Using publicly available information on the effects of nuclear weapons, com-monly used simulations of various scenarios of nuclear exchanges, the Ames group investigated how nuclear exchanges of from 100 megatons to 5000 megatons might affect global temperatures (the Mount St. Helens eruption, for comparison, was equivalent to about 10 megatons). Their one-dimensional radiative-convective model suggested that even the smallest exchange they considered could lead to subfreezing surface temperatures during the summer months. Larger exchanges could lead to fatal radiation doses from fallout, large-scale ozone depletion, and reduction of average light levels to 1 to 2 percent of clear-sky for a period of months. Their analysis suggested that the majority of the changes would not come from debris in the stratosphere, however, but from the smoke and dust in the troposphere.[22]

What happened in their model was that the smoke and dust absorbed the incoming visible light and re-radiated it in the infrared, causing heating of the upper troposphere while preventing sunlight from reaching the ground. Deprived of energy, the ground would cool rapidly. This produced a temperature inver-sion over much of the Northern Hemisphere, with the upper troposphere much warmer than the air near ground level. Rain forms when rising warm, moist air cools at higher altitudes; by reversing the temperature structure of the atmosphere, the smoke and soot would actually impair precipitation. In turn, this allowed the smoke and soot to remain airborne longer, prolonging the effect. While their one-dimensional model could not simulate changes to global circulation, that there would be such changes was obvious to the Ames group. Atmospheric circulation is nothing more than an energy transport process, and such a large-scale change in the atmosphere's thermal structure could not help but cause dramatic changes to the general circulation.[23]

The authors commented in their text that there were many uncertainties be-yond the lack of horizontal circulation in their model. The model did not have an ocean, for example, whose high heat capacity would tend to mitigate the cool-ing effect. Nor was it clear that data drawn from the Hiroshima and Nagasaki bombings and the aboveground testing programs of the 1950s adequately represented the blast and fire impacts of multiple detonations. How cities would burn after an explosion mattered, for example, and the two destroyed Japanese cities were not necessarily representative of how American and Soviet cities would burn. Forests and grasslands would burn with less intensity, and thus have still different impacts than urban conflagration, but it was again not clear to what extent nuclear weapons bursts would cause ex-urban fires. They concluded that while there were many effects that they could not quantify, "the first-order effects are so large, and the implications so serious, that we hope the scientific issues raised here will be vigorously and critically examined."[24]

In June 1982, as the Ames group was still examining their results, Carl Sagan had been approached by executives of the Rockefeller Family Fund, the Henry P. Kendall Foundation, and the National Audubon Society about organizing a public conference on the longer-term consequences of nuclear war. Sagan agreed to help assemble a steering committee to plan the conference. Paul Ehrlich of Stanford University, a biologist and author of the famous *Population Bomb*, agreed to help, as did Walter Orr Roberts, the founding director of University Corporation for Atmospheric Research (UCAR), and George Woodwell of the Marine Biological Laboratory at Woods Hole. The steering committee decided to base the conference around the Ames group's results.[25] First, they arranged a closed workshop at which the TTAPS paper would be subjected to peer review. (The acronym TTAPS is based on the initials of the paper's authors, beginning with Richard Turco and ending with Carl Sagan.) If the paper held up, it would become the basis for an assessment of its biological implications by a group of prominent biologists. The public conference would be scheduled if the TTAPS paper survived this first effort.

The workshop was held in Cambridge in April 1983, with about forty physical scientists attending to examine the TTAPS paper. This survived, at least according to Sagan, one of the co-authors, with only minor revisions, and then was examined by ten invited biologists. These scientists found the TTAPS paper compelling enough to draft a paper of their own on the biological consequences of the nuclear winter scenario, and the public conference was scheduled for 31 October 1983. Thirty-one scientific and environmental groups, including the Federation of American Scientists, the Union of Concerned Scientists, the Environmental Defense Fund, and the Sierra Club, as Sagan reports, funded the public conference.[26] This "Conference on the Long-Term Worldwide Biological Consequences of Nuclear War" included additional discussion of the two papers and the use of a satellite link to Moscow so that Soviet scientists could participate remotely.

Two more modeling efforts were discussed at the conference. In one, done at the National Center for Atmospheric Research (NCAR) by Curt Covey, Stephen Schneider, and Starley Thompson, the soot and smoke effects on general circulation of the atmosphere were examined through the use of a three-dimensional model. This group found that the major changes in general circulation that the TTAPS authors had assumed from their one-dimensional model did occur in their three-dimensional model. Three-dimensional flow also, again as the TTAPS authors had suspected, acted to mitigate the cooling effect of the smoke and soot. Their model found land surface temperatures dropping 20 degrees C in July, on average, and about half that in an April simulation—still quite dramatic enough to ensure massive crop failure. The Soviet effort had found similar results: large-scale changes to the general circulation and less, but still severe, surface cooling.[27]

At the conference, George Woodwell commented that what was most impressive in the TTAPS paper was not the result, but how obvious it was. He wondered publicly how it could be that no one had thought of it before, drawing comment from Tom Malone that apparently the community had needed a Paul Crutzen to stimulate their thinking.[28] From a qualitative standpoint, all of the processes made sense. Putting precise numbers on them was not necessarily possible without running the nuclear war scenario in the real world, which certainly no one among this group wished to see, but they were also not really necessary. The models, in this case, had served as digital thought experiments, forcing the assembled specialists to confront the ramifications of atmospheric processes they already understood. Over the next several years, there would be bitter arguments over the numbers due to the political and policy ramifications of these model studies, but the qualitative result—that the human and environmental consequences of nuclear war would be catastrophic—did not change.[29]

The TTAPS and Ehrlich papers were published together in the 23 December 1983 issue of *Science*, right behind an editorial penned by the magazine's publisher, William D. Carey. Carey congratulated the scientists who had prepared these articles for helping to bring to life the "conscience of science." In his eyes, their work represented a display of scientific responsibility. Scientists, had, after all created nuclear weapons in the first place, and therefore they had responsibility to "look squarely at the consequences of violence in the application of scientific knowledge."[30] Carey's argument was hardly new in 1983; indeed, many of the atomic bomb's own inventors had turned against their creation during the early 1950s. This had created a deep schism within nuclear physics between defenders of nuclear weapons and their opponents that never healed.[31] The late 1960s had seen similar demands from the American political left that scientists take responsibility for the consequences of science. But the period of détente had made the canyon between these factions irrelevant, and it had seemed to vanish. Reagan's revival of the Cold War had reopened this old wound, forcing the scientific community's members to take sides again.

Science's decision to openly praise what were, to a sizeable fraction of the scientific community as well as the larger American body politic, antinuclear, liberal/environmentalist tracts, was a defining moment. The journal was supposed to represent all of American science, not one ideological component of it, and its praise for the TTAPS and Ehrlich work, and indirectly the nuclear freeze effort itself, enraged conservatives. Paul Ehrlich had written one of the foundational works of the American environmental movement, and had served as president of Zero Population Growth and of the Conservation Society. This linked him indelibly to the environmental left. Sagan had been active in antinuclear circles as well, and he aggressively promoted the nuclear winter thesis, publishing, for example, in *Foreign Affairs* and *Parade*—far outside the scientific realm.[32] The conservative response, led by Robert Jastrow, who had retired from GISS to take a professorship at Dartmouth College, was to form the George C. Marshall Institute in Washington, a think tank aimed at supporting Reagan's Strategic Defense Initiative (SDI), nuclear power, and nuclear weaponry.[33]

Studying climate, then, led NASA scientists on both coasts into politically dangerous territory. In New York, Hansen had run afoul of the Reagan administration's anti-environmental, anti-regulation beliefs and been punished, while in California the Ames researchers had collided with foreign policy—although NASA, to be very clear, did not retaliate against them as the Department of Energy had against Hansen. Yet regardless of the political ramifications, climate, whether human-induced or volcanic (or solar, for that matter), was a legitimate scientific subject. It was also an important policy subject for officials interested in serving the public good, and not merely the ephemeral politics-of-the-moment. There would be substantial ramifications of either a dramatic warming or cooling of the Earth regardless of causation.

Hence, while these model and paleoclimate studies had been in progress, NASA officials had also turned to studies of an observing system to study extant climate processes. Vern Suomi, whose colleague Reid Bryson at the University of Wisconsin had been one of the first to argue that climate could change in very short periods of time, and who had deemed carbon dioxide–induced warming inevitable in his introduction to Charney's 1979 National Academy of Science study of the subject, helped orchestrate the first such study.

THE CLIMATE OBSERVING SYSTEM REPORT

After passage of the National Climate Program Act of 1978, officials from NASA, NOAA, and the other science agencies of the federal government had struggled for several years to formulate an interagency climate research program. Unlike the stratospheric ozone problem, climate research would involve virtually every Earth science discipline, and therefore the entire science establishment of the federal government. This made coordination very difficult. Disciplines were typically self-governing, without the directed nature that the stratospheric ozone research had taken on, and changing that would take time (and would also face considerable resistance). But NASA's role was relatively clear from the beginning. Its role would be the provision of global data, just as it had been in the old meteorology program. In fact, the meteorological satellite system formed the basis of the initial climate observing system concept.

In February 1980, meteorologist David Atlas, head of Goddard Space Flight Center's Laboratory for Atmospheres, convened a workshop to discuss how a climate observing system might be structured. His panelists were Vern Suomi, Thomas Vonder Haar of Colorado State University, and P. K. Rao of NOAA. During the first day of their two-day meeting, they heard briefings from various other specialists on the measurements needs for a climate observing system. Francis Bretherton from NCAR, for example, discussed the need for a climate-related oceanic monitoring system, while Tom Wilheit from Goddard explained the need for global precipitation measurement to understand ocean-atmosphere energy exchange. On the second day, the panelists and workshop attendees divided into working groups to begin drafting the sections of their report.[34]

In his opening remarks to the workshop, Suomi had argued for an evolutionary observing system based upon the existing operational meteorological satellite program, and this became the basis of the workshop's final report. Climate research required long-term data and repeated launching of the same sets of instruments, a quasi-operational mode of research that NASA was not interested in but which NOAA's satellite service had been designed for. At this time, NASA and NOAA were still linked via the Operational Satellite Improvement Program, and Suomi believed that moving instruments from NASA research satellites to NOAA operational ones via this program was the best way to get to a climate observing system. It would limit the number of new satellite development projects, saving money, while still permitting all of the relevant scientific research to be done.

The climate observing system this group postulated would start with improvements to the suite of operational meteorological satellite instruments, which already measured a number of key climate variables but not with sufficient accuracy or with reliable calibration capabilities. Table 7.1 suggests the complexity of the measurement problem. Atmospheric temperature profile instruments needed greater vertical resolution and reduced error limits, while the Advanced Very High Resolution Radiometer (AVHRR) needed improved resolution and more accurate measurement of sea and land surface temperatures. Additional vital oceanic measurements the group expected were to come from a then-planned National Ocean Surveillance Satellite (NOSS), which was being jointly developed by NASA, NOAA, and the U.S. Navy. NOSS was a descendant of the short-lived Seasat A mission, and was intended to be an operational satellite system like the weather satellites with a Shuttle launch tentatively scheduled for 1986. Its instruments included a scatterometer for measuring sea surface wind speeds, a precision altimeter for measuring sea surface height, the Coastal Zone Color scanner derived from Nimbus 7 for measuring ocean biological productivity, and a Large Antenna Multifrequency Microwave Radiometer, the only new instrument in the NOSS payload. This instrument would measure sea ice and sea surface temperatures, and, the group hoped, provide measurements of soil moisture content and oceanic precipitation, all of which were critical measures of climate processes that had not yet been made with decent accuracy.[35]

The group also drew on three other satellite missions that were already in progress for their climate system proposal, the Upper Atmosphere Research Satellite (UARS) project, the Earth Radiation Budget Experiment (ERBE), and Solar Max. The trace gases suspected of causing ozone depletion were also greenhouse gases, and therefore some of UARS's measurements were necessary for climate research. Some of UARS's instruments could be migrated to an operational satellite after UARS had proven them effective sometime in the mid-1980s. Only a new tropospheric chemistry instrument still needed to be developed for the climate observing system's chemistry satellite.

As the 1970 Study of Critical Environmental Problems (SCEP) group had argued, accurate and ongoing measurement of the Earth's albedo was also important to determining whether climate was changing. An ERBE was being developed at Langley Research Center for a Shuttle launch in 1984; unlike the previous efforts to measure the Earth's heat budget, this was a three-satellite mission including polar and equatorial orbits to permit understanding the diurnal variation in the Earth's energy flows. ERBE, in turn, drew on a new total solar irradiance instrument developed for the Solar Max mission, whose purpose was to measure a variety of solar parameters during the Sun's peak activity levels of the decade. Solar Max's total solar irradiance instrument, the Active Cavity Radiometer Irradiance Monitor (ACRIM), had been developed at Eppley Labs in Delaware and was slightly modified for the ERBE flights.[36]

This group also proposed two new development efforts. One critical set of measurements that climate researchers needed were of the cryosphere, the surfaces of the Earth that were completely or partially covered by snow or ice.[37] These were highly reflective and directly affected the Earth's albedo. Shrinkage of the Earth's polar ice caps, for example, would expose more of the dark ocean surface to sunlight, leading to more heat absorption and thus a warmer Earth (a positive feedback). Snow and ice extent were also excellent proxy measures of changing average temperatures. Temperature itself was extremely difficult to measure accurately, while extent of snow and ice cover was not. The two quantities were clearly linked, however. Worldwide shrinkage of ice fields, or gradual poleward retreat of the annual winter snow line, would clearly indicate a warming world even if a warming signal could not be found in error-ridden temperature records. Existing passive microwave radiometer technology could measure the spatial extent of snow and ice cover, while a new radar or, even better, lidar altimeter could measure the height of ice sheets above a baseline.

The other new development program the group supported was a new geosynchronous test platform.[38] The spin-stabilized Geosynchronous Operational Environmental Satellite (GOES) series of satellites was not suited to many of the new instruments that the group wished to mount for climate research, and they advocated the development of a three-axis stabilized geosynchronous platform that could support a wider array of instrumentation. They were particularly interested in measuring precipitation from geosynchronous orbit, as many convective systems had lifetimes shorter than the twelve-hour polar orbits used by the weather satellites. Their precipitation would not be adequately measured by the polar orbiters, but might be from geosynchronous orbit. Since the atmosphere's primary energy source was the latent heat of condensation, accurate measurement of precipitation was vital to untangling the process of energy exchange between surface and atmosphere. A new geosynchronous satellite seemed to be the best way to accomplish this measurement.

The incremental development plan the group had in mind would, by 1990, have resulted in a constellation of eight polar orbiters plus the new geosynchronous platform. Two of the polar orbiters would be improved versions of the Television-Infrared Observations Satellite (TIROS) N meteorological satellites for operational forecasting use (with ERBE as added payloads); two would be the NOSS operational oceanic observation system, each providing chemistry and cryospheric measurements; and the remaining two would be NASA research

TABLE 7.1
Climate Parameter Observational Requirements

Parameter	Desired Accuracy	Base Requirement	Horizontal Resolution	Vertical Resolution	Temporal Resolution	Index No.
Weather Variables (• Basic FGGE Meas.)						
• Temp Profile	1°C	2°C	500 km	200 mb	12–24 Hrs.	1
• Surface Pres	1 mb	3 mb	500 km	—	12–24 Hrs	2
• Wind Velocity	3 m/sec	3 m/sec	500 km	200 mb	12–24 Hrs.	3
• Sea Sfc. Temp	0.2°C	1°C	500 km	—	3 Days	4
• Humidity	7%	30%	500 km	400 mb	12–24 Hrs	5
Precipitation	10%	25%	500 km	—	12–24 Hrs.	6
Clouds			100 km	—	1 Day	7*
a. cloud cover	5%	20%				
b. cloud top temp.	2°C	4°C				
c. albedo	0.02	0.04				
d. total liq. H_2O Content	10 mg/cm²	50 mg/cm²				
Ocean Parameters						
Sea Sfc. Temp	0.2°C	1°C	500 km	—	1 Month	4a
Evaporation	10%	25%	500 km	—	1 Month	9
Sfc Sens. Heat Flux	10 W/m²	25 W/m²	500 km	—	1 Month	10
Wind Stress	0.1 Dyne/cm²	0.3 Dynes/cm²	500 km	—	1 Month	11
Radiation Budget						
Clouds (Effect on Radiation)			500 km		1 Month	7a
a. cloud cover	5%	20%				
b. cloud top temp	2°C	4°C				
c. albedo	0.02	0.04				
d. total liq. H_2O Content	10 mg/cm²	50 mg/cm²				
Regional Net Rad. Components	10 W/m²	25 W/m²	500 km	—	1 Month	16
Eq.-Pole Grad	2 W/m²	4 W/m²	1,000 km Zones	—	1 Month	17
Sfc Albedo	0.02	0.04	50 km	—	1 Month	18

Parameter						
Sfc. Rad Budget	10 W/m^2	25 W/m^2	500 km	—	1 Month	19
Solar Constant	1.5 W/m^2	1.5 W/m^2	—	—	1 Day	20
Solar UV Flux	10% per 50Å interval		—	—	1 Day	21
Land Hydrology and Vegetation						
Precipitation	10%	25%	500 km	—	1 Month	6a
Sfc. Albedo	0.02	0.04	500 km	—	1 Month	18a
Sfc. Soil Moist.	0.05 gm H_2O/cc Soil	4 levels	500 km	—	1 Month	22
Soil Moisture (Root Zone)	0.05 gm H_2O/cc Soil	4 levels	500 km	—	1 Month	23
Vegetation Cover	5%	5%	500 km	—	1 Month	24
Evapotranspiration	10%	25%	500 km	—	1 Month	25
Plant Water Stress	4 levels/2 levels	4 levels	500 km	—	1 Month	26
Cryosphere Parameters						
Sea ice (% Open Water)	3%	3%	50 km	—	3 Days	27
Snow (% Coverage)	5%	5%	50 km	—	1 Week	28
Snow (Water Content)	+1 cm	±3 vm	50 km	—	1 Week	29
Ocean Parameters						
Sea Sfc. Elevation	1 cm	10 cm	Variable	(As Indicated)	1 Week	12
Upper Ocean Heat Storage	1 KCal/cm^2	5 KCal/cm^2	500 km	—	1 Month	13
Temp. Profile	0.2°C	1.0°C	Variable	—	1 Month	14
Velocity Profile	2 cm/sec (near sfc)	10 cm/sec	Variable	—	1 Month	15
	0.2 cm/sec (at depth)	1 cm/sec	Variable	—	1 Year	15
Cryosphere Parameters						
Ice Sheet SFC. Elevation	10 cm	1 m	1–3 km	—	1 Year	30
Ice Sheet Horiz. Velocity	50 m/yr	100 m/yr	Point targets	—	1 Year	31
Ice Sheet Boundary	1 km	5 km	1–3 km	—	1 Year	32

Continued

TABLE 7.1
Continued

Parameter	Desired Accuracy	Base Requirement	Horizontal Resolution	Vertical Resolution	Temporal Resolution	Index No.
Variable Atmos. Composition						
Solar UV Flux	10% per 50Å	Interval	—	—	1 Day	21a
Stratos. Aerosol Opt. Depth	0.002	0.01	250 km N-S 1000 km E-W	3 km	1 Month	33
Tropos. Aerosol Opt. Depth	0.005	0.02	500 km	3 km	1 Month	34
Ozone	0.005	0.02 cm (total) 10% at effective alt.	250 km N-S 1000 km E-W	3 km	1 Month	35
Stratospheric H_2O	0.5 ppm	0.5 ppm	250 km N-S 1000 km E-W	3 km	1 Month	36
Reasonably Well-Mixed Tropospheric Gases (ground-based observations)						
N_2O	0.01 ppm	0.03 pm	—	—	1 Year	37
CO_2	0.5 ppm	10 ppm	—	—	1 Year	38
CFM's	0.03 ppb	0.1 ppb	—	—	1 Year	39
CH_4	0.05 ppm	0.15 ppm	—	—	1 Year	40

Note: All Climate B parameters are also required by Climate C & X.

*Under "Weather Variables" (Index No. 7), histograms of all four cloud parameters will be generated for 100 km × 100 km boxes.

flights to provide platforms for ongoing instrument development. These last two would replace the Nimbus line of research satellites. The polar orbiters could either be based upon the TIROS N satellite bus (designed for launch aboard expendable rockets), or upon the NOSS bus, which was being designed for Shuttle deployment. Six of the satellites would be operated by NOAA or the Defense Department as operational satellites, while the two NASA research satellites would remain under NASA control.[39]

This version of a climate observing system was largely forgotten over the next couple of years, however. The November 1980 election of Ronald Reagan dramatically altered the political landscape of any future Earth satellite system. Reagan favored shrinking the federal civilian government, and both NASA and NOAA were presented with budget rescissions—previously appropriated funds being taken back—during the spring of 1981 and took additional cuts in 1982. NOSS was dropped, the Operational Satellite Improvement Program was terminated, and the Commerce Department began planning to turn NOAA's weather satellites over to the Comsat Corporation—along with a $250 million subsidy to operate them. This closed off the instrument migration path from NASA research to NOAA operations without replacing it. There was no longer an institutional mechanism to bring about the evolution of the weather satellite system into a climate observing system. While the effort to transfer the operational weather satellites to Comsat failed in the face of congressional hostility, NOAA did not receive new funds to develop instruments on its own.[40]

Yet closing off the pathway to operational use of research instruments did not stop interest in a climate observing system within NASA or within the scientific community. Scientific leaders still wanted the research done, and NASA's Earth science and applications community wanted to maintain the vitality of their own research. As things stood after the Reagan cuts, only the Upper Atmosphere Research Satellite (UARS) and the ERBE were going concerns, but a number of the instrument programs on which David Atlas's group had based their proposal were continuing and their scientists continued to advocate for flight opportunities. In January 1982, the climate observing system effort began to be linked to NASA's human spaceflight program.

TOWARD GLOBAL HABITABILITY

The central focus of NASA Administrator James Beggs's efforts in the first half of the 1980s was promotion of a "permanent human presence in space," in the form of a permanent space station in low-Earth orbit.[41] The Soviet Union had maintained a string of temporary space stations during the 1970s and was building a permanent one in the 1980s; Beggs's effort in part was to counter Soviet space plans. But selling what would be a very expensive new project in the austere fiscal and political environments of the decade was difficult, and the agency's leadership needed to demonstrate its relevance to their funders. The best-known claim for the space station project is that it would foster new industries in space material processing. To gain scientists' support, NASA leadership also decided to link a climate observing system to the space station. This became the Global Habitability initiative.

A few months after Hansen's 1981 article was published, in January 1982 NASA Deputy Administrator Hans Mark had a brainstorming session with Harvard meteorologist Richard Goody and atmospheric chemist Michael McElroy on possible new initiatives in the Earth sciences. Mark, who had been director of the Ames Research Center during the 1970s, came away convinced that a Earth observing system aimed at global environmental change was both technologically and politically feasible. He recommended such an initiative to Administrator Beggs, and had Burt Edelson, the agency's associate administrator for space sciences, launch a two-track review of the idea. Edelson asked Goddard's David Atlas for an informal proposal for an observing system tied to the space station program, and he asked Richard Goody to head a panel composed of scientists to formulate a set of scientific objectives for the system.[42]

In early April 1982, Atlas forwarded a proposal assembled by himself, Harvey Melfi, and William Bandeen to use the space station for a long-term Global Environment and Ecology Mission (GLEEM). Its objective was to be understanding "both the natural and inadvertent affects [sic] on climate." The measurements Atlas proposed were largely the same as those proposed two years before, but some of the instruments to make them had grown considerably. The additional power and mass expected to become available from the space station project had caused researchers to propose much higher-resolution microwave and radar instruments, which in turn required larger antennae. A Multi-Channel Microwave Radiometer of 10 to 20 meters aperture replaced the advanced version of the TIROS N Microwave Sounding Unit, a 5- to 10-meter aperture High Power Microwave Radar replaced the precipitation radar proposal from the earlier study, and a 1- to 5-meter aperture High Power Lidar to measure wind velocities and aerosol densities were added to the manifest.[43]

Richard Goody's workshop convened 21 June 1982 at Woods Hole, Massachusetts. The study panel consisted of Robert Chase, Wesley Huntress, Moustafa Chahine, Michael McElroy, Ichtiaque Rasool, John Steele, and Shelby Tilford.

Forty-one additional attendees participated in the workshop's deliberations, including Wallace Broecker of the Lamont-Dougherty Geophysical Observatory, Paul Crutzen, James Hansen of GISS, former presidential science advisor Donald Hornig, Vern Suomi, and Robert Watson. The group's assignment was to examine "the viability of a major research initiative in the area of global habitability, specifically addressed to the question of changes, either natural or of human origin, affecting that habitability." Their charge included a timeframe, changes likely to occur in the next fifty to one hundred years, and a directive that it was to be a research program, which they interpreted as meaning "a concern with the foundations of knowledge needed for enlightened policy decisions."[44]

Goody's group believed that a research program into the Earth's continued habitability was urgent. Citing the ongoing ozone research and the measured increase of global atmospheric carbon dioxide levels, they contended that humanity's ability to cause global-scale changes was now obvious, but its ability to assess the impacts of those changes was poor. Scientists lacked a sufficient understanding of the physical, biological, and chemical processes involved. But assessment needed to be done relatively quickly, as the time lag between policy actions taken to mitigate damage and the improvement of an environmental problem could be very long. Again, they used the ozone problem to make their point. The many-decades residence time of chlorofluorocarbons (CFCs) in the atmosphere ensured that policy action taken in 1982 would not lead to an improvement in stratospheric ozone levels before the middle of the next century. There were other major environmental changes going on, and the committee listed widespread deforestation, chemical pollution of the oceans, and Soviet projects to alter the courses of three major rivers as examples of the kinds of things that needed to be better understood before policy actions could be formulated.[45]

Continued habitability of the Earth had a great deal to do with water, and the committee focused on the hydrologic cycle as the most obviously relevant study. Ocean circulation, air-sea exchange, and in particular the hydrological contribution of evapotranspiration were poorly understood. Cloud formation, and cloud effects on the Earth's energy flows, were other very important related unknowns. There were already efforts in the early stages of planning to study some elements of these processes. GARP's descendant, the Joint Steering Committee of the World Climate Research Program (WCRP), had proposed an International Satellite Cloud Climatology Project (ISCCP) to produce the first detailed climatological maps of global cloud coverage from the imagery returned by the meteorological satellites during the 1970s, and the committee recommended that NASA be a major participant in this. The committee also recommended immediate funding of an ocean topographic mission being studied at the Jet Propulsion Laboratory (JPL) called TOPEX (Ocean Topography Experiment). This was designed to measure small variations in the altitude of the ocean surface, which were related to circulation, temperature, ocean heat storage, and surface winds. As the oceans warmed, for example, thermal expansion would cause the average height of the ocean surface to rise. Finally, as the earlier Goddard workshop had, Goody's group argued that cryosphere measurements should be initiated quickly with existing microwave technologies in appropriate orbits, with resources committed to developing better instruments later. The technology already existed, they commented, but exploitation of it was lacking.[46]

The committee also reviewed the potential for space-based systems to study atmospheric pollution. NASA, of course, had already formulated the UARS mission, although it had not yet been approved as a new start, and the committee recommended that UARS be funded immediately. UARS was aimed at the stratosphere, however, and the committee found that NASA had no significant efforts going to devise instruments for measuring various tropospheric chemical species, either organic or those of human origin. Tropospheric pollution was well-known but poorly understood, and the committee recommended that NASA formulate a program similar to that of its UARP to investigate this area.[47] This became the tropospheric chemistry initiative examined in chapter 6.

The one major area the committee found that could not be subjected to study from space yet was land biomass. The Ocean Color experiment on Nimbus 7 had permitted the beginnings of an assessment of oceanic biological productivity, but nothing resembling that capability existed for land remote sensing. The imagery produced by Landsat could provide rough estimates of biological activity, but it gave no insight into soil moisture, nor did it help quantify plant respiration. Both were important measures of hydrologic activity. Very preliminary studies had been carried out of remote sensors that might be able to measure these variables at JPL, but these were a long way from maturity. Hence, Goody's group recommended that these efforts be accelerated, and that aircraft-based programs be implemented as a means of beginning the research effort earlier.[48]

Goody's committee, then, had recommended a relatively broad program of Earth science research aimed at understanding various components of environmental change over a decadal timescale. While this was not the timescale needed to understand the potential magnitude of anthropogenic climate change, which would require a research strategy designed around half-century scales or more, it offered a route to new instrument and measurement development that might eventually contribute to longer-term research. Further, there were decadal-scale climate phenomena known to exist, such as El Niño, which could be studied to begin parsing out the various climate feedback processes. This offered the potential to develop operational climate forecasting, as well as contributing to understanding of longer-term processes.

What Goody's panel had not done was make a case for linking this Global Habitability program to the space station project. This was not surprising, as the scientific community was quite skeptical of the utility of humans—astronauts— in space-based research. To this point, automated satellites had carried out most space-based Earth research and made all of the space program's major contributions to Earth science. The research program Goody and his colleagues were proposing was a larger effort in robotic data collection, with scientists on the ground carrying out the process of interpretation. It was not greatly different from the 1980 Climate Observing System Study. Goody's colleagues did not believe that having astronauts involved in this at any point would provide significant additional value. But Global Habitability was intended as part of NASA's justification for the cost of a permanent space station, so justification for astronauts was necessary.

The internal NASA process had been more accommodative to the astronaut corps. David Atlas had surveyed various instrument teams to glean their thoughts on what humans could contribute to a climate observing system. These responses had ranged from no need for humans to proposals for astronauts performing multi-day calibration studies in a "shirt-sleeve environment" within the instrument platform. The most common response, however, was that astronauts could repair or replace failed instruments, a capability that the agency was already building into several missions, including Solar Max and the Hubble Space Telescope. An occasional need for astronaut servicing became the primary justification for having humans involved with space-based research, making them expensive space mechanics.

Global Habitability's first public outing was in August 1982, at the second UN conference on the peaceful use of space, UNISPACE 82 (the first had been in 1968). NASA Administrator Beggs presented it as a new U.S. initiative during his opening remarks, positioning it as a NASA program that other nations, or experimenters from other nations, could participate in. He did not present it as a major *international* program, as GARP had been; the agency's leaders had specifically recommended against another GARP-like effort, arguing that it would be unwieldy and might restrict or dilute the program's scientific objectives. In addition, GARP's 1980 transformation into the WCRP would have made a new international climate initiative unwelcome—it would be a competitor to an existing effort. Instead, in addition to the announcement of opportunity process, which permitted any competent investigator to propose instruments almost regardless of nationality, NASA would welcome bilateral or multilateral agreements to provide instruments or additional platforms.[49]

The UNISPACE presentation, however, did not go over well with the meeting's delegates. In fact, it went over badly enough that members of Congress requested that the Office of Technology Assessment provide an analysis of why it had. Due to a dispute with the Soviet Union, the United States had started preparing for UNISPACE 82 in January 1982, which had not been enough time to make all of the arrangements for a smooth meeting. American officials also went to the meeting with the attitude that their job was to mitigate damage to U.S. interests, not to use it to promote U.S. interests. That negativity had left the delegation at a distinct disadvantage.

More important, however, substantive conflicts existed between the American position and that of large groups of states on several issues. One concerned the increasing militarization of space, which the American delegation tried to prevent discussion of at the meeting and thus sparked considerable unnecessary discontent. Another regarded land-imaging satellites, which would, of course, be part of any Global Habitability research program. Many nations did not want imagery being made available to their neighbors (and potential enemies), desiring veto rights over imagery of their territory. Many also wanted land-use satellites placed under UN control; they were opposed to the Reagan administration's plans to privatize the technology. Privatization was the worst possible outcome for poorer nations. In addition to the high cost of the imagery, governments would not be able to prevent it from going to potential enemies or to multinational corporations that were in a position to exploit imaged resources.[50] Hence, Shelby Tilford, one of the U.S. attendees at the meeting, characterized the response to Global Habitability at the meeting as openly hostile, not merely disinterested.[51]

But the Global Habitability program had not been intended as an international program, and lack of international support for it mattered very little in the short run. Hence, NASA kept working on the plan for it. The space-borne technological component of the research program was given the name System Z for the next round of planning, and Dixon Butler of Goddard Space Flight Center, a stratospheric chemist, was appointed to chair a concept study group to begin the long process of defining the system. By early September, the System Z working group's charter had been expanded to include all civil remote sensing needs, including all Earth sciences disciplines and applications and operational missions as well.[52] Operational missions had been added to the system to gain the support of NOAA, which had lost its pipeline to new instrument technology with the Operational Satellite Improvement Program's cancellation. System size expanded as well. Butler's instructions included a direction that no "arbitrary" size, mass, or power constraints should be applied in planning System Z. The only restraint to be applied to the committee's technologizing was a requirement for Shuttle launch into a polar orbit.

The basic concept behind System Z was integrated Earth observations. NASA had sent numerous probes to other planets that were equipped for comprehensive investigation of atmospheric winds, temperature, chemistry, and land surface (for those planets that had land), but had not done the same for Earth. This was largely because the Earth observations program had been in NASA's applications division, which was primarily interested in developing practical, commercializable uses for space technology. Its primary function was not the development of new knowledge. But new knowledge was precisely what Shelby Tilford sought in fostering System Z. It would utilize large satellites with many instruments boresighted to view the same scene simultaneously, so that accurate comparisons between instruments could be made. System Z would also require a commitment to extensive data analysis; to achieve what Tilford hoped for it, System Z's researchers would have to be able to examine processed data from many instruments.

CONSTRUCTING EARTH SYSTEM SCIENCE

President Reagan refused to launch the space station initiative during 1982 and again in 1983, and since System Z was tied to the station program, beyond conceptual planning nothing much happened until he finally gave his consent. He finally announced his Space Station Freedom initiative in the annual State of the Union address in January 1984. While System Z was not approved as part of the station program at this point, the agency increased its activity on System Z, now renamed the Earth Observing System (EOS), in response to the station approval. NASA assembled two more working groups to aid mission definition and system design. It asked Francis Bretherton, then at NCAR, in November 1983 to head an advisory committee composed of non-NASA scientists to prepare scientific strategies for EOS. And it assembled an internal panel early in 1984, the Science and Mission Requirements Working Group, to formulate a science strategy and tentative (known as "strawman") payloads.

Bretherton's committee took two years to produce its report, and in the interim the Dixon Butler's Science and Mission Requirements Working Group prepared the documents defining the desired instrument capabilities. In a report titled

From Pattern to Process: The Strategy of the Earth Observing System, Butler's group argued that scientists could use the comprehensive measurements available from the EOS sensor suite to investigate global-scale interactions by examining patterns of change. One example used in this report was the El Niño/Southern Oscillation phenomenon, or ENSO. Butler recalled later that El Niño had been on everyone's minds in 1982, as EOS planning began, because an El Niño event that year had been the most severe occurrence in at least the past century. Further, it had been entirely unexpected by the scientific community. A "canonical El Niño" had just recently been defined, and this one had not fit the definition; it had started later in the year and built up more rapidly. ENSO events were typically correlated to a rise in sea surface temperature of about 1 degree C in the equatorial Pacific, and generally evolved in a regular pattern. But only generally—each event differed slightly from its predecessors, with the 1982 El Niño appearing to be a very significant deviation from the pattern. Yet all El Niño events had certain things in common, and one hope for EOS was that analysis of its comprehensive data would permit scientists to identify the mechanism or process that triggered them.[53] Plate 4 shows a satellite view of the later 1997–98 El Niño.

But the EOS strategy was not to develop instruments specifically aimed at a given phenomenon. The committee did not advocate for an instrument or set of instruments aimed at ENSO, for example. Instead, the research strategy they advocated was based on the use of data from space-based instruments defined somewhat more generally, supplemented by appropriate in situ sensors. In the case of ENSO, because no near-term satellite instrument could measure temperatures or mass flows below the ocean surface, a network of moored buoys would be necessary to provide additional data. The general nature of the space-based component would permit its data to be used for other research agendas as well, increasing its usefulness and expanding, the committee hoped, the community interested in the program. This approach was controversial, however, because many scientists, particularly physicists, did not consider it a legitimate methodology. Drawing patterns out of large datasets was not a new approach to science. It extended back to at least as far as the work of Alexander von Humboldt in the early nineteenth century. But it was not the dominant methodology of the space age, providing a fertile ground for controversy.

The EOS Science Steering Committee also proposed a set of tentative payloads for the polar platforms, thirty instruments grouped into three packages. One set consisted of three surface imaging sensors and a lidar sounder: MODIS (Medium Resolution Imaging Spectrometer), which was essentially a substantially enhanced AVHRR instrument with much better stability and calibration;

HIRIS (High-Resolution Imaging Spectrometer); HMMR (High-Resolution Multifrequency Microwave Radiometer) for surface temperature and ice studies; and LASA (Lidar Atmospheric Sounder and Altimeter). The second package they named SAM (Sensing with Active Microwaves). This group consisted of three radar instruments, a Synthetic Aperture Radar operating in the L, C, and X bands; a radar altimeter for ice and ocean surface topography; and a scatterometer for measuring sea surface winds. The third package was aimed at atmospheric chemical and physical processes, but unlike the UARS, its instruments encompassed the troposphere. A Doppler Lidar would measure tropospheric winds for improved weather forecasting, while a set of interferometers would measure upper atmosphere winds. Two more instruments would measure chemical species in the troposphere and stratosphere, respectively. Finally, a group of instruments devised to measure solar radiation, radiation budget, and space weather rounded out this package.

The platforms needed to accommodate all these instruments were accordingly very large. The launch mass of the polar platforms was nearly twice that of the not-small UARS, and slightly higher than that of the Hubble Space Telescope. They would also obviously be very expensive, causing the committee to try to preempt that line of criticism. They argued that the large platforms were necessary to achieve simultaneity of measurements among the various instruments, which could not be done with instruments positioned on different satellites in different orbits. Further, some of the instruments, particularly the Synthetic Aperture Radar, needed more power than would be available from smaller satellites. Four of these platforms would ultimately be needed in sun-synchronous polar orbits, two "morning" platforms and two "afternoon" platforms to permit investigation of diurnal patterns.

A system of this scale required justification of similar scale, and Francis Bretherton's committee produced a study in 1987 that provided one: a thoroughgoing reconstruction of the Earth sciences themselves. Bretherton's group had started meeting in early 1984, forming working groups to help define individual subject areas and begin drafting reports, and met in a workshop in June 1985 to hammer out their inevitable differences. This proved a lengthy effort because Bretherton had a difficult time getting oceanographers, and particularly geologists, interested in the effort. As earlier reports, evaluations, and assessments had indicated, the climate problem was truly interdisciplinary. One of the reasons GARP had never made much headway with the climate element of its proposed program was that GARP's committees were largely composed of meteorologists, and climate research required, at a very minimum, extensive cooperation with physical oceanographers. The vast heat storage capacity of the world's oceans, and the fact that the atmosphere gained most of its energy from the latent heat of condensation, meant that understanding climate processes involved understanding the exchange of energy between ocean and atmosphere, and the movement of energy within the oceans. But GARP's meteorologists had not had extensive contacts with oceanographers, and they had little success at interesting oceanographers in their research during the 1970s. Meteorologists had become interested in the climate problem through model studies, and oceanographers had not yet adopted modeling as a research tool. This had to change before climate research could make any scientific progress.

There were other challenges resulting from disciplinary differences. Primary among these was timescale. The atmosphere changed on timescales of hours to weeks, the oceans on timescales of years to centuries, and the solid earth by timescales of millennia and longer. A satellite's lifetime was long compared to the ever-changing atmosphere, making satellites a reasonable research tool. But many ocean processes did not generate measurable change within a satellite's lifetime, limiting the usefulness of satellites to oceanography. Further, many ocean processes occurred far below the surface, where no conceivable satellite technology could detect them. Geologists, of course, faced similar problems with space research.

Bretherton's group took breaking these boundaries down as its real mission. This was necessary both to achieve broad interest in the climate program within the scientific community overall as well as to accomplish the science that they wanted to see done. Hence, they defined the goal of a new Earth System Science as "to obtain a scientific understanding of the entire Earth system on a global scale by describing how its component parts and their interactions have evolved, how they function, and how they may be expected to continue to evolve on all timescales."[54] They chose to present the Earth as a system composed of other systems, which could each be studied in relative isolation. This was, in essence, an engineer's view of the world. The Earth system could therefore be displayed, and analyzed, as a flow chart. The focus of Earth System Science was to be on the interactions between Earth's myriad physical, chemical, and biological processes. This approach was inherently interdisciplinary—one could not study ocean-atmosphere interaction without the presence of both meteorologists and physical oceanographers on the research teams.

Bretherton's committee resolved the timescale problem by separating Earth processes into two groups. "Solid Earth" studies would include the processes that operated on timescales of thousands to millions of years: plate tectonics, continental evolution, mantle structure and circulation, and magnetic field variations. And studies of the "Fluid and Biological Earth" would include processes taking place on scales of tens of years to centuries: biogeochemical cycles and the physical climate system. These were not perfectly separable, of course. Biological processes, for example, affected surface chemistry, bridging the two scales.[55] But some separation between the two scales seemed necessary to cope with the disciplinary challenges. It was also necessary to help define the structure of the observing system. The shorter-term climate and biogeochemical cycles required continuous monitoring, while the longer-term solid earth processes could be measured less frequently.

Finally, both the Bretherton committee and the EOS Science Steering Committee advocated proceeding with a set of pre-EOS missions that were already planned, but had not yet been approved: an ocean topography mission called TOPEX, and a rainfall measuring mission named TREM, for Tropical Rainfall Explorer Mission. In addition to providing some near term results, these missions were not suitable for the large polar-orbiting platforms. They needed specialized orbits tailored to their individual objectives. The TREM mission, for example, needed to be in an inclined equatorial orbit, not a polar orbit, in order to focus on tropical convective processes and energy exchange. Other missions proposed for EOS also required specialized orbits of various kinds, and this would become a source of tension as the large polar platforms began to absorb all of the agency's attention and most of its available resources.

In 1984, NASA had requested a survey of the needs of the Earth sciences for the period 1995–2015 from the National Academy of Sciences. When it was completed in 1988, it effectively ratified the EOS agenda. Subtitled "The Mission to Planet Earth," this called for integrated Earth observations to enable study of Earth as a system. It also proposed organization around four "grand themes": determination of the structure and dynamics of the Earth's interior and crust, including its evolution; understanding of the structure, dynamics, and chemistry of the atmosphere, oceans, and cryosphere and their interactions; characterization of the interactions of living organisms and the physical environment; and finally, an understanding of "the interaction of human activities with the natural environment."[56] This panel called for integrated, systematic space-based measurements by two to six large polar platforms; five new, large geosynchronous platforms; a series of smaller, specialized Explorer-type missions; and finally appropriate in situ measurements to complement the space assets. Finally, it envisioned a

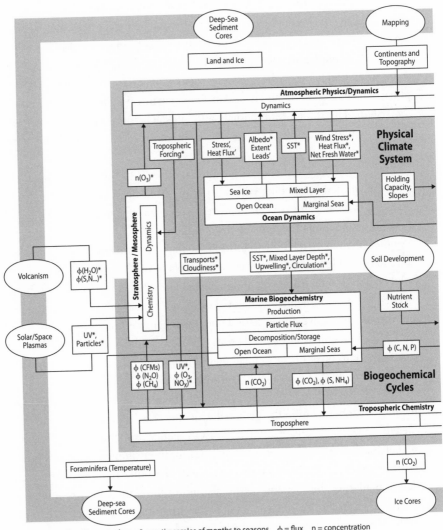

The "Bretherton Diagram," an explication of the interlinked biogeochemical processes that regulate the Earth's biosphere. Courtesy NASA.

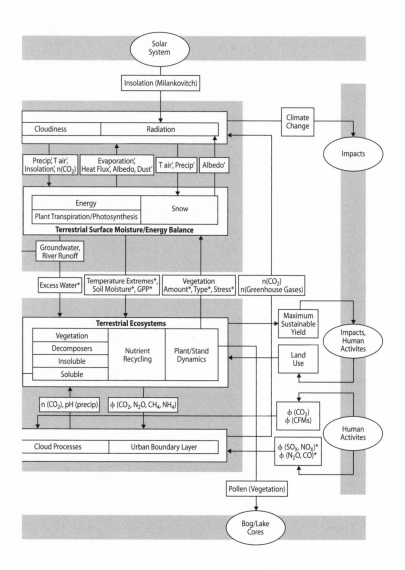

network of large in situ instrument installations they named PLATO, the Permanent Large Array of Terrestrial Observatories, which would serve as calibration and validation sites for the remote sensing instruments.

By 1988, then, the effort NASA had originally begun as a climate observing system had become a comprehensive Earth observing system aimed at a large variety of phenomena. The Bretherton committee had argued, and the National Academy of Sciences had accepted, that the global observing system's mission should be greatly expanded beyond the Global Habitability mission as defined by Richard Goody's 1982 panel. Goody had focused on processes that occurred on human timeframes, as these were of obvious relevance to policymakers. These became only one part of EOS's larger agenda. The expansion of the EOS mission generated the expansion of the polar platforms, which in turn raised their costs. Space Station Freedom was advertised as costing $8 billion, although as historian Howard McCurdy has documented, this was actually the funding level its promoters believed was the maximum cost it could be sold to Congress at—not what it would actually cost to build.[57] When NASA was done with all of the planning and preparatory work for EOS in 1989, the agency projected its cost at $17 billion through the year 2000.

CONTEXTS OF APPROVAL

Gaining funding for a project of EOS's magnitude was not a trivial activity. This was more money than NASA had received for anything, save Apollo and the Space Shuttle. The agency had just fought a political battle to gain approval of a space station that was half this cost. While EOS promised more direct human benefits, and certainly more obvious ones, than the station did, it was still a great deal of money. A complex series of processes took place to bring approval about. NASA leadership participated in all of these to one degree or another, but controlled none of them. When EOS was finally approved, it was as the NASA contribution to a larger national program, the U.S. Global Change Research Program, and a corresponding international program.[58]

Shortly after the 1982 failure of the Global Habitability initiative at UNISPACE, Burt Edelson, NASA's associate administrator for space science and applications, had approached Herbert Friedman, who chaired the National Academy of Science's Space Science Board, about the proposal. Friedman was an established space scientist whose own research had been in solar physics. He had responded by advocating an International Geophysical-Biophysical Program, planning for which was already in progress. Global Habitability might become a part of that program. The IGBP was to be a broad program of research into geodynamics, solar-terrestrial interactions, atmospheric chemistry, ocean circulation, and climate. Friedman expected IGBP planning to take five years, beginning with a meeting already scheduled for July 1983. Edelson, who wanted the Global Habitability program to move along faster than the IGBP was likely to, did not seek an explicit link between IGBP and Global Habitability, and therefore planning for the two proceeded semi-independently. The International Council of Scientific Unions (ICSU) approved the IGBP as the International Geosphere/Biosphere Program (IGBP) in 1986, and in 1988 the National Academy's Committee on Global Change published its recommendation for U.S. participation.[59]

The IGBP was designed to investigate the interaction of the Earth's biosphere, its ecosystems and the organisms they contained, with its physical climate processes. It was qualitatively obvious, for example, that organisms affected the hydrologic cycle, as they consumed and released water, but it was not yet possible to quantify their activities. Similarly, organisms removed carbon dioxide from the atmosphere and fixed some portion of it in their bodies. In the seas, much of this carbon wound up in the form of carboniferous sediments on the sea floor when the creatures died, effectively removing it from the atmosphere for millions of years. On land, much of it returned to the atmosphere via decomposition. But the fluxes, the difference between uptake of carbon dioxide and release back to the atmosphere, were not known. It appeared, for example, that only about 50 percent of the carbon dioxide released by human industrial processes remained in the atmosphere. Whether the remainder was removed primarily by phytoplankton in the oceans, or by land biomass, was a matter of some controversy. Numbers needed to be attached to these fluxes to permit future climate prediction.[60]

The IGBP's focus on the biosphere was paralleled at the international level by GARP's descendant, the WCRP. This was focused on physical processes, particularly ocean circulation and ocean-atmosphere interaction. The WCRP had a set of field programs in various stages of planning. A Tropical Oceans/Global Atmosphere program (TOGA) was being designed to examine air-sea interaction in the tropics and to help understand how tropical processes affected the atmosphere globally, while a World Ocean Circulation Experiment (WOCE) was evolving to better define heat and mass transport in the oceans. Both were critical to climate studies, and both were aimed at providing data for modeling efforts. Ocean circulation models were very primitive compared to weather forecasting models in the 1980s, largely due to lack of data. Meteorologists had been systematically collecting data on the atmosphere for forecasting purposes for more than a century, but as no American institution had as its mission ocean forecasting, no similar database of systematic measurements of the oceans existed. Nor, of course, had there been a systematic ocean model development effort for the same reason. WOCE, and to a lesser extent TOGA, were intended to foster ocean modeling at the international level the way GARP had fostered weather modeling.[61]

A series of congressional hearings called by Democratic leaders in 1987 and 1988 then produced additional pressure for approval of EOS. The most famous of these was held in June 1988. During 1987, a heat wave had begun in the eastern states, and late in the year what turned out to be a very severe drought descended across the midwestern farm belt. This environmental context played a major role in reviving popular interest in global warming, which received editorial comment from *The Washington Post* on 11 June and the *New York Times* on 23 June. The new set of hearings began the day of the *Times* editorial, with an opening statement by Senator J. Bennett Johnston of Louisiana (who had not attended similar hearings in 1987): "Today, as we experience 101° [F] temperatures in Washington, DC, and the soil moisture across the midwest is ruining the soybean crops, the corn crops, the cotton crops, when we're having emergency meetings of the Members of the Congress in order to figure out how to deal with this emergency, then the words of Dr. Manabe and other witnesses who told us about the greenhouse effect are becoming not just concern, but alarm."[62] The star witnesses at this hearing were Hansen and Manabe. Hansen spoke first, presenting the committee with three conclusions. First, 1988 was the warmest year in the instrumental record barring a substantial cooling later in the year. Second, "the global warming is now large enough that we can ascribe with a high degree of confidence a cause and effect relationship to the [enhanced] greenhouse effect." Third, climate simulations indicated that the anthropogenically enhanced greenhouse effect was now large enough to impact the probability of occurrence of extreme events.

He continued to explain that that the current warming of 0.4 degree C relative to the 1950–80 mean was not likely to be a chance occurrence. He placed the probability of a natural warming of that magnitude at only 1 percent. Further, the warming expected from the climate simulations was the same magnitude as this measured warming, reinforcing his argument that anthropogenic warming had been detected. He did not, however, attribute the 1988 drought to that warming. Instead, he argued that the warming would increase the probability of droughts, heat waves, and other events in the future.[63]

His group at GISS had taken a different approach to simulating the increase of atmospheric carbon dioxide than previous efforts had. His earlier simulations had run the Earth's climate at its normal carbon dioxide content to establish a control set of temperatures, and then the simulation was run again with doubled

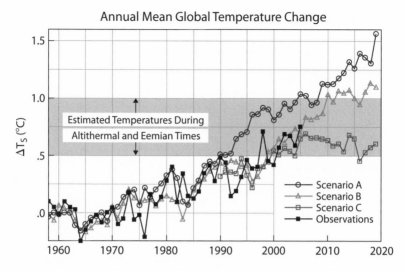

Annual Mean Global Temperature Change

Figure 1. Annual-mean global surface air temperature computed for scenarios A, B and C Observational data are an update of the analysis of Hansen and Lebedeff [J. Geophys. Res., 92, 13,345, 1987]. Shaded area is an estimate of the global temperature during the peak of the current interglacial period (the Altithermal, peaking about 6,000 to 10,000 years ago, when we estimate that global temperature was in the lower part of the shaded area) and the prior interglacial period (the Eemian period, about 120,000 years ago, when we estimate that global temperature probably peaked near the upper part of the shaded area). The temperature zero point is the 1951–1980 mean.

Hansen's 1988 projection for surface warming during the next several decades, updated with observations through 2005. The gray band represents the estimated peak temperatures during the past two interglacial periods, used to put the projected warming in historical perspective. The graph depicts the temperature impact of three different emissions scenarios he devised in an effort to demonstrate that policy action could be effective in curbing anthropogenic warming. Hansen, 1988, updated by Makiko Sato, 2005. Reproduced with the permission of the American Geophysical Union and the authors.

carbon dioxide. This produced two discrete climate regimes for comparison, but did not provide any information about what climate might be like during the transition between the two states. What humans were actually doing, of course, was increasing carbon dioxide content gradually. Hansen's group had tried to replicate that gradual increase by raising the carbon dioxide and other greenhouse trace gases content of the model atmosphere gradually, according to a set of emissions scenarios. Using observations for 1958 to the present, and projections of greenhouse gas forcings from three future emissions scenarios, Hansen's group had found that the model reproduced the observed warming to date. Further, under two of the emissions scenarios, within twenty years the Earth's global mean temperature would be higher than it had been "in the past few hundred thousand years."[64]

In this form, the GISS Model II used its mixed-layer ocean. This fictional sea represented an ocean confined to the mixed layer, about 150 meters, and instead of using an ocean circulation model, it prescribed the ocean's heat transport to reflect the current climate. This meant that instead of responding to atmospheric warming as the real ocean would, it would continue to act as it had in 1988. While this was physically unrealistic, it was an improvement over earlier models, which had not included ocean heat transport. Hansen referred to it as a "surprise-free" representation of the ocean. The parameterization of ocean heat transport meant that the simulation would not capture regional temperature shifts accurately, as small changes in ocean transport could make substantial changes in surface temperature and precipitation. Its value over previous models was that it could give a reasonable estimate of the time lag in the climate response to greenhouse gas forcings.[65] The timescale of the expected warming, after all, was a critical policy issue.

Unlike the other scientists at the hearing, Hansen did not speak as a representative of his institution. Instead, he appeared "in his capacity as a private citizen on the basis of [his] scientific credentials" to ensure his testimony was not taken as a reflection of either NASA or administration policy or approval. This was because his concluding testimony was highly controversial and did not support administration policy. After admitting that the models contained uncertainties, he argued that "the uncertainties in the nature and patterns of climate effects cannot be used as a basis for claiming that there may not be large climate changes. The scientific evidence for the greenhouse effect is overwhelming. The greenhouse effect is real, it is coming soon, and it will have major effects on all peoples."[66]

Hansen called for an immediate commitment to improvements in observations of climatological factors and in understanding of the climate system. In his written text, he pointed out that most of the warming already induced by human addition of radiative trace gases had not appeared yet due to the vast heat storage of the oceans, and it was impossible to determine how long the full warming would actually take to become apparent. Hence he argued for establishing a network of ocean temperature monitoring stations quickly, so that ocean warming could be measured during the 1990s, when he expected surface warming to become apparent to "the man in the street." He also endorsed the measurements called for by the Earth System Science committee.[67]

The only new scientific information presented at this hearing was Hansen's analysis of global temperatures, and these were as yet unpublished.[68] Yet the lack of substantiated new scientific findings did not matter. Due to the environmental context of 1988, the hearings resulted in numerous articles in high-profile popular media. Included in the resulting hearing transcript were a series of reprints commenting on the severity of the drought and its potential linkage to global warming, including articles and editorials from the *New York Times*, *The Washington Post*, and *Fortune*.[69] In its 4 July issue, for example, *Fortune* writer Anthony Ramirez had placed global warming in the context of its inverse, the "Little Ice Age" of the fourteenth century via historian Barbara Tuchman's *Distant Mirror*. He had then explained the physics of the warming, cited NCAR founder Walter Orr Roberts on its inevitability, and discussed corporate concern about it. Alaska and Siberia had, as expected by the models, warmed about 2.7 degrees F in the preceding decades, causing oil companies with extensive infrastructure investments there to worry about the impact of melting permafrost and shippers to wonder about the impact of increasing numbers of icebergs from a potentially disintegrating arctic ice cap.[70]

Hansen's testimony became the focal point of most of the articles that followed the hearings, and he became the de facto leading advocate for research on global warming and for mitigation of it. Because of his advocacy, he was confronted by his colleagues at a meeting later that year over his certainty regarding the detection of greenhouse warming when they were still highly uncertain. But as an extensive article by *Science* reporter Richard Kerr made clear the following June, the climate community was very split about Hansen. No one in the corridors of power had paid much attention to them since 1982, and the attention they were starting to receive because of his declaration was welcome. Further, most of the community actually did believe that greenhouse warming would occur based simply on what Stephen Schneider called "physical intuition." One could not endlessly raise atmospheric concentration of radiatively active gases without a climatic response. But Hansen's claims of *detection* they largely believed were unsound. There was not yet a sufficiency of evidence to support him, in their collective opinion.[71]

The year 1988 also witnessed action in the international arena, further increasing pressure for action on global warming. The United Nations Environment Program and the World Meteorological Organization (WMO) had scheduled a conference in Toronto to discuss climate change and begin to devise a strategy to cope with it. This meeting resulted in the creation of a science assessment group called the Intergovernmental Panel on Climate Change (IPCC). This was patterned after the 1985 international ozone assessment NASA's Watson had orchestrated, to serve the same purpose: to provide a single, thoroughly reviewed, and

authoritative scientific statement on climate. Bert Bolin of the Stockholm Meteorological Institute was appointed chairman, and the panel was subdivided into
three working groups. The first group's task was producing a report reflecting the
state of the science of climate change, while the second group's was assessment
of the potential environmental and socioeconomic impacts of climate change.
The third group's task was to formulate a set of possible response strategies. The
IPCC's three reports were due in 1990, a very short time given their intent to
involve more than three hundred scientists from twenty-five nations in the writing
and review processes.[72]

Such was the political pressure generated by the 1988 hearings, the increasing international momentum toward climate research programs, and the lousy
weather, that during his campaign to succeed Reagan in the White House, Vice
President George H. W. Bush had promised to counter the "greenhouse effect
with the White House effect" and take climate change seriously. After his inau-
guration in January 1989, he started out well enough, having his secretary of state,
James Baker, attend the first meeting of the IPCC and call for action to counter
the threat of global warming. He also had the Federal Coordinating Council for
Science, Engineering, and Technology's Committee on the Earth Sciences prepare and submit a brief report outlining a proposed U.S. Global Climate Research
initiative with his proposed fiscal year 1990 budget. At a scant, booklet-sized thirty-
eight pages, the document contained little detail and did not look beyond fiscal
year 1991.[73] This had quickly been welcomed in the U.S. Senate, however, where
the Committee on Commerce, Science, and Transportation had prepared a similar bill, the National Global Change Research Act of 1989.[74]

The global change research proposal did not gain the administration much
credit, however. It promptly began backing away from its commitments to the
international policy process, influenced by a report from the George C. Marshall
Institute. With the rapidly ending Cold War removing the group's primary reason
for existence, it had changed its emphasis to combating environmentalism
instead. Their climate report was designed to cast doubt on greenhouse warming,
arguing that the current warming trend was probably due to solar activity and that
the twenty-first century would likely be cooler, as solar activity receded. As the
Marshall Institute was not a scientific research organization, they provided no
evidence to back their claims, and their clear purpose in writing it was to prevent
regulation of greenhouse gases (they supported climate research). Their Washington office had contacted the White House to request the opportunity to brief staff
on it. Physicist William Nierenberg, one of the Institute's founders who had
recently retired as head of the Scripps Institution of Oceanography, did the brief-

ing in response to a White House request. This intercession provided a faction within the White House that was opposed to regulation to seize the opportunity to block efforts by Environmental Protection Agency (EPA) Administrator William K. Reilly to participate in the international framework convention process.[75]

The administration did not have a science advisor during its first six months, and the person finally appointed to the job, Yale nuclear physicist D. Allan Bromley, was thrust into this controversy relatively unprepared. He also initially accepted the Marshall Institute's analysis, and he received rough handling in hearings conducted by the Senate subcommittee for science and space, particularly from Senator Albert Gore, Jr. He defended the administration's stance against regulation, this time using the work of British modeler John F. B. Mitchell, who had seen large changes in his model's climate predictions occur when he included the effect of ice. Bromley had interpreted this as showing the immaturity of the field of climate modeling, justifying a no-regulation stance.[76]

Bromley also recognized that he did not have a full grasp of the field, and he began asking for briefings from other scientists, including Warren Washington, one of the founders of NCAR's climate modeling division, Bob Watson, Daniel Albritton, and Stephen Schneider. Bromley eventually concluded, as had all of the post-1975 National Academy assessments, that carbon dioxide–induced warming was inevitable. He also believed that greenhouse gas mitigation was necessary, but that it had to be done in an economically rational way. As he explained in his 1994 memoir, this meant trying to improve the regional fidelity of the climate models. Economically sound responses to the greenhouse problem required knowledge of regional impact so that appropriate mitigation strategies could be selected. In turn, this made investment in much more powerful computers and in NASA's proposed EOS vital.[77] Promoting research, instead of mitigation action, of course, also allowed the administration to present itself as doing something about climate change, a deliberate strategy of avoiding the political costs of regulatory action.[78]

The administration therefore greatly expanded its Global Climate Research Program in its fiscal year 1991 budget proposal, asking for a 57 percent increase over 1990. It requested full funding of NASA's EOS, with a first launch of one of the large polar platforms scheduled for 1998. The first year increment for EOS was $132 million, with a run-out cost through fiscal year 2000 of $17 billion. NASA planned to construct fourteen facility instruments for the polar platforms, and it had selected an additional twenty-four instruments during its 1988 competition. The facility instruments were assigned to NASA centers for construction, with

TABLE 7.2
EOS Polar Platform Instruments, 1984

ALT	Radar Altimeter
AMRIR	Advanced medium-resolution imaging radiometer
AMSR	Advanced microwave scanning radiometer
AMSU	Advanced microwave sounding unit
APACM	Atmospheric physical and chemical monitor
ARGOS+	enhanced ARGOS data collection and platform location system
ATLID	Atmospheric lidar
ATSR	Along-track scanning radiometer
CR	Correlation radiometer
ERBI	Earth Radiation Budget Instrument
ESTAR	Electronically Scanned Thinned Array Radiometer
F/P-INT	Fabry-Perot Interferometer
GLRS	Geodynamics Laser Ranging System
GOMR	Global Ozone Monitoring Spectrometer
HIRIS	High-resolution Imaging Spectrometer
HRIS	High-Resolution Imaging Spectrometer
IR-RAD	Infrared Radiometer
LASA	Lidar Atmospheric Sounder and Altimeter
LAWS	Laser Atmospheric Wind Sounder
ITIR	Imaging Thermal Infrared
MAG	Magnetosphere Currents/Fields
MERIS	Medium-Resolution Imaging Spectrometer
MLS	Microwave Limb Sounder
MODIS	Moderate-Resolution Imaging Spectrometer
MPD	Magnetosphere Particle Detector
NCIS	Nadir Climate Interferometer Spectrometer
PEM	Particle Environment Monitor
SAM	Sensing with Active Microwaves
S&R	Search and Rescue
SAR	Synthetic Aperture Radar
SCAT	Scatterometer
SEM	Space Environment Monitor
SISP	Surface Imaging and Sounding Package
SUB-MM	Submillimeter Spectrometer
SUSIM	Solar Ultraviolet Spectral Irradiance Monitor
TIMS	Thermal Infrared Multispectral Scanner
VIS-UV	Visible-Ultraviolet Spectrometer

Source: Adapted from "From Pattern to Process: The Strategy of the Earth Observing System, vol. II," 1984, pp. 14–17.

Note: These were the instrument concepts, often called "strawman payloads," initially intended for the Earth Observing System polar platforms. In 1989, formal instrument proposals were accepted, often under different names. The GOMR concept instrument became MOPITT, for example.

their science teams chosen by NASA headquarters.[79] Table 7.2 gives the EOS instrument concepts as of 1984.

By this time, NASA had also signed agreements with the European Space Agency to supply a second polar platform, so that the observation system would consist of the U.S. platform in a sun-synchronous "afternoon" orbit and the European platform in a sun-synchronous "morning" orbit, and also with Japan, to provide a third platform.[80] A large part of EOS's high cost was in its data system, the Earth Observing System Data and Information System (EOSDIS), which would have to ingest about 12 terabytes of data per day. This was critical to the success of the research program. If the data could not be ingested, processed, and made available to researchers relatively quickly, the hoped-for rapid progress in climate science would not happen.

Finally, by the time it was approved, EOS had been separated from the Space Shuttle. Originally conceived as a set of large, astronaut-tended platforms about the same mass as the Hubble Space Telescope, by 1989 it had become clear that this was unrealistic. As built, the Shuttle could not place a payload of more than about 500 kilograms into polar orbit, far less than the necessary 11,000 kilograms. An enhanced Shuttle or, alternatively, a cargo-only Shuttle, such as the Shuttle C design being pursued at the Marshall Space Flight Center, would have been necessary to launch them. These had not been funded. Further, the polar launch facility for the Shuttle at Vandenberg Air Force Base had been built incorrectly, and would have required expensive reconstruction. While waiting for an air force decision about whether it would fix and activate the Vandenberg facility, the administrator had ordered the platforms be made dual compatible with the Titan 4 expendable launcher and with the Shuttle, effectively eliminating the need for Shuttle launch or servicing.[81] The air force ultimately decided not to repair and activate the Vandenberg launch pad, forcing EOS on to the Titan 4.

A number of threads converged to bring about approval of the EOS. First among these was an environmental context that drew media attention—and therefore public and congressional attention—back to the subject of climate change. The second was activism by scientists at the national and international levels. The sense of urgency the leaders of the research community had about climate had led them to propose a variety of research programs in the early 1980s, and as they worked out the details of their programs they also advocated for them at the national and international levels. Their promotion of research helped keep the subject of climate in the public eye. The controversies surrounding the Antarctic ozone hole also helped, of course, by keeping the fact of global-scale, anthropo-genic environmental damage clearly, and almost constantly, in Americans' minds. Finally, NASA's own advocacy of climate research helped unify the disparate national and international efforts. NASA's leaders during the period believed that climate research was a high-priority subject, and the network of scientists the agency recruited to help define the climate observing system's mission tied it directly to independent scientists who were also backing other climate research programs.

NASA's effort to launch its observing system also led it to propose a radical agenda for American science: large-scale interdisciplinary research, or better yet, unification of the disparate Earth science disciplines into a new research field. Interdisciplinary research was problematic for scientists (indeed, for all scholars). A physical oceanographer, for example, could only bring expertise to bear on part of a grant proposal for a study of air-sea exchange. Similarly, a highly qualified meteorologist would only be partly qualified to review a paper on that subject for a scientific journal. This made it difficult to maintain high scientific standards through the scientific profession's traditional mechanism of peer review, and it led many scientists to be skeptical about the value of interdisciplinary research. Yet every new scientific discipline faces the same problem—no expertise to provide cogent criticism until the field begins to mature. Hence while NASA's Earth System Science agenda was controversial, it was not at all illegitimate. The space agency was trying to create a new scientific discipline, as it effectively had in planetary science, and resistance was to be expected.

Finally, it took NASA ten years to win approval of the EOS despite the reality that its leaders believed it to be a high priority. The dissolution of the obvious, incremental evolutionary path to a climate observing system in the 1981–83 weather satellite privatization drive had forced the concept into the human space flight side of the agency. There it suffered from the gigantism that infected the Shuttle program and grew enormously. By becoming a very expensive new start, it was further delayed in the austere Reagan (civilian) science budgets. Yet removal of EOS from the human spaceflight program just before its congressional approval did not cure it of these ills. Instead, placement of the system back into the robotic science program permitted critics to demand that it be made far less expensive.

CHAPTER EIGHT

Missions to Planet Earth: Architectural Warfare

Management turbulence [is] defined as continual changes in cost, schedule, goals, etc. At each step, contracts must be renegotiated, people reassigned, designs changed and schedules revised. Soon a disproportionate amount of time is spent in the pursuit of these change practices instead of producing the end product itself.

—*Augustine Report, 1990*

The Earth Observing System (EOS) was subjected to cuts every year through 1995. This, in the words of a harsh National Academy of Sciences assessment, resulted in management turbulence.[1] Seemingly unending reviews, redefinitions, rescopings, and rebaselinings consumed time, squandered project resources, and demoralized staff. By the time the system definition finally stabilized in 1996, the project had been shrunk from $17 billion to $7 billion through the beginning of the new century. The resulting delays pushed the first launch into 1999, twenty-one years after the original National Climate Program Act had gotten planning for EOS started.

EOS's fiscal troubles stemmed from a number of causes. The sudden collapse of the Soviet Union in 1991 undermined the American space program overall, as it had always been primarily a tool of Cold War propaganda.[2] Space Station Freedom suffered the same descoping, and in fact survived an effort to terminate it entirely in 1993 by a single vote in the Senate. Large flagship science projects outside NASA, like the Superconducting Supercollider, were attacked.[3] The early 1990s also witnessed an unusual bidding war over the federal deficit, with the two political parties trying to cut areas of the budget that largely benefited the other's principal constituencies. While this party competition eventually produced the first balanced budget in a generation, the result for NASA was catastrophic. The agency's budget declined 18 percent between 1991 and 1999, when it became obvious that further cuts would require terminating major agency functions.[4] Further, the Republican Revolution of 1994 placed anti-environmental leaders firmly in charge of Congress, making EOS an obvious target.[5] Finally, NASA administrator Daniel Goldin, appointed in early 1992, believed that everything the agency did could be done far more cheaply, and therefore did not resist the cuts. He had also directly attacked EOS's "Battlestar Galactica" approach in a private White House briefing.[6] He welcomed the fiscal discipline being imposed as a way to force his preferred "faster-better-cheaper" approach to space science on the agency.[7]

In the process of shrinking EOS, NASA leaders abandoned the large satellite approach. They also abandoned the idea of maintaining the primary EOS measurements for fifteen years. In the early years of the downsizing, the range of scientific questions was narrowed, but expanded again later as the savings from the smaller satellite approach became clearer. NASA also began to try to migrate some of the key measurements to a new, joint air force–National Oceanic and Atmospheric Administration (NOAA)–NASA operational weather satellite program in hopes of maintaining continuity of measurements; finally, it sought to make the climate observing system international as it had once tried to make the weather satellite system international. It found international partners for some missions, and in some cases dropped planned measurements entirely when other countries promised to take them over. And yet, by the first years of the twenty-first century, NASA had fully abandoned Shelby Tilford's original conception of an integrated, long-term, well-calibrated set of Earth observations. Instead, the nation's strategy was back where it had been in 1980, and the successor to EOS, a new generation of meteorological satellites, seemed to be firmly on the twenty-years-to-space track.

DEMISE OF THE POLAR PLATFORMS

The revolt against EOS had started even before the program had achieved White House approval in 1989. There were three major streams of criticism developing even as Shelby Tilford and Dixon Butler tried to shepherd EOS through the approval process. A vocal contingent within the scientific community opposed EOS because of its inflexibility, slow pace, and gigantism. Second, from a technological direction, particularly the clandestine world of the Strategic Defense Initiative (SDI), came criticism of EOS for its use of "old" early 1980s technology instead of new and untried concepts developed in (mostly) military labs. Third, at $17 billion, there was immediate criticism of the program's cost. Many scientists believed that it would swallow all the resources available for Earth science research for the next two decades, and that therefore scientists who had not been part of EOS in its formative phase would be shut out of NASA funding effectively forever. These three lines of criticism converged in a demand for a smaller, less expensive, and more flexible system architecture.

As it stood in 1990, EOS consisted of a set of four Hubble Space Telescope–sized polar orbiting platforms, two built by NASA and one each provided by the European Space Agency and the National Space Development Agency of Japan (NASDA). These were to be launched between 1996 and 1998, and repeated three times each to achieve the desired fifteen-year time span. The first NASA platform was assigned instruments for surface studies, atmospheric temperature soundings, and ocean and land surface altimetry. This platform also contained instruments intended to replace those on the NOAA series of operational weather satellites. The second NASA payload carried a very large synthetic aperture radar and a suite of atmospheric chemistry and physics instruments. These included instruments aimed at studies of the thermosphere and mesosphere, and of the Earth's magnetic field. The European and Japanese platforms were similarly equipped. In its original guise, EOS was meant to envelop the full range of the Earth sciences. It was not focused on any particular scientific question despite its origins in the early climate observing system planning exercises of the late 1970s. It also contained no provision for smaller missions, for missions requiring non-polar orbits, or for missions responsive to new scientific questions that might arise over the fifteen-year period. This basic system architecture, not the science EOS was supposed to do, was the initial point of conflict over EOS.

The Space Studies Board of the National Research Council (NRC) had issued the first formal statement of scientific discontent over EOS in a 1988 study, *Strategy for Earth Explorers in the Global Earth Sciences*. It had been prepared at the request of NASA's Space and Earth Science Advisory Committee, which was composed primarily of university-based researchers. This group, whose chairman was geodicist Byron Tapley of the University of Texas, sought to supplement EOS with a series of smaller satellites that would take less time and money to develop.

Artists' conception of the EOS large platform. This version carries a large radar, the panel in the lower left. Courtesy NASA/Caltech/JPL.

Tapley's group believed that smaller satellites provided a number of benefits. One was that the large polar platforms could not make some of the desired measurements. Structurally, large platforms would be too flexible for use in precision altimetry measurements, for example, which were necessary for oceanographic and cryospheric research. Another was that large platforms carried large risk. Loss of a small, inexpensive satellite in the event of a launch vehicle accident, or from a spacecraft failure, was much easier to overcome financially than loss of one of the large platforms. Finally, smaller satellites also offered greater programmatic flexibility. New scientific demands could be accommodated more easily if NASA funded a line of small satellites in addition to the large platforms. Hence, Tapley's committee argued for a line of Earth Explorers within the Mission to Planet Earth that would be similar in nature to the Explorer series of satellites used by NASA's physics and astronomy community. They sought sufficient funding for two small missions per year or one medium mission every three years.[8]

Tapley's committee reflected a growing unhappiness within the scientific community over the sheer length of time NASA was taking to fly new Earth science missions. Much of the work done in American science was done by graduate students and postdoctoral fellows, who had short time horizons (four to six years) and could not participate effectively in missions taking ten to twenty years to develop and launch. This undermined space-based science in general by drying up the pool of future talent. The possibility that EOS could take decades was made real for the committee by the Upper Atmosphere Research Satellite (UARS), which by 1988 had been in development since 1978 and, because of the Challenger accident, would not fly for several more years. It had thus far taken ten years to develop this critical group of allegedly high-priority measurements, and as a result, the scientific limelight had been stolen by the ground- and aircraft-based instruments NASA and NOAA had deployed in the National Ozone Expedition (NOZE) and the Airborne Antarctic Ozone Experiment (AAOE). EOS, planning for which could also be traced back to 1978, seemed to be taking an even slower road to orbit, and since it would be the only Earth science program for the next twenty years, the committee found its large platform approach deeply troubling.[9] EOS seemed to be an example of Big Engineering getting in the way of scientific research.

During 1990, the Goddard Institute for Space Science's (GISS) James Hansen began advocating a set of small satellites called CLIMSATs (Climate Satellites) that were also about the size of the proposed Earth Explorers. His CLIMSAT proposal involved two small satellites, each with three instruments: improved versions of the Langley Stratospheric Aerosol and Gas Experiment (SAGE) and Earth Radiation Budget Experiment (ERBE) instruments and an Earth Observing Scanning Polarimeter (EOSP), to measure cloud radiative properties. EOSP was a derivative of instruments previously sent to Venus and Jupiter. All three had been selected for EOS during the 1988 announcement of opportunity process, and represented relatively mature technologies with previous flight experience. Hansen argued that their technical maturity meant that they could provide critical measurements sooner than would the large EOS platforms.

Hansen's scientific motivation for the CLIMSAT proposal was a 1989 article by influential radiative transfer specialists, who had argued that water vapor feedback would enhance the carbon dioxide greenhouse enough to make it detectable by an ERBE-like instrument within the next couple of decades.[10] Hence, they had concluded, ERBE or something very much like it, needed to be flown for the next several decades. Hansen added SAGE to monitor both ozone, which was also a greenhouse gas, and aerosols, which tended to cool the Earth. The scanning polarimeter was to study the interaction of aerosols and clouds. These three instruments, Hansen contended, were the ones necessary to detect anthropogenic climate change, and needed to be flown as quickly as possible to establish a continuous record. He also believed that they would be inexpensive enough to be reflown and maintained in orbit for long-term monitoring; the hugely expensive EOS constellation would almost certainly not be.[11]

Hansen's proposed CLIMSAT system was aimed at studying the changing climate. EOS, however, had evolved for process studies, not climate monitoring. Hence, CLIMSAT would not fulfill EOS's other science goals, such as better forecasting of El Niño. EOS's perceived ability to contribute to better regional prediction had been an important reason for its approval by the White House, so CLIMSAT was not really an alternative to EOS. It was complementary. But it was badly timed. Hansen's public advocacy of it while NASA was attempting to get EOS through Congress tended to make it appear as a smaller, cheaper competitor to EOS. Hansen himself argued that CLIMSAT was complementary to EOS, and should be funded via an expanded Earth Probes program. But it nonetheless caused difficulties for Lennard Fisk, Shelby Tilford, and Dixon Butler, who had to defend EOS's mission anew.[12]

Yet a third line of criticism emerged from the National Space Council. Veterans of the Defense Department's SDI contended that it had developed a variety of microsat technologies that could make EOS's measurements earlier and much less expensively. As ballistic missile defense was being conceived in the late 1980s, dozens of low-altitude, infrared scanning, micro-satellites called Brilliant Eyes were supposed to provide guidance information to ground-based interception missiles. These became the source of a third challenge to EOS, layered on top of the Earth Probes and the CLIMSATS.[13]

The collapse of East Germany in 1989 and the subsequent end of the Cold War, finally, led to reexamination of the federal budget overall, and NASA's in particular. NASA's role as a vehicle for technocratic competition and propaganda victories left the agency without its primary political function. Space Station Freedom was immediately threatened with termination; it survived but was repeatedly descoped until (temporarily) stabilizing in 1994 as the International Space Station. EOS was also subjected to budget reductions almost immediately. At hearings on EOS in April 1991, Senator Albert Gore, Jr., of Tennessee, warned Lennard Fisk, NASA's associate administrator for space science, that the fiscal year 1992 budget then being debated in the Senate would not fully fund EOS; the conference committee report that emerged placed a run-out cap on EOS of $11 billion through fiscal year 2000.[14]

The conference committee report also required NASA to empanel an independent review of EOS. NASA Administrator Richard M. Truly asked Edward Frieman, who had succeeded William Nierenberg as director of the Scripps Institution of Oceanography in 1986, to assemble a panel to help the agency restructure EOS. This group became known as the EOS Engineering Review Committee. Frieman was invited to Washington at the end of April to meet with senior administration officials and get their views on EOS; his interviews included Shelby Tilford; oceanographer W. Stanley Wilson, now EOS program scientist; NASA Administrator Truly; Mark Albrecht, the Space Council's director; and Vice President Dan Quayle. He reported to his committee members that "there seems to be strong support within the Administration for EOS."

But he also found three common concerns in all of his interviews: EOS was too expensive, it did not make use of newer small satellite technology, and "[it] is perceived by some as being too distant to help solve the critical near-term global change policy concerns."[15] As a result, Frieman recalled later, there was great hostility at the Space Council and at the White House toward NASA's approach to EOS. In a letter to Frieman in mid-May, Representative George E. Brown, chairman of the House Committee on Science, Space, and Technology, put it slightly differently: "despite my own conviction and that of many of my colleagues that increased investments will be needed to better define future environmental policy options, major scientific undertakings such as EOS, for which a clear technical consensus does not yet exist, is difficult to sustain in this budgetary environment [sic]."[16]

Frieman's committee recommended several approaches to bringing down EOS's cost. The first of these was reducing the program's scope. As it existed in 1990, the EOS concept included sensors aimed at all possible Earth science disciplines, from solid Earth geophysics to Sun-Earth interaction. Frieman's committee advocated descoping EOS back to the Global Habitability initiative's focus on climate change.[17] It made this recommendation based upon the scientific priorities laid out by the Intergovernmental Panel on Climate Change (IPCC), whose 1990 report had called for an increased emphasis on "the various climate-related processes, particularly those associated with clouds, oceans, and the carbon cycle."[18] The IPCC's three priorities, in turn, became one basis of EOS's reconstruction.

The IPCC had chosen to emphasize clouds, oceans, and the carbon cycle because these were the sources of the largest uncertainties in predicting the rate of future warming. Ocean modeling was in its infancy compared to atmospheric modeling, with substantial model development not starting until the early 1980s.

Ocean remote sensing was also in its infancy. Seasat A's ninety-nine-day mission had been the first dedicated oceanographic flight, and to date the only one. NASA, the European Space Agency, and the Japanese space agency all had ocean-sensing instruments planned for flight during the 1990s, but with the exception of a modified version of the Coastal Zone Color scanner, these were all essentially first-generation instruments. They would require time to be fully understood, just as had the atmosphere sensing instruments. And because the oceans served as Earth's primary heat storage system, improving ocean circulation data and ocean circulation models was fundamental to gaining better climate forecasting.

The IPCC's focus on clouds grew out of the Langley Research Center's ERBE. In January 1989, the project's science team had published a paper in *Science* that weighed in on a long-standing controversy over whether clouds produced a net warming of the Earth by absorbing outgoing infrared energy, or whether they produced a net cooling by reflecting away incoming sunlight. The ERBE data clearly showed that clouds produced a net cooling effect, which the team chose to define in terms of forcing. Cloud reflectance of incoming sunlight was shortwave forcing, while cloud absorption of outgoing infrared, or longwave, radiation was longwave forcing. Defining their terms in this fashion allowed them to separate and analyze independently the two different effects clouds had.[19]

The ERBE data also allowed the team to assess how cloud forcings varied with latitude. Shortwave forcing was highest at high latitudes, while in the tropics shortwave and longwave forcings nearly cancelled each other out, as shown in Plate 5. And the magnitudes of the individual shortwave and longwave forcings were about 10 times the radiative forcing that would be produced by a doubling of carbon dioxide concentration. Because of this, the authors pointed out, a shift of mid-latitude cloudiness patterns toward the equator during the last ice age would have dramatically reinforced cooling and enhanced glaciation.[20] An expansion of tropical cloudiness patterns out of the tropics would, alternatively, have a warming effect. Clouds, their data made clear, were the dominant factor in regulation of the Earth's climate.

The ERBE team also found from a brief survey of six existing global climate models that they had predicted this variation in cloud forcing qualitatively, but they had not done so quantitatively. The six models had shown wide variation in their analyses of cloud forcings, with shortwave forcing, for example, differing from the ERBE data by as much as 50 percent. Because the cloud effect on climate was so much more powerful than a carbon dioxide doubling, the inconsistent treatment of clouds by leading climate models introduced a large measure of uncertainty to forecasts of future climate.

Writing for *Physics Today* later that year, ERBE team members Veerabhadran Ramanathan, Bruce Barkstrom, and Edwin Harrison pointed out that the scientific community did not know why the cooling effect should be dominant.[21] The physics of clouds were not well understood, and therefore the models could not be programmed to generate them from physical principles. Instead, clouds were incorporated into the models via parameterization, which allowed their known effects to be described within the models but largely prevented cloudiness patterns from changing as the model climates changed. This significantly impaired the models' ability to forecast climate. This view was reinforced the following year, when Robert Cess published a paper surveying fourteen climate models, finding that while they relatively accurately represented clear-sky radiative transfer through the atmosphere, they diverged considerably in their treatment of cloud impacts.[22] Hence, the IPCC had made cloud forcings their top research priority in their 1990 assessment.

Taking the independent IPCC's report as its cue, Frieman's committee had argued that EOS be refocused on climate change, and particularly on cloud and aerosol forcings and ocean measurements. They also argued for a greater internationalization of EOS to reduce the cost to the United States. While NASA had gotten agreements for European and Japanese polar platforms in the late 1980s, it had not attempted to create an integrated international global observing system. Instead, it had pursued international efforts somewhat piecemeal. It had gotten French support for TOPEX (Ocean Topography Experiment)/Poseidon without making arrangements for a successor; gained a space for a NASA-built scatterometer on the first Japanese environmental research satellite, ADEOS, scheduled for 1996; and forged a partnership with Japan for the Tropical Rainfall Measuring Mission. Foreign experimenters also had a number of instruments chosen for EOS via the standard announcement of opportunity process. But NASA had not tried to negotiate an overall division of labor among the other spacefaring governments over who would support which measurements over the two decades EOS was supposed to run.

Finally, Frieman's reengineering committee recommended a substantial restructuring of EOS. EOS had been planned around large polar platforms that required the Titan 4 launch vehicle, which was the most expensive by far (except, of course, for the Shuttle). Delinking EOS from the Shuttle program had not changed that, partly because Tilford, Butler, and their planners believed that simultaneity of measurements was scientifically necessary, and therefore the instruments needed to be on the same platforms. This policy had also been based on the lack of a mid-sized launch vehicle, with the only rockets available for the Vandenberg Air Force Base site (the only site available for polar launches) being the small Delta and the large Titan. However, for its own reasons, the U.S. Air Force had decided in 1991 to improve its Vandenberg facility to take the mid-sized Atlas rocket, and this had been the opening wedge for Frieman to recommend repackaging EOS onto satellites sized for Atlases. Atlas-sized satellites would permit some, but not all, of the desired EOS sensors to be clustered together for simultaneity.

But Frieman had also been under a great deal of pressure to shrink EOS further, by splitting more of the sensors off onto small, single-instrument payloads. After accepting the responsibility of preparing his review, Frieman had been brought to Washington to interview senior members of the administration about EOS; after his return, he had told his panel members that this subject came up in several different meetings.[23] Advocates of this approach, many of whom had experience in classified programs related to Reagan's SDI, argued that cost was an exponential function of complexity. A five-instrument satellite would cost far more than five one-instrument satellites. Hence, one could achieve large cost reductions by building more, simpler satellites.

These advocates also argued that the simultaneity claim of EOS's supporters was overblown. Satellites could, they believed, be maneuvered to fly in formation close enough together so that their measurements would be essentially simultaneous. Formation-flying had not been demonstrated in any program known to the Space Studies Board of the National Academy of Sciences as late as 1995; nonetheless, it became part of Frieman's recommendations. As a potential future technology, formation-flying offered a great deal of promise; a major cost of satellite building was the cost of trying to integrate many different instruments onto the same satellite bus. Limiting satellites to one or two instruments could permit many more instruments to be flown for the same amount of money—if the formation-flying idea worked out. Finally, the smaller satellite approach promised to produce some research results sooner, which would help mollify scientists and policymakers who were becoming outspoken in their criticism of the slowness of EOS, while also responding to the scientific advocates of the Earth Probes concept.[24]

In February 1992, the Senate subcommittee on Science, Technology, and Space reviewed the reengineered EOS. As NASA's Lennard Fisk presented it, EOS now consisted of six payloads: three Atlas-sized satellites, one Delta-class satellite, and two sized for a new small launch vehicle intended to replace the dis-continued Scout rocket. The three Atlas packages were named EOS-AM, a polar, "morning"-orbit satellite instrumented primarily for surface studies; EOS-PM, a polar, "afternoon"-orbit satellite instrumented for weather and cloud studies; and EOS-CHEM, a polar, "afternoon"-orbit chemistry satellite to succeed UARS. The Delta payload was EOS-ALT, an altimetry satellite for ocean circulation and ice sheet studies, while the two smallest were EOS-COLOR and EOS-AERO, for ocean color and aerosols, respectively. These were to be launched between June 1998 and 2002. They would then be repeated to achieve the required fifteen-year coverage, so there would ultimately be three EOS-AM flights, three EOS-PM flights, and so on.

Fisk also explained that to reduce a gap in coverage of the Earth's radiation budget, a very high priority for the IPCC, a Clouds and the Earth's Radiant En-ergy System (CERES) instrument would be added to the Tropical Rainfall Mea-suring Mission, scheduled for launch in 1997, while a Stratospheric Aerosol and Gas Experiment (SAGE) III would be added to a "flight of opportunity," a Russian Meteor weather satellite, as things would turn out.[25] Finally, in the process of repackaging EOS, NASA also shrank it, deleting instruments aimed at geomag-netism, upper atmosphere research, and solid Earth geophysics. The largest instruments, a space-based large-aperture lidar and synthetic aperture radar, were also dropped, as they could not be accommodated on the much smaller Atlas-sized satellites.

This reengineering of EOS did not silence public criticism by the scientific community. There were several points of contention, including the continued slow pace of the program, its lack of competition, and its apparent consumption of all available funding for Earth science missions for the foreseeable future, but one will suffice for discussion: launch order. The AM package, to be launched first, was primarily aimed at land and ocean surface studies, and was not directly relevant to the central question of *detecting* anthropogenic climate change. The PM package contained instruments aimed at that question. Climate models al-most universally predicted that as the troposphere warmed, the tropopause, the boundary between the troposphere and the stratosphere, would change its aver-age altitude, while the stratosphere would cool significantly. Due to the large errors and poor calibration records of the 1970s era satellite temperature sounders still in use, a definitive detection of this effect could not be made using their data. Radiosondes had similar problems, with inherent biases that had to be identified and corrected.[26] As the launch order controversy played out in the first half of the 1990s, it was clear to the climate research community that instruments capable of

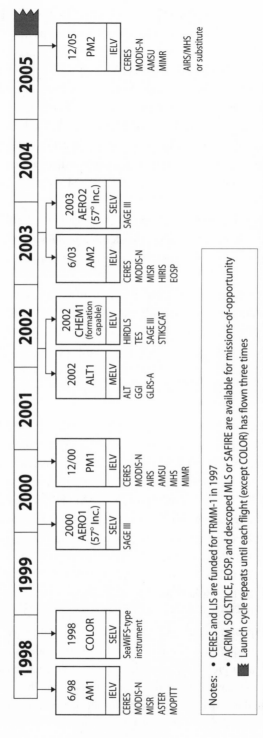

| 1998 | 1999 | 2000 | 2001 | 2002 | 2003 | 2004 | 2005 |

6/98 AM1		
IELV		

CERES
MODIS-N
MISR
ASTER
MOPITT

1998 COLOR		
SELV		

SeaWiFS-type
instrument

2000 AERO1 (57° Inc.)		
SELV		

SAGE III

12/00 PM1		
IELV		

CERES
MODIS-N
AIRS
AMSU
MHS
MIMR

2002 ALT1		
MELV		

ALT
GGI
GLRS-A

2002 CHEM1 (formation capable)		
IELV		

HIRDLS
TES
SAGE III
STIKSCAT

6/03 AM2		
IELV		

CERES
MODIS-N
MISR
HIRIS
EOSP

2003 AERO2 (57° Inc.)		
SELV		

SAGE III

12/05 PM2		
IELV		

CERES
MODIS-N
AMSU
MIMR

AIRS/MHS
or substitute

Notes:
- CERES and LIS are funded for TRMM-1 in 1997
- ACRIM, SOLSTICE, EOSP, and descoped MLS or SAFIRE are available for missions-of-opportunity
- ▰ Launch cycle repeats until each flight (except COLOR) has flown three times

EOS as of 1992. From Ghassem Asrar and David Jon Dokken (eds.), *EOS Reference Handbook*, NASA Earth Science Support Office, 1993, p. 10.

measuring tropospheric temperature with much better calibration capabilities were vital to the question of anthropogenic climate change.

The Jet Propulsion Laboratory's (JPL) Atmospheric Infrared Sounder (AIRS) instrument, which was expected to produce improved vertical resolution and reduced error compared to the 1970s generation of temperature sounders on the operational weather satellites, was supposed to be able to measure the predicted changes. The AIRS principal investigator, Moustafa Chahine, intended to overcome the cloud-clearance problem infrared instruments had by pairing AIRS with the Advanced Microwave Sounding Unit (AMSU) and using the microwave data in his temperature retrieval algorithms. Similarly, the atmosphere was expected to become wetter as it warmed, and a PM instrument to be provided by NOAA initially, and ultimately by Brazil, called the Measurements of Humidity Sounder, could determine whether this was happening. The humidity sounder's data was also to be used to improve AIRS's retrievals, and the three instruments effectively formed a single package aimed at the classical meteorological values of temperature and moisture. They also, of course, had direct climate relevance. These three instruments were to be flown with the Moderate Resolution Imaging Spectroradiometer (MODIS), which would provide cloud-top temperatures and altitudes, also likely to change in a warming world; the Clouds and the Earth's Radiant Energy System (CERES), whose radiation budget measurements nearly everyone agreed were of the highest scientific priority and directly relevant to anthropogenic climate change; and a new microwave imager. Because this PM package was the central climate satellite, many members of the Frieman's engineering review panel thought that it should be the first launch.[27]

The counterargument that NASA officials had made was that the AM satellite, whose contractor, General Electric, had originally been assigned to build the large polar platforms, was already under construction and delaying it would only raise overall costs without providing any benefits. Further, its most expensive instrument, ASTER, a high-resolution land imager, was being provided by Japan and thus the AM satellite, despite having more instruments, was less expensive to NASA. It was also less technologically challenging. The AIRS instrument, EOS-PM's centerpiece, was the most difficult U.S. instrument, and none of NASA's leaders believed that it could be accelerated successfully. Finally, the PM satellite contract award had not been made yet. Nothing would be gained by trying to rearrange the flights except raising costs. As of late 1991, it was already too late to reprioritize.[28]

While NASA's decision not to change the launch order stood, it did not sit well outside the agency. In June 1992, Pierre Morel, by now head of the planning staff for the World Climate Research Program (WCRP), argued that scientific priority still clearly belonged to the PM satellite. In early 1993, *Science* reporter Gary Taubes reported that many members of Frieman's committee did not think NASA had taken their recommendations seriously, and while EOS was now much smaller, it was still not focused on the correct set of questions.[29] Hence the larger scientific community continued criticizing EOS for its misplaced priorities and lack of a tight focus on detection of anthropogenic climate change; here the mid-1980s strategy of trying to produce long-term, continuous, and comprehensive datasets for use by all parts of the Earth science community was working against EOS. Instead, with budgets shrinking, the community sought the opposite: an observing system aimed at a narrowly drawn, specific set of questions.

RESCOPING AND REBASELINING EOS

Shortly after the Senate review of the reconfigured EOS, NASA received a new administrator, Daniel S. Goldin. Vice President Dan Quayle had had visions of building his own political future on a vastly expanded space program aimed at Mars colonization, and he had found Administrator Truly to be unsupportive. Truly was too wedded to the high-cost Space Shuttle to undertake reforms Quayle thought necessary. Hence, Quayle had attempted to arrange a bypass around Truly by issuing a directive specifying that NASA's assistant administrator for exploration communicate directly with the Space Council, which Quayle chaired, instead of going through Truly's office.[30] When this workaround was exposed, it was not seen favorably either in Congress or in the White House. Truly was then removed. After a prolonged search, Quayle chose Goldin. Goldin was an advocate of what he called "faster-better-cheaper," an effort to do more small missions with less money.[31]

Goldin then embarked on what was known as the "Red team/Blue team" study of all the agency's programs. In essence, all of the agency's programs were reviewed by two teams, a Blue team drawn from the program in question and a Red team composed of people not associated with it. In EOS's case, the Blue team was led by Goddard Space Flight Center's Chris Scolese, while the Red team was chaired by JPL's John Casani. The Blue team reviewed its own program and then defended it before the Red team; the Red team did an independent review and critiqued the Blue team. In EOS's case, Goldin also required the teams to find $3 billion in additional savings. This initiated what insiders called the rescoping of EOS to separate it from the previous years' reengineering, and which was not entirely complete before the EOS budget was cut again—a $750 million slice—this time by Congress. This cut provoked a rebaselining.[32] It also led to the departure of Shelby Tilford.

This time, there was no new outside review committee; instead, EOS's reconstruction was carried out largely internally, with Ed Frieman and some of his committee members serving as informal consultants. The first rescoping stage of the cuts was carried out by forcing all but one of the EOS payloads onto Delta-class rockets or smaller; to protect the 1998 launch schedule, EOS-AM remained an Atlas-class payload. Several instruments were cut from the EOS-PM and EOS-Chem satellites, and other instruments were descoped. But the most significant cuts in the rescoping exercise were made to the data system, the Earth Observing System Data and Information System (EOSDIS). This forced, or enabled, depending on one's point of view, a fundamental redirection of the system.

EOSDIS had originally been conceived in accordance with a centralized data processing model that derived from large-scale Defense Department computing projects, such as the SAGE air defense system.[33] This model was based upon the high cost and large size of high-powered computers, in company with the very limited supply of programming talent for them during the early years of digital computing. This was known as the MITRE model due to its historical linkage with the MITRE Corporation, which had been the system engineer for SAGE and other projects like it. For EOSDIS, this centralized model meant a single, large-scale installation would develop the retrieval algorithms for the instruments, process, and then validate the datasets before providing them to offsite Distributed Active Archive Centers (DAACs).[34] EOSDIS would also provide satellite command and control.

By 1993, as the rescoping was taking place, criticisms of this approach had appeared. From inside the agency, a number of experienced investigators opposed the centralized model. Langley engineer Bruce Barkstrom, for example, who had been the architect for the ERBE and CERES data processing systems, believed that the programmers doing algorithm development had to be very familiar with an instrument and the data it produced to get useful results; because of the historical evolution of NASA's various Earth science components, that expertise was not all located in one place. Langley had specialized in aerosols and atmospheric radiation, while JPL had specialized in physical oceanography and stratospheric chemistry. It made little sense to redevelop that expertise elsewhere. Further, he argued that the centralized model, like the original EOS large platform model, was unnecessarily expensive. In software development, the interfaces between modules of various programs had to be carefully controlled to ensure they remained compatible. Increasing the number of programs a given computer system had to run increased the number of interfaces exponentially, not linearly, and hence cost was also an exponential function of complexity. Finally, computing power had become dramatically less expensive, and programming expertise more widespread, in the decades since the formulation of the MITRE model. Centralization of computing resources was no longer necessary. In his view, a distributed system built around the extant DAACs made much more sense.[35]

Similar arguments appeared in an NRC review of EOSDIS that was presented to NASA in September 1993. Shelby Tilford had requested the review late in 1991, and NRC chairman Bruce Alberts had asked Charles Zraket of Harvard University to chair the new study. Zraket's group argued that EOSDIS's centralized design did not have the flexibility necessary to achieve the EOS's desired results. It was "simply an automated data distribution system" that would provide a set of standardized products to users. Researchers would not be "able to combine data from different sensors, alter the nature of the products to meet new scientific needs, or revise the algorithms used to process data for different purposes."[36] The ability to do this sort of interdisciplinary investigation was one of EOS's major selling points; Zraket's group did not believe EOSDIS would support it effectively. Further, the growing availability of computing power made a distributed architecture both possible and desirable. This new distributed architecture model descended from the Advanced Research Projects Agency's (ARPA) ARPANET, which by 1993 had become publicly available as the Internet.[37] This had demonstrated that a distributed computer network could change with extraordinary speed as users and user demands changed.

Zraket's panel therefore recommended redesigning EOSDIS around the extant DAACs, and adding new ones if additional subject area expertise was necessary. The DAACs would be the system's interface with EOS's user communities, while an EOSDIS Core System would provide the network infrastructure the archive centers would require. Their vision was of EOSDIS evolving into "UserDIS," with science users able to access datasets from multiple sources and integrate them into new products. This meant, in turn, relying on the "entrepreneurial spirit of the DAACS and other interested organizations."[38] But the resulting data system would be much more flexible, able to adapt to new uses and new user-generated products as EOS itself evolved.

The redesign cost money, however, which, like the actual budget cuts, had to be paid for in capability. Dixon Butler recalls that the only really firm cost figures available were attached to processing capacity, and this was where most of the cuts were taken. EOSDIS's processing capacity was reduced by more than half. The number of data products was reduced even further, from 809 to 128.[39] To a degree, the reduction in data products would be made up by researchers who could be expected to use the available datasets to create new ones. User innovation was, after all, the hoped-for outcome of the conversion of EOSDIS to a decentralized network. Lost, however, was the ability to process the huge amounts of data in real time. Instead, data releases would be made months after receipt. While this violated EOS's original intent, it was not perceived as a great loss at the working level. Many researchers had seen the real-time goal as unrealistic, because of the time necessary to evaluate the quality of the data. Very few of NASA's missions had succeeded at producing immediate, high-quality data outputs right after launch, a fact that had made the research community skeptical of Butler's real-time goal.

Tilford, who was the architect of the original EOS concept of long-term, comprehensive Earth observations, by late 1992 had begun to fight the redirection of EOS toward smaller, less capable satellites and a less-than-real time data system. Tilford did not believe Goldin's goal of shrinking all satellites down to single-instrument payloads would permit the achievement of simultaneity in related measurements; while Tilford had not liked the large platform approach, he considered its antithesis, the micro-satellite approach, to be at least equally bad. Hence, Tilford's and Goldin's goals were not the same.

Goldin started looking for a replacement for Tilford in mid-1993. Ed Frieman recommended one of his former students, Charles Kennel, who was an astrophysicist at UCLA. He had recently completed service on an NRC panel that had drafted a decadal survey of the field, which set priorities for astrophysical research for the 1990s and early 2000s. This had strongly supported the use of small satellites just as the 1988 Earth science report had, and of course helped make Kennel an appealing choice. Kennel flew to Washington to meet Goldin for an interview at his apartment in the Watergate hotel. Kennel remembers that this went long into the night. Goldin explained the political problems that EOS faced, both from the scientific community's criticism of EOS and from congressional opponents of climate research, and that he was looking for an associate administrator for it who was not connected to the Earth sciences at all in order to immunize the program from criticism. Reflecting the ongoing scientific criticism of EOS, Goldin made clear that he wanted *science* to be in charge of the Mission to Planet Earth. He also wanted EOS refashioned to use smaller satellites, made more flexible, and finally, kept within a very restrictive budget.[40]

Kennel accepted the position, intending to stay two years before returning home to California. Goldin announced the choice on 6 January 1994.[41] This produced some difficulties for Kennel, as Tilford was well-known and popular in Washington science circles. Tilford was widely regarded as the developer of the concept of integrated Earth observations and had built one of the most successful research programs in the government. Criticism of the decision therefore erupted, channeled toward Kennel's lack of experience in the Earth sciences. He was, after all, to be responsible for the largest single Earth science program ever carried out. This was resolved when he was allowed to choose Robert C. Harriss, who had left Langley Research Center for the University of New Hampshire in 1988, to be the director of the Mission to Planet Earth's Science Division. This made Harriss responsible for decisions about what science proposals were funded. Kennel and Harriss came aboard in January 1994, initiating the next phase in EOS's transformation.

RESHAPING EOS

Kennel joined NASA at an odd moment in NASA history. The 1992 election had put William Jefferson Clinton, a centrist Democrat, in the White House with former Senator Albert Gore, Jr., as his vice president. Gore was a self-proclaimed environmentalist and had written a book titled *Earth in the Balance* about a litany of environmental crises.[42] His ascension had seemed to promise stability for EOS and a revitalized set of environmental policies. That isn't what happened. Gore had not supported EOS's original large platform approach, preferring a more rapid deployment of smaller satellites.[43] In his own interview with Kennel, he had made clear that EOS had to shrink while its science had to be protected. Gore also appointed NASA's Robert Watson to head the environmental section of the Office of Science and Technology Policy, and Watson reiterated that theme. The presence of an environmentalist vice president did not signal safety for EOS, particularly as Gore was directly involved in the international negotiations over greenhouse gas emissions reductions that descended from the 1992 Framework Convention on Climate Change. Instead, EOS remained under pressure from the administration until the Republican Party took over the House of Representatives in the 1994 election, when its association with the activist vice president made it a target of political retaliation.

Kennel set in motion an initiative to redefine EOS shortly after arriving in Washington that resulted in the reshaping plan. Its core objective was to separate the definition of the system from its hardware configuration and instead define it by the science it would do. Kennel did not believe that it was possible to fly three identical sets of instruments, as had been EOS's original design. Components went out of production, and no two instruments ever built were truly identical.

Instead, what made the data from instruments comparable was the establishment of a calibration record, preferably by flying an old instrument and its replacement side by side to intercalibrate, although there were other possible approaches. This meant that new technology could be infused between instrument generations. It also meant that hardware no longer needed to be the defining factor in the system.

Within a week after arriving in Washington, Kennel was approached by Michael Luther, who was in charge of mission development for the Mission to Planet Earth, and asked whether he would support a proposal for a line of small satellites to complement the larger observatory missions. He agreed, and this became the Earth System Science Pathfinder (ESSP) series of missions. These missions were to be chosen by the competitive announcement of opportunity process. This permitted nearly anyone to propose new missions, with mission prioritization and selection done by a panel of NASA and non-NASA researchers. Kennel hoped that the smaller, less expensive ESSP missions would allow him to maintain, or perhaps expand, the science content of EOS while adapting to the imposed budget cuts. Since one major goal of the ESSP program was to speed up the process of getting new missions into space, the first announcement of opportunity was scheduled for early 1996.

Kennel also formed a science team to examine EOS again. The team was chaired by Michael King, the EOS project scientist at Goddard Space Flight Center and one of the MODIS instrument principal investigators. King had also been a prominent advocate of redefining EOS by its science, not its system architecture, and his team constructed a set of twenty-four measurements that were to represent EOS. These are given in Table 8.1. For the atmosphere, these were clouds and radiation, precipitation, chemistry, aerosols and volcanic effects, and atmospheric structure. Other measurements included surface temperatures, ocean circulation, ice and snow coverage, soil moisture, and solar irradiance.[44] The task of his science panel was to find a way to maintain this set of measurements past 2000, when EOS was likely to be held within an annual cost cap of $1 billion, and to suggest ways to migrate some of the measurements to the next-generation operational weather satellites.

The next-generation weather satellite program was called NPOESS: National Polar Orbiting Environmental Satellite System. In 1993, Vice President Gore had initiated a "national performance review" aimed at streamlining the federal government. One of the resulting recommendations was merging of the separate military and civilian weather satellite programs, and in May 1994 President Clinton had issued a Presidential Decision Directive ordering their convergence. The

TABLE 8.1.
EOS Measurements

Atmosphere	Cloud Properties
	Radiative Energy Fluxes
	Tropospheric Chemistry
	Stratospheric Chemistry
	Aerosol Properties
	Atmospheric Temperature
	Atmospheric Humidity
	Lighting
Solar Radiation	Total Solar Irradiance
	Solar Spectral Irradiance
Land	Land Cover / Land Use Change
	Vegetation Dynamics
	Surface Temperature
	Fire Occurrence
	Volcanic Effects
	Surface Wetness
Ocean	Surface Temperature
	Phytoplankton and Dissolved Organic Matter
	Surface Wind Fields
	Ocean Surface Typography
Cryosphere	Land Ice
	Sea Ice
	Snow Cover

Source: Adapted from *EOS Science Plan: The State of Science in the EOS Program,* National Aeronautics and Space Administration, January 1999, table 1.2, 10.

directive ordered NASA, NOAA, and the Defense Department to form an Integrated Project Office (IPO) to manage the program. NASA's primary responsibility within the IPO was the provision of new technology. Since several of the EOS instruments had been intended as testbeds for the next-generation weather satellites, the formation of the NPOESS project presented an opportunity to migrate these to an operational status. Moustafa Chahine's AIRS was one obvious candidate. Another was MODIS, which contained heritage channels from the Advanced Very High Resolution Radiometer (AVHRR) imagers on the current weather satellites but with better resolution and calibration capability. A third was the AMSU. These instruments were on the EOS-PM payload, and thus migrating these and some other relevant instruments to the post-2004 NPOESS constellation would eliminate the need for NASA to replace the EOS-PM mission, and therefore reduce its post-2004 financial needs.[45] NPOESS was not a complete solution to the *climate* observing system problem, which required oceanic, cryospheric, and chemical measurements as well, but it was a start.

The Republican victory in the fall 1994 congressional elections gave Kennel still another challenge. The minority Republicans on the science committees became the majority members, and opponents of climate science took over the committee chairs in January 1995. The new Republican leadership immediately targeted the U.S. Global Climate Research Program, and NASA's Mission to Planet Earth, for termination. This anti-environmental revolt had been brewing since the early Reagan administration, when the so-called "wise use" movement began. This had sought the rollback of land-use restrictions in the American West, where most of the nation's public lands were located.[46] Later in the decade, and well into the early 1990s, conservative propagandists had attacked the anthropogenic ozone depletion hypothesis in the public arena, as discussed in chapter 6. In this highly charged context, the new chairman of the House Science committee, Robert S. Walker, relied on claims by members of the George C. Marshall Institute that solar irradiance changes would be responsible for whatever warming might happen in calling for a cut of $2.7 billion from the Mission to Planet Earth 1996–2000 budgets. He also scheduled hearings designed to showcase the arguments of fringe scientist-contrarians on both ozone depletion and climate change.[47]

But both Bob Harriss and Ed Frieman recall that Walker had other concerns beyond the nakedly partisan. The EOS concept had been based on the collection of long-term datasets whose analysis would provide EOS's scientific return. This meant spending billions of dollars for an unpredictable return in the distant future. EOS had not been structured to provide shorter-term results that would help to justify it. It also had not been primarily aimed at developing applications that might have economic benefit or aid in the provision of some public service. There were no shorter-term results to help convince congressional leaders that the public's money was being well-spent. Harriss had already set out to change this, carving out a small amount of money to establish an applications section within his science division and trying to recruit scientists interested in developing near-term results. Yet this was controversial within the EOS science community, which thought an applications emphasis diluted the Earth System Science concept, and in any case the strategy was just getting started.[48]

For several reasons, then, Congressman Walker wrote to the National Academy of Sciences president, Bruce Alberts, in early April and requested another review of EOS and of the larger U.S. Global Change Research Program by its Board of Sustainable Development. The chairman of the board was Ed Frieman, who had chaired the reengineering study of four years before. Frieman accepted the charter, and his review took place in July 1995.[49]

The EOS principal investigators met to prepare for the questioning they would face, in Santa Fe, New Mexico, at the end of June. Kennel explained the situation to the group, pointing out that the proposed $2.7 billion cut would allow completion of the first mid-sized satellite, AM 1, plus Landsat 7 and the Tropical Rainfall Measuring Mission, but would not permit completion of the remaining constellation. The basic concept of an Earth observation system would die from a cut this size. The only bright spot was that the Senate science committee was not inclined to go along with the House's cut, and there was therefore a good chance that it would not happen. But NASA and the EOS investigators needed to convince Frieman that EOS could not tolerate any further cutting, and that it was finally in a sustainable form aimed at appropriate scientific objectives.

Frieman's Board on Sustainable Development met to investigate EOS at the Scripps Institution of Oceanography on 19–28 July 1995. Kennel, Robert Harriss, Michael King, Joe McNeal, Ghassem Asrar, and Claire Parkinson, the EOS-PM project scientist, presented the results of the reshaping initiative to the board. The reshape plan kept the first three intermediate EOS missions, AM, PM, and Chem, as multi-instrument flights while using smaller satellites to fly the altimetry and aerosols missions. New technology developed under EOS would allow future flights of the primary EOS instruments to be lighter and thus less expensive, and so there would be no need to fly three identical sets of 1980s era instruments. Further, the ESSPs, they argued, would permit expanding the range of scientific investigations while also allowing new investigators to participate in the program. Finally, they argued that construction of the AM and PM satellites was far enough along that changing them would not result in any savings. Instead, future savings would have to come from technology infusion and migration of some measurements to NPOESS.

The NASA group also argued that EOS was well-focused on a thoroughly reviewed body of scientific questions. Harriss reminded the board members that EOS and the U.S. Global Change Research Program had been formulated together and were strongly linked. NASA scientists and EOS investigators had been involved in the National Research Council's planning for the U.S. Global Change Research Program, and EOS was the primary observational tool for the program. In their presentation, Harriss and Asrar gave thirteen areas in which EOS measurements contributed to important climate change questions, beginning with the cloud feedback problem. This they described as the largest source of uncertainty in climate model predictions, echoing the IPCC.[50]

After the NASA presentations, the board members engaged in a somewhat acrimonious debate over their evaluation that continued through the report-writing and reviewing phase. Two principal points of conflict were over a perception some members of the board had that the Mission to Planet Earth was still too focused on remote sensing to the detriment of in situ measurements, and that the reformulated EOSDIS did not go far enough in decentralizing data processing and analysis. The first complaint descended from the second-class status sub-orbital science often seemed to have at NASA, a problem reinforced by Administrator Goldin's announced intent to make NASA an orbit-only agency by 2000. This goal obviously jeopardized the in-atmosphere research programs that had allowed the agency to respond quickly to the Antarctic ozone hole and that was also the source of many of its new instrument concepts for future space-borne use. EOSDIS, finally, came under criticism for not placing enough control over data products into the hands of the investigators; it still appeared to be tied too much to centralized Big Engineering concepts.

After the panel settled its differences, it produced a report that called for implementing the AM, PM, Landsat, and Tropical Rainfall Measuring missions as planned and modifying the chemistry mission's tropospheric instruments to focus on ozone and its precursors. It also called for expansion of in situ, process, and modeling studies in Mission to Planet Earth, and advocated re-reconfiguring EOSDIS to transfer responsibility for data product generation to a "federation of partners," which could be composed of universities, research corporations, or other entities.[51] Finally, they explicitly argued that the 1995 reshape exercise Kennel had initiated had achieved all the savings possible at that point in the satellite infrastructure. Further cuts would merely produce delays, result in loss of data continuity, or eliminate the technology development that offered the primary potential for future cost reductions.[52]

Frieman's group also sought to deflect the political critics of climate research by mounting a strong defense of the science. Their first bullet point in the executive summary stated "*science* is the fundamental basis for the USGCRP [U.S. Global Change Research Program] and its component projects, and that fundamental basis is scientifically sound" (emphasis in original). To reinforce their point, they included a series of science working group reports on atmospheric chemistry, ecosystems, decadal to centennial scale climate processes, and seasonal to interannual climate processes. These provided details on the scientific needs in each area. The decadal to centennial scale climate report, for example, written by Eric J. Barron, an oceanographer at Pennsylvania State University, argued that the past decade of research had shown that the Earth's climate was variable on short as well as long timescales, and that human activities had the potential to cause further changes. He listed a series of specific accomplishments, including ERBE contributions to improving climate models and improved ability to assess the impact of volcanic eruptions. The same research had demonstrated that the Earth's climate processes were highly complex, and the U.S. Global Change Research Program needed to establish solid understandings of both natural climate variability and anthropogenic forcings.[53]

Ed Frieman then took the draft report to Washington and briefed it to the congressional committees. It was favorably received in the Senate, where Harriss had been actively working with western state senators interested in possible land-use and wildfire detection capabilities offered by EOS, but Walker, while reportedly impressed by the report, still attempted to impose the first installment of his desired cut, $323.9 million.[54] Walker and some of his political allies then staged what journalist Ross Gelbspan has called a "book burning" in late 1995, producing a forum in which the political critics of ozone and climate science were allowed to proclaim against Robert Watson, Daniel Albritton, and climate researcher Jerry Mahlman, among others, in advance of the release of the IPCC's 1995 assessment.[55]

The IPCC's report was thus immediately controversial when it was released. It contained the first consensus statement by a major scientific group that human influence on climate had been detected. Their phrasing in the summary for policymakers, that "the balance of evidence suggests a discernible human influence on global climate," was weak, but had been carefully chosen to reflect the limited current knowledge and of ongoing controversies over, for example, why tropospheric warming was clear in the radiosonde record but not in the data provided by David Staelin's Microwave Sounding Unit aboard the NOAA series of polar orbiters.[56] Too, the summary for policymakers was subjected to a line-by-line vetting and approval process at a plenary session of the scientific working group, which included members from nations whose major export product was oil, and thus had to be as uncontroversial as possible while still getting the point across.

Nonetheless, the IPCC report was attacked shortly after its publication in 1996, this time by the fossil fuel industry–funded group Global Climate Coalition and by the Marshall Institute, whose scientists launched attacks in the *Wall Street Journal* on Benjamin Santer, the lead author of the assessment's chapter on "Detection of Climate Change and Attribution of Causes."[57] These groups accused Santer of having altered the text to deemphasize uncertainties on the basis of rumors that they never sourced, at least in public; this caused a furious exchange of electronic mail messages between William Nierenberg, one of the authors of the Marshall Institute letter, and Thomas Wigley of the University of East Anglia in Britain, co-author of the chapter, and provoked the Executive Council of the American Meteorological Society to publish an open letter in support of Santer in the *Bulletin of the American Meteorological Society*. They also published the complete text of the letters sent to the *Wall Street Journal*, and indicated how the *Journal* had edited them.[58]

The purpose of these public attacks on the scientists involved in the IPCC process was to generate public doubt over climate science, to stifle a growing international drive to devise a new treaty that would mandate carbon dioxide emission reductions and perhaps limitations on other greenhouse gases as well.[59] They were not directly aimed at EOS. However, because EOS was part of the national climate science program it was affected by these political controversies as well. In this context of overt, extremely partisan attacks on climate science, the U.S. Senate largely blocked the House drive to eliminate EOS, agreeing to a cut of only $91 million.

For the next several years, the system architecture for EOS stabilized. A $150 million cut in the fiscal year 1997 budget affected the launch dates for the PM and chemistry missions, but did not affect the core measurements beyond the resulting delay. Kennel then found money to implement the ESSP missions in fiscal year 1996 despite the cuts by assuming that the larger EOS satellites would not need to be replaced at the end of their official five-year life spans. He explained later that by the mid-1990s most spacecraft far outlived their expected terms. The UARS was already a year past its designed life, for example, with all but its cryogen-limited instruments still operating, while the ERBE was five years past its intended life span in 1995, with the SAGE II instrument still fully functional and the ERBE instrument partly functional. Hence he had some confidence that the satellites would outlive their design lives quite significantly.[60] The advantage of assuming longer-than-expected life spans was that it meant the replacement satellites did not have to be built as quickly, freeing resources for other uses.

From the first ESSP announcement of opportunity, NASA selected two missions for development, a Vegetation Canopy Lidar proposal to measure forest canopy density, and the Gravity Recovery and Climate Experiment (GRACE) to make improved measurements of the Earth's gravity field. These were cost-capped at $60 million and $86 million, respectively, and were scheduled for launch in 2000 and 2001. Each mission represented a science area that had been left out of the post-1991 EOS design. GRACE, for example, was chosen because its measurements would be beneficial to the investigators on the EOS altimetry missions, improving the quality of their ocean circulation measurements. It could also play a role in ice sheet studies. A satellite pair, finally, it would demonstrate formation-flying, with high-precision spacing between the two spacecraft provided by laser ranging.[61]

Kennel returned to UCLA as the university executive vice chancellor at the end of 1996, and his deputy for engineering, William F. Townsend, succeeded him as acting associate administrator for Mission to Planet Earth. Goldin tried for about a year to find a university scientist to replace Kennel; discovering that no one wanted to move to Washington to take the highly controversial job, he appointed Ghassem Asrar associate administrator, and Townsend became deputy director of Goddard Space Flight Center. Harriss left soon after, moving to Texas A&M for a few years before finally settling at the National Center for Atmospheric Research (NCAR) in Colorado.

As it stood when Kennel left, EOS consisted of a set of twenty-four measurements, which were packaged into the (unchanged) AM payload, a reduced six-instrument PM payload, a still smaller four-instrument Chem payload, Landsat 7, and a set of smaller satellites: IceSat, a laser altimetry satellite for ice sheet measurement; Jason 1, a replacement for the highly successful TOPEX-POSEIDON ocean altimetry satellite; and SAGE III, which was to go into space on a Russian weather satellite. Ocean surface wind measurements were to be made by SeaWinds, a scatterometer derived from SeaSat A, on the Japanese satellite ADEOS 2, and QuikScat, a rapid replacement for the failed ADEOS 1. Finally, negotiations for an integrated, global climate observation system patterned after the international weather satellite system were just beginning. These would take more than a decade to bear fruit.

TRIANA AND THE POLITICS OF CLIMATE SCIENCE

One morning in March 1998, Vice President Gore woke up with a vision of a satellite in the L-1 position between Earth and Sun. At L-1, the gravity of the Earth and Sun exactly balance, permitting a satellite to remain there indefinitely.[62] His satellite would continuously stream an image of the sunlight Earth onto the Internet, and Gore had suggested the idea to NASA administrator Goldin. Gore thought such a satellite would have both scientific and "mystical" value. Considering the proposal a challenge to demonstrate the agency's newfound flexibility, Goldin had then had the newly renamed Earth Science Enterprise issue an announcement of opportunity for a science mission to use L-1 for Earth observation. This mission became "Triana" to its supporters, "GoreSat" to its detractors, and a metallic symbol of the intense politicization of Earth science at the turn of the century.[63]

NASA received nine proposals for the mission in response to the announcement, and chose one from the Scripps Institution of Oceanography. Francisco Valero had proposed a satellite with three instruments: a narrowband spectrometer with channels chosen to replicate MODIS and Total Ozone Mapping Spectrometer (TOMS) channels for aerosol, cloud, and ozone study; a set of broadband radiometers for albedo and radiation budget studies; and a solar plasma instrument. The advantage of these choices was that they would provide the same data as MODIS, TOMS, and CERES, but for the entire sunlight side of the Earth simultaneously. From their positions in low Earth orbit, these instruments saw swaths of the Earth, which then had to be stitched together mathematically to arrive at a full "Earth" of data. This introduced errors that data from simultaneous viewing would not have, and thus the L-1 satellite's data could be used to check and correct the data produced by the other satellites. Further, the Moon would occasionally occlude the satellite's view of Earth, and since the satellite would only see the same part of the unchanging lunar surface, it made an excellent calibration target—several, but not all, EOS instruments already used it for that purpose.[64] On the strength of its benefits to the other EOS satellites, the Scripps proposal had been accepted.

"GoreSat" was immediately challenged by congressional Republicans led by Dave Weldon of Florida, who inserted an amendment to the fiscal year 2000 budget canceling it. This passed the House but not the Senate; the resulting conference committee report barred NASA from spending money on Triana until the National Academy of Sciences passed judgment on its scientific merits. It also barred NASA from launching the satellite prior to 1 January 2001, after the 2000 presidential election, exposing the political relevance of the mission: Gore was the expected Democratic contender in the race to succeed President Clinton. On 14 October 1999 Ghassem Asrar wrote to Bruce Alberts, president of the National Research Council, requesting the study; on 3 March 2000, the review panel, chaired by James J. Duderstadt of the University of Michigan, responded with a letter report. This was positive, although not without significant technical caveats, and the satellite development restarted.[65]

This was not the last criticism of Triana, however. In a letter to *Science*, science policy scholar Roger Pielke, Jr. and former science director for Mission to Planet Earth Robert C. Harriss criticized NRC for failing to adequately carry out its tasking. The review panel had specifically refused to examine Triana's probability

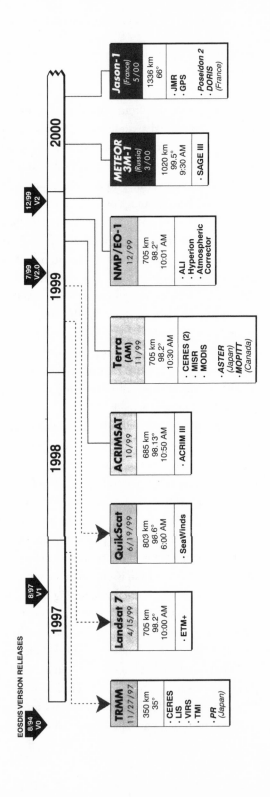

EOSDIS VERSION RELEASES

8/94 V0

8/97 V1

7/99 V2.0

12/99 V2

1997   1998   1999   2000

TRMM 11/27/97
350 km
35°
· CERES
· LIS
· VIRS
· TMI
· *PR* *(Japan)*

Landsat 7 4/15/99
705 km
98.2°
10:00 AM
· ETM+

QuikScat 6/19/99
803 km
98.6°
6:00 AM
· SeaWinds

ACRIMSAT 10/99
685 km
98.13°
10:50 AM
· ACRIM III

Terra (AM) 11/99
705 km
98.2°
10:30 AM
· CERES (2)
· MISR
· MODIS
· *ASTER* *(Japan)*
· *MOPITT* *(Canada)*

NMP/EO-1 12/99
705 km
98.2°
10:01 AM
· ALI
· Hyperion
· Atmospheric Corrector

METEOR 3M-1 *(Russia)* 3/00
1020 km
99.5°
9:30 AM
· SAGE III

Jason-1 *(France)* 5/00
1336 km
66°
· JMR
· GPS
· *Poseidon 2*
· *DORIS* *(France)*

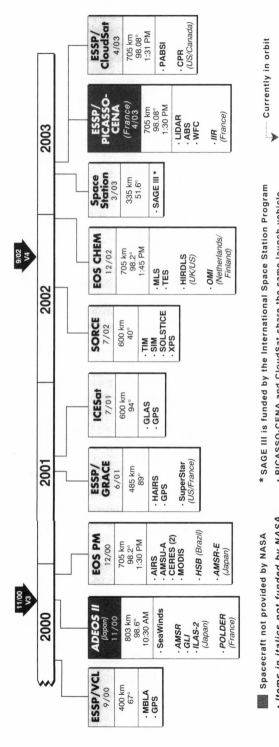

Instruments, platforms, and expected launch dates for EOS as of 1999. From Michael D. King and Reynold Greenstone, eds., *1999 EOS Reference Handbook*, p. 20. Courtesy NASA.

of success, or, more important, its priority relative to other possible Earth science missions. NASA typically based the research topics proposed in an announcement of opportunity on the relevant National Academy of Science decadal survey to ensure a given mission had the support of the scientific community. In Triana's case, it had not done so. The Earth science survey of 1988 had contained no suggestion that L-1 might be good for Earth viewing (although L-1 was already in use by several Sun-viewing missions). While this was possibly an oversight, or lack of imagination on the Earth science community's part, the fact that Triana had skipped the usual scientific vetting and prioritization without drawing condemnation from the National Academies was very troubling to Pielke and Harriss. It was not at all clear to these two that Triana would have been rated by the community above a synthetic aperture radar mission, for example, which evidence from the European and Canadian Synthetic Aperture Radar (SAR) satellites, as well as JPL's DC-8 based SAR, suggested could provide earthquake warnings and measure the flow velocities of large glaciers. Hence, they criticized NRC for not carrying out what they perceived as its responsibilities to provide adequate advice to Congress.[66]

NASA's bypassing of the policy process in Triana's case left it open to attack by opponents of climate research; similarly, it left NRC and its parent organization, the National Academy of Science, open to attack. Further, their refusal to criticize Triana, EOS, or the larger agenda of climate research was taken as evidence of partisan intent by members of the American political right. By this time, bizarre theories had circulated in right-wing circles that the National Academies and the editors of the major science journals, including the British journal *Nature*, were members of a pro-Gore conspiracy.[67] Hence despite Kennel's and Harriss's efforts in the mid-1990s to make sure *science* drove EOS, by 2000 it was clear to right-wing political activists and the politicians they communicated with that precisely the opposite was the case. They believed politics drove EOS and climate science.

TRANSITIONS

During the Triana conflict of 1999, the newly renamed Earth Science Enterprise's leadership also formalized the first part of its transition strategy to move some of the EOS measurements to the NPOESS. The NPOESS project office had selected a somewhat different set of sensors for its weather satellites than had been chosen for EOS-PM, the satellite containing the primary EOS weather measurements, and hence the EOS-PM instruments would not serve as the prototypes for

NPOESS. Further, the differences meant that the NPOESS instruments would not have had their scientific qualities and calibration capabilities demonstrated by EOS, or their algorithms validated. In a sense, they represented starting over, redoing for new instruments what was already being done for EOS. However, the NPOESS project office was not a scientific research organization, and it was also not required to produce a climate-quality, as opposed to a weather-quality, dataset. To fulfill the demands of climate science, Ghassem Asrar and Bill Townsend formulated one more new, medium-sized EOS mission, inelegantly named the NPOESS Preparatory Project (NPP).[68]

The year before, the National Academy of Science's Committee on Global Change Research had acknowledged the fiscal improbability of maintaining the EOS constellation indefinitely and had called for a single additional medium mission to serve as a bridge between the EOS-PM mission and the first NPOESS launch in 2009 to ensure data continuity.[69] The NPP satellite would carry copies of the instruments chosen for NPOESS, and assuming everything went according to schedule, it would overlap the life spans of the EOS-PM satellite and the first NPOESS satellite. This overlap would permit cross-calibration between the EOS-PM and NPP satellite, then the NPP and NPOESS satellite, effectively moving the calibration forward in time. This was considered essential by the climate science community to ensure comparability of measurements made by the three satellites. Further, the NPP's science teams would generate the data reduction algorithms for the NPOESS sensors and they would also define and produce the climate-quality data products for the new sensors. Indeed, this was one of the central weaknesses of the NPOESS transition, as far as the climate science community was concerned. It was not within NPOESS's managerial responsibility to support climate-quality data production and archiving, let alone research using the data. NPP, of course, was a project limited in time span, not a permanent research organization designed to sustain climate research for the foreseeable future. NPP and NPOESS were therefore at best partial solutions to the problem of how to maintain climate-quality data production in a post-EOS era.[70]

The National Academy of Science's 1998 study had also recommended that EOS be restructured into smaller, more focused missions along the Earth Probe and ESSP lines.[71] A series of planning workshops took place during 1998 and 1999 to map out a new direction for NASA's Earth science research. The strategy that emerged was to use NPOESS to maintain key long-term measurements and use smaller Earth Probe and ESSPs to develop new measurement capabilities that might be migrated to NPOESS, or maybe NOAA, or perhaps some as-yet-undefined climate agency later. These might include carbon cycle measurements or soil moisture measurements; the new strategy also called for new technology investment to reduce the cost of maintaining the observation system in the future. It left unclear, however, how the new measurements would be transitioned to operational use or to what agency they would be sent. By 2000, therefore, NASA no longer planned to replace the four EOS mid-sized satellites when they reached the end of their lives. The budget cuts, congressional hostility to climate science, and scientific hostility to the EOS approach left the agency without the political support necessary to sustain the program through its intended fifteen-year span.

EOS-AM was finally launched in 1999, two decades after planning for a climate observing system had started. It was renamed Terra. This very long lead time for space hardware limited the agency's ability to respond to important policy questions. As the global warming controversy heated up during the 1990s, NASA was not in a position to respond with new global observations; it could not even accelerate measurements that it had in progress. To the larger scientific community, it had been clear that the PM mission had contained the highest-priority instruments for detection of climate change, but having started down a different path, the agency was effectively trapped by its prior history of decisions. This fact had left EOS vulnerable; as Representative George Brown had pointed out in his letter, the clear lack of consensus, either technological or scientific, of how EOS should be structured made it difficult to defend even when the basic concept of a comprehensive observing system was politically favored. EOS thus shrank, first to a set of four medium satellites plus a flurry of small satellites, then to one that was essentially Vern Suomi's concept of 1980—a climate observing system based upon the weather satellites plus research satellites.

When Shelby Tilford had launched the EOS effort in the early 1980s, the Reagan administration had reverted to the grandiose space engineering dreams of the early space race, encouraged by unsubstantiated, and unrealized, claims of inexpensive, reliable, and routine space access via the Space Shuttle. This had bound the project to the larger Space Station Freedom initiative, which itself did not survive without severe descoping. The EOS architecture had originally been a means of promoting the use of humans in space. It had not been chosen as the most efficient means to accomplish a given set of scientific goals. Nor had it been constrained within a specified budget. Instead, "no arbitrary constraints" had been applied to the project during its earliest phases, leading to an architecture that became politically unsustainable as soon as the Cold War demand for space spectacles vanished. Big Engineering was EOS's Achilles heel.

Further, because the initial EOS conception was so expensive and long-term, it was widely seen in the scientific community as locking in funding to a select group of researchers and locking out everyone else. To many of the scientists who were not selected at the 1988 announcement of opportunity, there would be no more opportunities within their working lives. This violated scientific norms of fairness and competition and left no means of introducing graduate students and new researchers to space science, creating resentment that further undermined EOS. There were also sound technical reasons for having alternatives to the giant platforms. The polar platforms' Sun-synchronous orbit was not useful for certain science missions, such as ocean topography, and the lack of alternatives within the original architecture meant no means of carrying out these other missions. NRC's expressed demand for smaller, competed missions—eventually the ESSPs—reflected these frustrations.

Finally, the fact that environmentally relevant science had become an issue associated with only one political party during the 1980s began to harm NASA's atmospheric science programs. The brief 1970s, when both parties had supported environmental improvement while disagreeing over regulatory methodology, was long over. Instead, environmentalism, and any branch of science that touched on the natural or human environment, became merely another partisan issue. NASA's decision to make itself the lead agency for atmospheric and climate sciences in the late 1970s had brought the wrath of the new majority party down on it in the 1990s. This would only get worse as EOS began to fly.

CHAPTER NINE

Atmospheric Science
in the Mission to Planet Earth

Humans have enjoyed the fruits of the industrial revolution and avoided a large cost in climate change, as aerosol cooling has mitigated greenhouse warming. Payment
comes due when humanity realizes that it cannot toler-
ate the further exponential growth of air pollution that
would be needed for continued mitigation of global
warming.

—*James E. Hansen, 2004*

The seemingly endless arguments over the appropriate architecture for the Earth
Observing System (EOS), and the lack of new space hardware between the Upper
Atmosphere Research Satellite's (UARS) launch in 1991 and the Tropical Rainfall
Measuring Mission (TRMM) in 1997, did not impair the activities of NASA's
atmospheric science programs. The Global Tropospheric Experiment (GTE) carried out a set of Pacific Exploratory Missions (PEMs) that investigated chemical
outflow from Asia to determine relative contributions from biomass burning and
industrial emissions. The Upper Atmosphere Research Office carried out a series
of expeditions to further examine stratospheric ozone production, transport, and
loss processes, expanding their efforts into the tropics and mid-latitudes. Under a
new program jointly funded with NASA's Office of Aeronautics, they revisited the
subject of aircraft impact on stratospheric ozone in a program named Atmo-

spheric Effects of Stratospheric Aircraft.[1] But the agency's field scientists also began to examine the radiative effects of aerosols and clouds in detail. Necessary to resolve critical scientific questions involving human impacts on climate as well as for the more prosaic need to validate satellite sensors, these expeditions marked another new direction for NASA's atmospheric research.

Between 1997 and 2004, NASA launched the four surviving observatory-class missions of its EOS and the first few Earth System Science Pathfinders (ESSPs). As it had for the UARS, it sponsored field experiments for calibration and validation purposes, while also supporting expeditions for tropospheric chemistry and atmospheric radiation studies. The catastrophic eruption of Mount Pinatubo provided a real-world geophysical experiment against which to test climate model treatment of stratospheric aerosols, while the climate impact of tropospheric aerosols remained a contentious issue.

With the launch of Landsat 7, EOS-AM, and EOS-PM, NASA's atmospheric scientists were able to begin the long process of validating the space-borne instruments and their data products. This was done using both surface observations and airborne experiments, in much the same way the GTE program had conducted its field experiments in the 1980s. Surface data, which could be collected continuously, could be linked to airborne measurements, which were larger in scale but severely limited in time, through to the global satellite data. The advantage of this methodology for satellite operations was that continuously operated surface stations would provide twice-daily checks on satellite instrument performance during overpasses while also gaining the full range of diurnal effects that the Sun-synchronous polar orbiters would not. The relatively inexpensive (at least compared to satellite costs) surface stations therefore expanded the range of science that could be done while providing independent data for verification.

The expansion of NASA's efforts occurred against a backdrop of increasing scientific concern about human-induced climate change. During the late 1980s, the National Science Foundation's Office of Polar Programs and a separate European consortium had engaged in major drilling projects on the Greenland ice sheet, seeking additional evidence of past climate shifts. Russian scientists also completed a major drilling project in Antarctica. All these cores showed unmistakably that Earth's climate tended to shift rapidly and nonlinearly in response to forcings, not slowly and gradually as most scientists had expected. Severe climatological consequences of human emissions would appear in a few decades, not centuries. This and other evidence led the National Academy of Sciences to publish a study of "abrupt climate change," as this possibility came to be called, in 2002.[2] Further, the threshold at which change would begin to become irreversible seemed to be low. By the time the National Academy of Sciences published its study, some scientists thought humans had already passed it. By 2004, NASA's Hansen thought significant policy action had to take place by the end of the first decade of the twenty-first century. Delay of another decade, he said (wearing his guise of private citizen from Kintnersville, Pennsylvania) in a speech at the University of Iowa, "was a colossal risk."[3]

Yet while the scientific community saw global warming as a certainty, and its consequences in much starker terms than it had at the beginning of the 1990s, policy action became politically impossible. Former oil industry executive and Texas governor George W. Bush ran his 2000 election campaign against Vice President Al Gore with a promise to take action on global warming; he then followed in his father's footsteps and reneged. Instead, he questioned the legitimacy of the Intergovernmental Panel on Climate Change's (IPCC) Third Assessment Report and asked the National Academy of Science to review it. The Academy, with a few quibbles, affirmed it. He took the unusual step of unsigning the Kyoto Protocol, and administration diplomats actively blocked further international discussion of climate mitigation actions until late in 2007.[4] Widespread suppression of government climate scientists by political appointees took place across all science agencies, including NASA.[5]

The political climate in Washington became extremely hostile to climate science.

ON VOLCANOES, AEROSOLS, AND CLIMATE MODELS

After his 1988 study of greenhouse gas–induced warming, NASA's Hansen had turned to using his model to explore the role of solar and aerosol forcings in climate dynamics. While it had been clear to most climate scientists by the late 1970s that long-lived greenhouse gases building up in the atmosphere would eventually overwhelm the cooling effects of much shorter-lived aerosols, the relative contributions were not well-known. Tropospheric aerosols were a particularly difficult challenge, as their global distribution was not known at all. The scientific community knew that they varied in time and space, but had never quantified them. A number of the EOS instruments were aimed at studying aerosol distribution for this reason.

For a different reason, Hansen was also interested in reinvestigating solar contributions to climate. The Marshall Institute's claims that the twentieth-century warming was solar in origin needed to be put to the test. Solar irradiance data existed from satellites, astronomical observatories, and proxy measurements (mostly radioisotopic in nature), and these could be used to show what Earth's climate would have been given only solar irradiance changes. One could simply keep the greenhouse gas levels in the climate model at the pre-industrial level and vary the solar irradiance in accordance with the observational data to find this out.

The results of his first efforts to quantify these effects were published in a 1990 review article, "Sun and Dust in the Greenhouse." The well-calibrated solar irradiance measurements made by the Solar Max and Nimbus 6 and 7 satellites had shown a variation of 0.1 percent during the solar cycle, while anthropogenic greenhouse gas forcing was already the equivalent of 10 times that variation. So Hansen accepted that solar variation of a few tenths of a degree probably had happened in the past, but he rejected the Marshall Institute's 1989 claim that a cooling sun would soon reverse the warming trend Hansen had identified in 1988. There was no extant evidence for decadal-scale solar variations of more than 1 percent.[6]

The aerosol issues were not nearly as clear. While the previous couple of decades of research on stratospheric aerosols left some confidence their impacts on climate were fairly well understood, the same was not true of tropospheric aerosols. They had been the basis of the conflict between Will Kellogg's climate warmers and climate coolers in the early 1970s, and their true impact had still not been quantified. Hansen's group at the Goddard Institute of Space Science (GISS) thought that tropospheric aerosols imposed a cooling equivalent to about a quarter of the anthropogenic greenhouse gas forcing. But Hansen also pointed out that Robert J. Charlson, an aerosol specialist at the University of Washington, thought that the effect could be half to three-quarters of the anthropogenic greenhouse effect.[7] So Hansen concluded that the net impact of tropospheric aerosols could not yet be determined, and represented the greatest source of uncertainty about climate forcings.

Aerosols also impacted the radiative properties of clouds. Hansen's colleagues referred to this as the "secondary aerosol effect," to separate it from the aerosols' direct impact on radiative transfer. Aerosols changed the size of cloud particles, which in turn impacted their reflectivity. Michael King, the EOS project scientist at the Goddard Space Flight Center, had spent his early career studying the impact of shipboard diesel engine emissions on clouds. These were clearly visible in weather satellite imagery as bright tracks in the clouds above a ship, shown in Plate 6.[8] Sulfate aerosols, at least at the sizes emitted by industrial power sources, had the effect of raising cloud shortwave reflectivity. In other words, they produced a negative climate forcing (a cooling), but one that had not been quantified either regionally or globally.

Hansen and his team of modelers at GISS set out to improve their climate model's treatment of various forcings and feedback effects in this period, hoping to develop a better understanding of the relative importance of various processes. They needed to be able to make many simulations, permitting them to change a wide range of aerosol-related factors one by one. This would help determine which factors had the largest effects on climate, and were thus the most important ones to try to measure in the atmosphere. GISS did not have ready access to the latest supercomputers, however. So they first modified the Model II into what they called the "Wonderland model," which covered only a 120 degree range of longitude and deliberately did not have accurate topography. This saved enough computational cycles to allow dozens of simulations, but still allowed the radiative effects of aerosols to be studied.

Hansen first used the Wonderland model to explore the potential mechanisms behind an apparent reduction in the diurnal cycle of surface air temperature. Average nighttime temperatures over land, particularly in the Northern Hemisphere, had increased during the preceding decades significantly more than average daily temperatures had. This had occurred in early studies carried out with the Wonderland model, too, and Hansen had wanted to study the phenomenon systematically. The great advantage of climate models, of course, was that one could manipulate the world one forcing (or feedback) at a time, narrowing the range of possible mechanisms. Hansen estimated the individual forcings acting on climate between 1850 and 1998. In this case, what Hansen found as he worked through the possible combinations was that the spatial pattern of the diurnal suppression could only be reproduced by a combination of the global effects of greenhouse gas emissions, slightly higher cloudiness overland, and changes to tropospheric aerosols. The aerosol changes, he thought, were probably due to increasing amounts of tropospheric sulfate aerosols, produced by both industrial emissions and biomass burning; these would have the same cooling effect in the troposphere as in the stratosphere.[9]

His findings also had implications for the trend of future warming. In the journal *Atmospheric Research*, Hansen pointed out these provided "quantitative confirmation of the widely held suspicion that anthropogenic greenhouse gas warming has been substantially counterbalanced by a forced cooling."[10] They also meant that the rate of greenhouse warming was likely to accelerate in the near future. Sulfate emissions regulations enacted during the 1980s in many

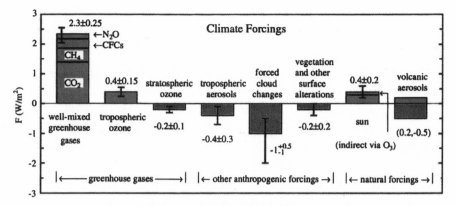

FIG. 2.   Estimated radiative forcings between 1850 and the present.

Climate forcings, as estimated in 1998. Forced cloud changes are those induced by aerosols (also called the secondary aerosol effect). Land cover changes affect the Earth's albedo, or reflectivity, by changing the color of the surface. From Hansen et al., "Climate Forcings in the Industrial Era," *Proceedings of the National Academy of Sciences* 95 (1998), fig 2.

nations would cause tropospheric aerosols to decline, while the long-lived greenhouse gases would continue to build up.

The radiative changes to clouds that Hansen's model studies suggested were necessary to account for the diurnal suppression, finally, called for observational study to confirm. He advocated flight of an instrument like Rudy Hanel's old Infrared Interferometer Spectrometer (IRIS) from the Nimbus and planetary programs, which could provide some of the desired cloud property measurements even in multilayered cloud environments, and Larry Travis's aerosol polarimeter, initially selected for EOS but later dropped in the early 1990s downsizing. The polarimeter would provide aerosol and cloud microphysical measurements necessary to evaluate cloud changes. This instrument, too, was a derivative of a planetary instrument, in this case one sent to Venus.

The eruption of Mount Pinatubo in 1991 also gave Hansen the opportunity to test the Model II against a real-world experiment. Pollack's group at Ames Research Center had studied volcanic effects on climate in the 1970s and early 1980s, of course, but Pinatubo was a much larger explosion. NASA and National Center for Atmospheric Research (NCAR) responded to Pinatubo's explosion by measuring the volcano's ejecta plume with in situ and remote sensing techniques developed over the preceding two decades. Both the still-functioning Total Ozone Mapping Spectrometer (TOMS) instrument on Nimbus 7 and the UARS provided information on the evolution of the plume over time, making Pinatubo the most thoroughly documented volcanic eruption to date (at least from an atmospheric science perspective).

Pinatubo's eruption caused the largest aerosol injection into the stratosphere of the twentieth century, and was estimated to be the third largest perturbation of the industrial era (behind Tambora in 1815 and Krakatau in 1883). Of the estimated 30 teragrams of mass shot into the stratosphere, the old TOMS instrument's data suggested about two-thirds was sulfur dioxide. This transformed into radiatively active sulfate as it aged. Stratospheric Aerosol and Gas Experiment (SAGE) II and TOMS showed the plume moving around the world in twenty-two days, spreading relatively quickly southward to 10S latitude, then more slowly dispersing to higher latitudes. UARS measured the rapid warming of the stratosphere that followed, and the heating effect also lofted the aerosols, moving them higher in the stratosphere. But despite the large increase in sulfate, there was no measurable increase in active chlorine immediately after the eruption. Instead, as the plume spread to the poles, the aerosols appeared to increase the surface area available for the wintertime production of chlorine dioxide (the inactive precursor species involved in ozone depletion), leading to significantly larger ozone loss than in prior years. The UARS's scientists were able to characterize the stratosphere's chemical response to the volcano and its recovery. Two and a half years later, the aerosol loading had diminished to about one-sixth of the original amount.[11]

Almost immediately after the eruption, NASA headquarters had called an interagency meeting in Washington to discuss preliminary information gathered about the eruption; after this, using data gleaned from some of his colleagues, Hansen used the Model II to make a forecast of the volcano's climate impact. It had predicted an immediate low-latitude cooling, becoming essentially global by mid-1992 and peaking at about minus 0.5 degree C (globally averaged, of course) late in 1992. In his 1993 paper, he argued the volcano had provided an "acid test for global climate models." The expected cooling was about 3 times the standard deviation of global mean temperature; this should, he thought, be measurable despite the apparent onset of an El Niño (which tends to warm the troposphere).[12]

The surviving Earth Radiation Budget Experiment (ERBE) showed that the eruption increased the Earth's albedo, as expected, by a significant amount. This in turn caused cooling of the troposphere, which despite the El Niño that year, resulted in the year being measurably cooler than the twenty-six-year mean. In fact, the weather satellite data showed almost exactly the amount of cooling predicted by the GISS Model II, and in a spatial and temporal pattern that was highly consistent with the model's, too. (Though Hansen was quick to point out in a 2006 interview that the forecast hadn't been perfect.) A 1999 reviewer called the consistency between the prediction and the independent analyses "highly significant and very striking"; they had led, he argued, to increased confidence in the models' representation of climate processes.[13]

ATMOSPHERIC CHEMISTRY

During the 1980s, the Tropospheric Chemistry Program's field studies had focused on the chemistry of the boundary layer, the lowest portion of the atmosphere. Their goal had been to develop knowledge about the exchange of gases between the biosphere and the atmosphere, as well as developing new techniques for making reliable measurements. After the Atlantic Boundary Layer Expedition (ABLE) 3 missions to the Arctic, however, the GTE team started to shift their efforts toward study of transport between the boundary layer and the upper troposphere as well as toward the investigation of long-range chemical transport. They also shifted most of their operations to the Pacific basin, where rapid economic development in Asia promised to provide a rapidly changing atmosphere to study. These changes in direction were made partly because GTE had accomplished its initial set of objectives, to examine biosphere/atmosphere fluxes, and the obvious next set of questions involved how surface emissions were lofted higher in the atmosphere to be transported great distances.

The 1990s GTE missions had two focuses: establishing the extent and impact of pollutant outflow from the Asian continent on the remote Pacific atmosphere's chemistry, and further characterizing the impact of biomass burning. Three sets of missions took place during the decade: PEM-West, which extended from Alaska down the East Asian coast and then eastward to Hawaii; PEM-Tropics, which were largely carried out in the Southern Hemisphere tropics; and finally TRACE-P, designed to examine the impact of biomass burning in Asia on the Pacific basin.

In 1991, the International Geosphere/Biosphere Program (IGBP) had launched a tropospheric chemistry subprogram known as the International Global Atmospheric Chemistry Program (IGAC). This established a set of regional experiments, and NASA's PEM-West missions were designed in collaboration with the larger East Asia/North Pacific Regional Study, known as APARE. As had been the case for ABLE-2 and ABLE-3, the PEM-West experiments took place in two field phases, intended to coincide with the dominant meteorological conditions. For East Asia, the periods of interest were characterized by minimum and maximum outflow from the continent, February–March and September–October, respectively. Due to the intervening TRACE-A experiment in 1992 and the Upper Atmosphere Research Program's (UARP) Arctic expedition in 1993, the PEM-West studies were carried out in the fall of 1991 and the spring of 1994.

The principal focus of PEM-West was on tropospheric ozone chemistry, and the NASA DC-8 was equipped with instrumentation very similar to its payload in ABLE-2 and ABLE-3. The mission scientists sought to improve understanding of the photochemistry of industrial and biomass burning emissions as well as their transport into the Pacific. The Pacific basin was the largest region of the Earth in which there were no significant direct human impacts, and this made it a valuable place in which to study atmospheric chemistry that was relatively unperturbed. Yet it was also obvious by 1991 that it was not pristine. Long-range transport from continental areas ensured that to some extent, pollutants must have had an impact. Determining what those impacts were, and generating a database against which future changes could be measured, were primary goals.[14]

The PEM-West expeditions were followed by the PEM-Tropics expeditions in 1996 and 1999. These added some new measurements, including instruments aimed at the elusive hydroxyl radical. The GTE deployed two aircraft for these missions: the DC-8, which remained focused on ozone photochemistry, and the NASA P-3B, which was equipped primarily for sulfur chemistry. The P-3B's outfitting reflected both the availability of new instruments as well as the increasing scientific importance of the sulfur cycle. In addition to the strong radiative impact of sulfate aerosols known from volcanic eruptions, sulfate aerosols were also suspected of being a major trigger for cloud formation. Cloud droplets required a surface to form around, and sulfate particles were common in the atmosphere. Sulfur aerosols had a number of sources beyond industrial emissions and biomass burning, and GTE's scientists were interested in characterizing the relative importance of two of these, dimethyl sulfide from the ocean surface and sulfur dioxide from volcanoes.

The PEM-Tropics expeditions were carried out during September 1996 and March 1999, with the aircraft operating out of Easter and Christmas Islands, Tahiti, Ecuador, and Christchurch, New Zealand. These flights turned out to be full of surprises for the science teams. During the 1996 expedition in the dry season, what they termed a "pollution river" streamed southwest from Asia across the Pacific, remaining confined within the marine boundary layer until it reached Fiji, where it was broken up by convection in the South Pacific Convergence Zone.[15] Ozone concentrations in the South Pacific during the wet season of 1999 were half those of the dry season in 1996, reflecting the profound impact of biomass burning on the Pacific troposphere. Ed Browell's aerosol lidar clearly demonstrated that layers of pollutants from burning emerged from Africa and South America and were able to maintain their structure across vast distances. Further, the P-3B's measurements suggested that while volcanoes were the dominant source of tropospheric sulfur dioxide in the Southern Hemisphere, industrial emissions were dominant in the Northern Hemisphere. And while dimethyl sulfide was a major source of sulfur within the marine boundary layer, it turned out not to be a major source of sulfur to the troposphere overall.

GTE's research of the preceding fifteen years had shown that tropospheric ozone concentrations were the result of a balance between photochemical production and loss within the troposphere. The view from 1978 had been that the troposphere received its ozone from the stratosphere, while the succession of missions to Brazil, to the Arctic, and to the Pacific clearly showed otherwise. Stratospheric inflow, while occasionally large on a local scale, was a minor impact on the troposphere considered globally. Because of this, and because human activities affected the concentrations of ozone source species in the troposphere, changing industrial and biomass burning emissions had the ability to change the atmosphere's ability to remove ozone and its source species on a global scale.[16]

In parallel with its tropospheric expeditions, NASA conducted a series of field experiments to further examine stratospheric ozone processes during the 1990s. These were funded jointly by the office of Mission to Planet Earth and the Office of Aeronautics, and were named STRAT, POLARIS, ASHOE/MAESA, and SOLVE.[17] The last of these expeditions, whose name was a "nested" acronym, the SAGE III Ozone Loss and Validation Experiment (with SAGE being, of course, Stratospheric Aerosols and Gas Experiment), was conducted in parallel with the Third European Stratospheric Experiment on Ozone (THESEO 2000). As implied by its name, SOLVE was to be in part a validation experiment for the first copy of Pat McCormick's SAGE III instrument. Three copies had been built, with the first to be launched in 1999 on a Russian Meteor satellite. Another copy was supposed to be attached to the International Space Station, to provide diurnal coverage similar to that provided by the ERBE satellite, but that never occurred. The Meteor was delayed until December 2001, however, but due to the investment made in organizing it, SOLVE was carried out on schedule. Table 9.1 lists the stratospheric chemistry expeditions of the 1990s and early 2000s.

SOLVE took place in three phases beginning in November 1999 and ending in March 2000, after the breakup of the Arctic polar vortex. The schedule was set by the science teams' desire to observe the complete lifecycle of the polar vortex and the evolution of its chemistry. As in the early 1990s, NASA provided the DC-8

TABLE 9.1.
Polar Stratospheric Airborne Ozone Research Expeditions

Antartic Airborne Ozone Expedition	August–October 1987
Airborne Arctic Stratospheric Expedition	January–February 1989
Airborne Arctic Stratospheric Expedition II	January–March 1992
Airborne Southern Hemisphere Ozone Experiment/ Measurement for Assessing the Effects of Stratospheric Aircraft (ASHOE/MAESA)	March–October 1993
Stratospheric Tracers of Atmospheric Transport	October 1995–October 1996
Photochemistry of Ozone Loss in the Arctic Region in Summer (POLARIS)	April–September 1997
SAGE III Ozone Loss and Validation Experiment (SOLVE)	September 1999–March 2000
SAGE III Ozone Loss and Validation Experiment II (SOLVE II)	January 2003–February 2003
Polar Aura Validation Experiment (Polar-AVE)	January 2005

and ER-2 for the mission, with a series of European aircraft participating in the parallel THESEO effort. Despite the non-arrival of SAGE III in orbit, a series of satellite sensors also contributed to the field experiment: McCormick's SAGE II, still operating on the extremely long-lived ERBS; HALOE, and MLS on the UARS; a recently launched TOMS instrument that replaced Nimbus 7; and two European instruments. The expedition also included a number of balloon pay-loads, including the now-venerable Jet Propulsion Lab (JPL) Mark IV interferom-eter and a new JPL instrument, the Submillimeterwave Limb Sounder, designed to demonstrate a new technique for measuring chlorine monoxide, ozone, and hydrogen chloride.

The expedition flew out of Kiruna, Sweden, housed in a new set of hangars built expressly for supporting Arctic airborne science. The winter of 1999–2000 proved to be unusually cold in the stratosphere, and widespread formation of Polar Stratospheric Clouds (PSCs) took place. These had been only intermit-tently sampled in previous field campaigns, and the scale of their appearance during this campaign gave the expedition the ability to substantially increase knowledge of their composition using data from the ER-2 and from a special bal-loon payload designed specifically to measure profiles from within the clouds.

About five hundred scientists from twenty nations were involved in various aspects of SOLVE/THESEO. David Fahey found large particles of condensed nitric acid in the polar vortex, providing new evidence regarding the process of denitrification. Removal of nitrogen species from the stratosphere was a key pro-cess in the ozone depletion reaction but was not well-understood; Fahey's parti-cles were large enough to settle out of the stratosphere under the force of gravity.

These were dry particles, containing very little water, which finally explained why extensive nitrate removal took place without extensive dehydration, a mystery left unanswered by all previous field expeditions. Very low levels of nitrates, combined with the extensive PSC presence, produced ozone losses of 60 percent at the ER-2's 20 kilometer cruise altitude by mid-February. An ozone hole of the magnitude found in Antarctica did not form, however, due to the Arctic's differing meteorology. Extensive downwelling of ozone from higher altitudes, driven by more energetic vertical circulation within the vortex, partially replaced the ozone being chemically destroyed.

There were other significant results from the expedition, some of which reflected improving model capabilities as well as the ability to take models, and modelers, into the field with the experimental teams. New meso-scale meteorological models were used to forecast the formation of mountain waves that influenced PSC formation as well as producing turbulence that was dangerous to the fragile ER-2. Modelers accompanying the expedition also used composite meteorological/chemical forecast models to aid in the flight planning for all aircraft to ensure they encountered interesting phenomena, and for balloon releases to ensure both that they would drift into areas of interest and that their payloads landed in accessible areas for recovery.

The expedition also expanded confidence in the ability of space-borne instruments to characterize the ozone destruction process. The satellite, aircraft, balloon, and surface instruments all generally agreed in their measurements of ozone loss, as did the primary chemical models. The agreement was not perfect, as every instrument and model had its own unique error structure, but in the words of the project leaders, the results provided "great confidence in our ability to quantify ozone loss inside the polar vortex."[18] With the non-launch of SAGE III and the aging status of the UARS Microwave Limb Sounder (MLS) instrument, which was no longer operating reliably, European instruments would be the only space-borne instruments capable of providing ozone profiles in the Arctic and Antarctic stratosphere until the launch of NASA's EOS-Chem satellite, and validation of their capabilities meant that there would not be a data gap if Joe Water's nine-year-old MLS finally failed.

CLOUDS, AEROSOLS, AND TROPOSPHERIC RADIATION

Chemistry, NASA's focus for the previous twenty years, began to be more closely linked to atmospheric radiation during the mid-1990s, as the PEM missions were proceeding. This reflected the changing priorities of NASA and of the larger scientific community as interest in ozone declined due to the Montreal Protocol and the resolution of the intense stratospheric ozone controversy and, perhaps more important, as the initial deployment of the EOS and its focus on climate processes neared. New aircraft-based instruments, often intended as testbeds for similar EOS instruments, became available in 1995, and these became the basis for field experiments aimed more directly at radiative transfer through the atmosphere.

One of the first of these field experiments was SCAR-B, the Smoke, Clouds, and Radiation-Brazil experiment. Yoram Kaufman, a specialist in aerosol remote sensing at Goddard Space Flight Center, had organized this experiment to begin validating the products expected from the center's Moderate Resolution Imaging Spectrometer (MODIS). Its testbed sensor, known as MAS (for MODIS Aircraft Simulator), was designed as a payload for the NASA ER-2. Kaufman had become interested in the radiative effects of aerosols from biomass burning from the work of Sean Twomey, who had first identified the indirect aerosol effect in 1974, and Robert Charlson. He had started his career at Goddard developing a method to extract aerosol optical depth from the Geosynchronous Operational Environmental Satellite (GOES) satellite imagery, which was made difficult by the uncalibrated nature of that instrument. One of MODIS's selling points had been its great attention to spectral calibration, overcoming that weakness. So Kaufman had proposed to develop a fire data product for MODIS. To prepare for the spaceborne instrument he and Mike King, the EOS project scientist, had convinced NASA headquarters to fund a set of three field experiments aimed at the effects of smoke on clouds and radiative transfer. Two were carried out in the United States in 1993 and 1994 in preparation of the larger South American expedition; in 1995, they went to Brazil during the fire season. Like the GTE expeditions to Brazil, it was organized with the help of the Instituto de Pesquisas Espaciais.[19]

One reason for the choice of Brazil as the location for the field experiment was that Brazil had installed a set of ground-based aerosol monitoring stations. Originally developed in France, these instruments were part of a new global aerosol monitoring network called AERONET (Aerosol Network). The network was based on a federation concept; a project office managed by Brent Holben at Goddard Space Flight Center coordinated the network and maintained the sensor calibration program, but the sites overseas were owned and operated by the international partners. The AERONET instrument itself was a solar-powered, fully automated Sun-tracking radiometer designed at the University of Lille that transmitted its data back to Goddard via the geosynchronous weather satellites. Processed at Goddard, the data then went into a public database; any of the partners (indeed, anyone willing to take the time to understand the system) could access and use the data. Mike King thought this was one of the federated network's great advantages. It gave local users the ability to make whatever use of the data they wanted. It gave the Brazilian government a fairly sophisticated ability to monitor pollution around São Paulo, for example, while also serving NASA's more global interests. The large number of potential users and resulting products increased the value of the network substantially, expanding political support for it in the process.[20]

SCAR-B was carried out during August and September 1995, using the ER-2 and two other aircraft, the University of Washington's instrumented C-131 and an instrumented Bandeirante provided by the Brazilian space agency. SCAR-B drew on other sources of information about regional fires as well. The Cooperative Institute for Meteorological Satellite Studies at the University of Wisconsin had developed an experimental automated algorithm to extract fire information from the GOES satellites, and for SCAR-B the Institute had provided thirty-minute updates to the field experiment. The regional scale available from the geosynchronous satellite imagers enabled demonstration of the scale of biomass burning effects, and also helped indicate how many fires there actually were. For one day in September, for example, the GOES fire count exceeded four thousand.

A major scientific goal of the SCAR field experiments had been to examine the consistency of the measurements produced by various instruments. The methodology developed by the science teams was to fly the C-131 over an AERONET station to gain simultaneous measurements; the C-131 could provide both remote sensing and in situ measurements that could be checked against the AERONET instrument's data. Similarly, underflying the ER-2 with the C-131 established independent datasets reflecting the same parcel of air. The ER-2 data and AERONET datasets could also be compared. The multiple comparisons available between datasets enabled the expedition's scientists to put some firmer limits on the range of possible values for aerosol effects, while also enabling improvements to the data processing algorithms needed to turn the measurements into knowledge.

The following year, NASA scientists also began to look at the impact of aircraft contrails on radiative transfer in a field study called SUCCESS. (SUCCESS stood for SUbsonic aircraft Contrail and Cloud Effect Special Study.) Carried out in 1996 in Salina, Kansas, and Ames Research Center in California, this study was financed by a new NASA program to assess the impact of subsonic (i.e., tropospheric) aircraft on the atmosphere. Michael Prather, who had left GISS in 1991 for the University of California, Irvine, had been asked by the Congressional Office of Technology Assessment to arrange a workshop aimed at helping the Federal Aviation Administration devise new emission rules for aircraft. Prather recalled that Jack Kaye and Michael Kurylo, Bob Watson's longtime assistant in the UARP, as well as Robert Whitehead, the director of NASA's Aeronautics programs, attended. There were a lot of unknowns about aircraft emissions, but one of the most significant for radiative transmission was the impact of water vapor and particularly of contrails. These suggested to Whitehead that an assessment program for subsonic aircraft would be useful to establish a knowledge base in advance of future environmental regulation. This was the origin of the subsonic assessment program.[21]

Contrails' impact on climate was not really a new issue. The 1970 Study of Critical Environmental Problems (SCEP) had discussed the possibility that aircraft water vapor emissions could affect climate. These findings had not sparked significant research programs, however. In one sense, the 1970 SCEP study had been too early. At the time, commercial jet aircraft were relatively new and only a few hundred were in service. The lower-altitude propeller-driven aircraft that had preceded them had been far less likely to produce contrails. But by the 1990s, the total number of commercial jet aircraft approached ten thousand. Many thousands of flights per day took place over North America, making the potential impact much more significant.

William L. "Bill" Smith, who moved from the University of Wisconsin to Langley Research Center in 1996, recalled that the revived interest in examining the radiative impact of aircraft contrails grew out of a 1993 paper on cirrus clouds. Researchers had reanalyzed data from Rudolf Hanel's Infrared Interferometer Spectrometer (IRIS) instrument on the Nimbus 4 satellite, and found that there was unexpectedly strong infrared absorption above the Pacific Warm Pool that appeared to come from cirrus clouds that were "optically thin," having little impact on visible light transmission and therefore invisible to humans and visible-light satellite imagers. They concluded that high-altitude cirrus made a major contribution to maintaining the high surface temperatures of the Warm Pool.[22]

The advent of commercial jet aviation was widely suspected of having increased cirrus cloudiness in the Northern Hemisphere, but there was little rigorous data to back that suspicion. Walter Orr Roberts, NCAR's founder, had made this observation to a New York Times reporter back in 1963, for example. He could see the aircraft contrails gradually spread out, thin, and merge into an indistinct cirrus-like layer.[23] It was less clear how often this phenomenon happened. Gaining a quantitative estimate of the phenomenon's occurrence would be extremely difficult. The existing weather satellites' imagers could not see these thin, high-level clouds reliably. The net radiative impact also was not well-known, particularly as it appeared that even when cirrus clouds were too thin to visible, they still had infrared opacity.[24] SUCCESS's scientists, including Smith, Brian Toon, and Patrick Minnis from the Earth Radiation Budget team, along with many others, intended to try to study the cirrus/contrail nexus with the new instrumentation that had been evolving for the past decade.

The science team chose Salinas, Kansas, for the experiment because the Department of Energy had established an experimental ground site there for its own Atmospheric Radiation Program that would supplement the aircraft-based measurements. The Atmospheric Radiation Program was part of the Energy Department's contribution to the U.S. Global Change Research Program, consisting of a set of instrumented ground sites, support for model development, and support for field programs.[25] The ground site instrumentation was designed to provide high spectral resolution profiles of the radiative characteristics of the atmospheric column above the site; it was very similar to the data output for the aircraft instruments intended for the SUCCESS experiment. The instrument suite included cloud lidars and radars in addition to a variety of passive sensors. The radars included traditional precipitation radars and millimeter wavelength radars that could also measure non-precipitating water and ice, providing the fullest characterization of cloud properties currently possible. Overlap with the aircraft instruments would permit comparison between ground and airborne instrumentation to help resolve errors.

The airborne instruments for SUCCESS included in situ sensors on a T-39 and on the DC-8, and remote sensors on the ER-2. Many of the DC-8's instruments derived from the GTE and UARP's development efforts, and were chosen to study the chemistry of aircraft emissions. Cloud and contrail ice crystals and water droplets needed nuclei of some kind to form around, and the particulate emissions of aircraft engines were suspected of fostering their condensation. The science teams were interested in whether certain kinds of nuclei were more efficient at precipitating crystals or droplets. Because of the interest in sulfate aerosols, the Langley Research Center's 757, which was used as the source of the experiment's contrails, burned both high- and low-sulfur fuel during the experiment.

The ER-2's remote sensors were chosen to represent prospective satellite instruments that were planned for the EOS. These were the MAS, an airplane-based version of the Goddard Space Flight Center's MODIS; the High Resolution Interferometer Spectrometer (HRIS), from the University of Wisconsin's Space Science and Engineering Center; and the Langley Research Center's Cloud and Aerosol Lidar System. Bill Smith had proposed the interferometer instrument in 1979, and it had begun flying on the ER-2 in 1986; its purpose was to produce very high-resolution temperature and humidity profiles of the atmosphere as well as provide spectral information on aerosols. The lidar could provide information on aerosol and cloud particle density and shape, microphysical measurements necessary to evaluate the adequacy of cloud physics models used by other instrument algorithms. MODIS, which contained channels for land and ocean surface sensors as well as for atmospheric studies, was expected to provide data representing cloud particle size and distribution, but from infrared radiance retrievals. The retrievals used a model of infrared transmission based on particle size; the accuracy of the model needed to be verified by other means. These included the use of the ER-2's lidar instrument as well as in situ particulate sampling by the other expedition aircraft, cloud radars on the ground, and the Wisconsin interferometer.

Initially, SUCCESS looked like a failure. Project scientist Pat Minnis remembers that the weather over Kansas refused to cooperate and provide conditions suitable for contrail formation, so eventually the team moved to Ames Research Center, where meteorological conditions off the Oregon coast were reported to be perfect. The expedition's results from the Ames flights, however, were striking. In their eighteen research flights, carried out during April and May 1996, meteorological conditions in the upper troposphere were often cold and humid enough for cirrus formation, but cirrus clouds did not form until aircraft flew through the area, changing the chemical environment. Yet the science teams could not find a chemical difference between aircraft-induced cirrus and natural cirrus that formed elsewhere.

Further, contrails sometimes persisted for more than seven hours and spread to cover more than 10,000 square kilometers, forming optically invisible cirrus that were nonetheless effective infrared absorbers.[26] This evidence strongly suggested that aircraft-induced contrails had significant climate effects, but was hardly conclusive. The true extent of the contrail impact could not be determined without reanalysis of the older satellite and ground observations to discern how common this effect really was. Minnis spent the next several years trying to quantify it using data from the International Satellite Cloud Climatology Project (ISCCP) organized at GISS; his work was eventually incorporated in a special IPCC study, *Aviation and the Global Atmosphere*.

Ames's ER-2 deployed almost immediately after SUCCESS to a very similar field experiment called TARFOX, conducted over the Atlantic Ocean. TARFOX, which stood for Tropospheric Aerosol Radiative Forcing Observational Experiment, was aimed at discerning the radiative impact of pollutant outflow from North America. Like the Tropospheric Aerosols and Chemistry Expedition (TRACE) mission in 1992, this was part of an international experiment carried out under the umbrella of the IGAC experiment and was focused largely on aerosols. Flown from Wallops Island, Virginia, TARFOX was aimed at the East Coast aerosol plume because the radiative impact of each of the major continental pollutant plumes was not well known, but needed to be characterized for future improvement of climate models. Averaged globally, aerosol radiative impact seemed to be between one-quarter and one-half that of the anthropogenic carbon dioxide content of the atmosphere, but in the regional pollutant plumes, the radiative forcing from aerosols would be much larger. But even the sign of the regional impact was not known, because it depended on the composition of the plume. Some aerosol types, such as volcanic sulfates, produced net cooling, while other aerosols produced warming.[27]

In TARFOX, the most interesting findings were that the East Coast aerosol plume contained an unexpectedly large amount of organic material, and that water condensed on the aerosol surfaces was a major influence on the aerosols' optical effects. Comparison of data from the MODIS simulator on the ER-2 with that from the sunphotometers also provided important verification that the retrieval algorithms used by the MODIS science teams produced reliable aerosol information, a key goal for NASA. Similarly, the experiment provided validation of algorithms used to produce aerosol data products by the new European environmental satellite ERS 2, launched the year before.[28]

After the extensive activity of 1996, there was a nearly two-year lull in atmospheric radiation field experiments while the science teams analyzed and processed their data, published, and planned the next mission. This was the FIRE Arctic Clouds Experiment in 1998. This was intended to be part of the validation process for the first EOS satellite, EOS-AM. However, it did not launch on time. EOS-AM was completed on schedule but the EOSDIS Core System was not ready for it; the Command and Data System that was supposed to control the satellites failed compatibility testing with the satellite, triggering a frantic nine-month effort to rebuild its software. But FIRE was carried out as scheduled due to the investment made to organize it as well as the continued relevance of its science objectives.[29]

FIRE stood for the First International Satellite Cloud Climatology Project's Regional Experiment, and dated from 1983. The ISCCP had originally been organized to develop a cloud climatology for the Earth from the database of satellite imagery available from the global network of weather satellites. William Rossow at GISS was ISCCP's lead scientist. The program's principal goal was to develop algorithms that would allow characterization of clouds from the data produced by the Advanced Very High Resolution Radiometer (AVHRR) imager on the weather satellites. FIRE was formed as a subprogram somewhat later, when the group's scientific leaders decided that they needed observational data to help them understand what the satellite sensors were actually recording. Hence, FIRE supported cloud modeling and satellite cloud retrieval activities, and carried out cloud-related field experiments.

Even before the striking ERBE results in 1989, cloud parameterizations within weather and climate models had been understood to be at best weak representations of real cloudiness patterns, and FIRE's goal was to provide observational data to improve them. FIRE's project manager, David S. McDougal, was at Langley Research Center; NASA was the lead agency for FIRE, but the National Science Foundation, Office of Naval Research, the National Oceanic and Atmospheric Administration (NOAA), the Department of Energy, the U.K. Meteorological Office, and France's Centre de la Recherche Scientifique were all collaborators. The small size of the program had led it to focus on marine stratocumulous and cirrus clouds to make the best use of its limited resources; the ERBE team's results in 1989 had produced increasing interest in FIRE and a consequent expansion of its resources. It remained focused on stratocumulous and cirrus, but began to deploy to more exotic locations than Madison, Wisconsin, and San Diego, California, its field sites during the 1980s.[30]

Many of the scientists involved in FIRE were part of the SUCCESS expedition of 1996, which marked the largest cloud experiment to date although not formally a FIRE experiment; the FIRE Arctic Clouds Experiment two years later marked the first FIRE deployment outside the mid-latitudes. The FIRE science team, led by Patrick Minnis of Langley Research Center, had chosen the Arctic because polar cloud patterns were significantly different than those of the mid-latitudes. High-altitude, optically thin clouds were very common during certain parts of the year, and there were kinds of clouds that did not exist elsewhere on Earth, including "diamond dust," ground-level ice clouds. Further, arctic cloud patterns were often multilayered, and the radiative impact of multilayered clouds was different than that of single layers.

FIRE's Arctic experiment was organized around two other Arctic experiments, the Surface Heat Budget of the Arctic Ocean (SHEBA) project and the Atmospheric Radiation Measurement program's new ground facility in Barrow, Alaska. SHEBA, funded by NOAA, had established a ground station within the Arctic sea ice by trapping an icebreaker in the ice pack and letting it drift for a year. This was aimed at developing a detailed dataset regarding the Arctic surface energy and ice balance using instrumentation provided by the Atmospheric Radiation

Measurement (ARM) program. ARM also established a permanent ground station at Barrow, Alaska, like that used by SUCCESS in Salinas, Kansas; it was part of growing ground infrastructure for the generation of long-term, high-resolution, and relatively inexpensive atmospheric radiation datasets. As had been true with SUCCESS, the purpose for gathering overlapping datasets from ground and airborne sensors was to ensure reliable measurements via cross-calibration.[31]

The FIRE Arctic experiment involved four aircraft: the Ames Research Center ER-2, a C-130 from NCAR, and Convair 580s from the University of Washington and from the Canadian Institute for Aerospace Research. The ER-2 simulated a satellite again, carrying a payload of instruments intended for future space applications: the MODIS simulator, the Wisconsin HRIS, a new Cloud Lidar System, and a new JPL instrument, the Airborne Multiangle Scanning Radiometer (AirMISR). The Multiangle Scanning Radiometer shared the EOS-AM satellite with MODIS and was essentially set of nine imagers designed to improve the angular sampling of cloud properties. The lower-altitude aircraft, as in SUCCESS, were equipped to measure cloud microphysical characteristics via both direct sampling and remote sensing, including particle size, composition, and concentration. They also carried remote sensing instruments for measuring upwelling radiation; the Canadian Convair 580 carried an airborne version of the Landsat Thematic Mapper, for example.[32]

The FIRE expedition flew from Fairbanks and Barrow, Alaska, and Inuvik in the Canadian Northwest Territories, during the spring and summer of 1998, using several bases in order to overfly the drifting SHEBA ship.[33] The science team had gone to the Arctic expecting complex cloud structures but had been surprised at just how complex clouds and the related surface energy budget was; in the Arctic, it turned out that clouds cooled the surface during the winter but warmed it in summer. The most significant outcome of the expedition for NASA was its determination that the combination of sensors being assembled for EOS would permit better discrimination of cloud types and properties than the older generation of satellite instruments. It also provided data for use in improving both satellite retrieval algorithms and cloud and climate models. Minnis, for example, remembers that the expedition's data enabled a big improvement in the algorithms used to generate cloud information by the MODIS and Clouds and the Earth's Radiant Energy System (CERES) instruments when they finally reached space.[34] They also found direct evidence that anthropogenic aerosols changed the characteristics of higher cloud layers, further illustrating that emissions would change cloud properties and thus impact Earth's climate.

During the FIRE Arctic experiment, NASA was evaluating proposals for new missions under its second ESSP announcement of opportunity, and in December it announced the selection of proposal by David Winker of Langley Research Center to fly a cloud lidar in space. Initially known as PICASSO-CENA, this was a joint mission with the French space agency, and included a number of other partners: Ball Aerospace Corporation; the Institut Pierre Simon Laplace, a coalition of French environmental research laboratories; and Hampton University, a historically black college in Hampton, Virginia, to which Langley scientists Pat McCormick and James Russell had retired in 1996. The PICASSO-CENA team eventually changed the mission name to CALIPSO, the Cloud-Aerosol Lidar and Infrared Pathfinder Satellite Observation satellite.

Langley's Pat McCormick and Ed Browell had advocated putting lidars into space for cloud and aerosol studies for many years, but faced the innate conservatism of the agency—technology that hadn't flown in space was too risky to fly in space. In 1988, however, they had gotten Shelby Tilford to accept a proposal to fly a lidar demonstration experiment on the Space Shuttle. Named LITE, for the Laser In-space Technology Experiment, it flew on Discovery in 1994. This flight confirmed that a space-based lidar could provide cloud and aerosol details, even in multilayered cloud environments. It also clearly identified the sub-visible cirrus that eluded the AVHRR. LITE's success eliminated doubts about the scientific merits of a lidar. Browell remembers that there was still doubt that a lidar would have a usefully long lifetime in space, however, as LITE only flew for two weeks. The Goddard-built Mars Observer Lidar Altimeter, which operated in Mars orbit from 1997 to 2001, eliminated those doubts by far exceeding its expected lifetime.[35]

Another ESSP, Cloudsat, was selected for flight the following year. Cloudsat's principal investigator was Graeme Stephens of Colorado State University, and like CALIPSO the mission involved a partnership of several institutions. Stephens's research specialty was cloud radiative effects. Cloudsat's principal instrument was a millimeter wavelength radar that was similar to that used in the ARM stations, and Stephens hoped its ability to measure both precipitating and non-precipitating regions of cloud would lead to improved model treatment of clouds. He also believed that it would uncover previously unknown forms of convection. The Cloudsat and CALIPSO instruments had originally been packaged as a single mission proposal in 1993, but the EOS program structure at the time left no way to get this medium-class mission into space. Hence, Stephens and CALIPSO's principal investigator, Winker, had wound up competing with each other for an ESSP selection several years later. Cloudsat was to be developed as a partnership between JPL, Colorado State, the Canadian Space Agency, the Department of Energy, Ball Aerospace, and the U.S. Air Force.[36]

CALIPSO and Cloudsat completed a sequence of cloud- and aerosol-related satellites that came to be called the Afternoon Constellation, or "A-Train." When completed, it became the first atmospheric science experiment carried out by a formation of satellites. The constellation placed five satellites — EOS-PM, Cloudsat, CALIPSO, a French aerosol polarimetry satellite named PARASOL, and finally EOS-CHEM — into a single line, with the spacecraft separated by at most a few minutes. Cloudsat and CALIPSO, whose spacing was the most critical, were put only fifteen seconds apart so that their instruments would see the same clouds at the same time. This way, NASA could still achieve the original intent of the huge EOS platform approach but with far less risk. This formation was supposed to be completed in 2003, however, and in fact was not until 2006 due to a combination of technical problems with the CALIPSO lidar and to a workers' strike against the launch vehicle manufacturer.

AN EARTH OBSERVING SYSTEM — FINALLY!

The first two EOS satellites, Landsat 7 and EOS-AM, were finally launched during 1999, into the same orbital plane but forty-five minutes apart. EOS-AM was renamed Terra, a name proposed by Sasha Jones, a student in St. Louis, Missouri, after its orbit insertion.[37] The field experiment originally intended as part of Terra's validation, FIRE Arctic Clouds, was well over by this time, but a second, larger, expedition had been planned for Africa's fire season. Named Southern African Fire-Atmosphere Research Initiative (SAFARI) 2000, it marked the beginning of NASA's effort to demonstrate that EOS would do the things it had been promising for two decades.

EOS project scientist Michael King had been a major proponent of returning to Africa to validate the Terra sensors, which included instruments to study clouds, aerosols, and carbon monoxide distribution in addition to land surface sensing. King had made presentations to the governments and legislatures of several African nations seeking their support, and explaining the potential benefits to them from the EOS data products that would eventually result. One data product the MODIS instrument was expected to produce was one revealing fire outbreaks, a product of obvious value in a region of annual, and very large-scale, biomass burning (and also one popular with western state representatives). The Canadian Measurements of Pollution in the Troposphere (MOPITT) instrument would also measure carbon monoxide, a primary product of burning and an ozone precursor.

And while SAFARI 2000's purpose was not directly to study fire, the 2000–2001 fire season turned out to be the most intense the mission scientists had ever seen.[38]

SAFARI 2000 was patterned on the SAFARI experiment of 1992, as a regional-scale examination of biogeochemical cycling between land and atmosphere, and its experiment design was similar to that devised for the GTE's operations in Brazil during the 1980s and SCAR-B in 1995. SAFARI scientists erected measurement stations in Zaire, Botswana, and South Africa to provide local measurements that could be linked to the regional-scale aircraft measurements; in addition to NASA's ER-2, the University of Washington's Convair 580, the U.K. Meteorological Office's C-130, and a pair of South African Aerocommanders were included. And the AERONET project office at Goddard Space Flight Center placed fifteen AERONET units in the experiment region to provide further correlative measurements.[39]

The ER-2 was instrumented with the aircraft versions of Terra's instruments plus three additional sensors: the Cloud Physics Lidar, a new version of the Wisconsin HRIS that could scan across the aircraft's flight track called Scanning HIS, and a very early testbed for a new hyperspectral imager. The lidar provided the ability to examine aerosol density and cloud particle properties at very high spatial resolution, while the interferometer could characterize them at very high spectral resolution. These were important components of improving scientists' knowledge of the information content of the data that Terra gathered. The interferometer also represented the next generation of infrared temperature sounders; an instrument similar to Scanning HIS had been chosen to succeed the Atmospheric Infrared Sounder (AIRS) instrument on the National Polar Orbiting Environmental Satellite System (NPOESS) series of next-generation weather satellites. Hence, as it had in the previous cloud- and aerosol-related field experiments, the ER-2 represented a simulated satellite that could check and improve upon the real satellite's measurements.[40]

Due to strong La Niña conditions in 1999, the colder, wetter inverse of El Niño for Southern Africa, a greater abundance of biomass existed in Africa during 2000 for burning than was typical, and the result was a fire season so extensive that the project's principal scientists titled their review article on the expedition "Africa Burning." They encountered vast layers of aerosols produced by biomass burning, and the dominant meteorology during the period had them exiting the continent southeastward in a "river of smoke" clearly visible in the imagery from several different satellite and aircraft sensors.[41] The extreme conditions provided the expedition scientists the unique opportunity to demonstrate the ability of the satellite instruments to produce quality data even under unusual atmospheric conditions. The MODIS aerosol product had to be modified after the data from the sun photometers and lidars showed that over land it consistently underestimated aerosol density while being quite accurate over the ocean.

The SAFARI expedition's data also allowed assessment of the regional radiative forcing produced by the aerosol plumes, which turned out to be considerable. The expedition's measurements of radiative fluxes, and the radiative transfer models these were compared to, found that aerosol masses imposed a negative forcing, or cooling, an order of magnitude larger than that imposed by the doubling of carbon dioxide, although, of course, the aerosol impact was temporary and regional. This affected weather patterns and also cloud structures, with the biomass-generated aerosols reducing cloud particle size. Smaller particle size suppressed precipitation; particles had to grow large enough to settle out of the clouds for rain to happen. Changing particle size also affected the radiative impact of clouds. Perhaps more interesting, the radiative impact of aerosols depended upon the presence or absence of clouds; in cloud-free conditions the aerosol layers produced a cooling effect, while in the presence of underlying clouds they had the opposite impact.[42]

NASA launched the second EOS satellite, the Delta rocket-sized EOS-PM, in May 2002. It became Aqua under the agency's new naming scheme, reflecting its primarily cloud- and water-oriented mission. Aqua's six instruments were the AIRS infrared temperature sounder; the Advanced Microwave Sounding Unit (AMSU); a new scanning microwave radiometer (AMSR-E); MODIS; CERES; and the Humidity Sounder from Brazil, which replaced the NOAA humidity sounder after NOAA did not receive the funds to complete the instrument. Like Terra, Aqua's launch was followed by an extensive field campaign, this one in Florida and similar in design to the FIRE Arctic Clouds Experiment. Called CRYSTAL-FACE, for Cirrus Regional Study of Tropical Anvils and Cirrus Layers-Florida Area Cirrus Experiment, this expedition also presented the first physical realization of the strategy to transition the Aqua measurements to the follow-on NPOESS. Several of the participating aircraft were equipped with copies of instruments intended for the NPOESS satellites. The CRYSTAL experiment had started its life as FIRE IV, reflecting the beginning of a fourth phase of FIRE-sponsored expeditions aimed at the radiative and meteorological effects of active tropical convection features, but the consensus of the planning team had been that it was time for a new name.

As carried out in July 2002, the Florida phase of CRYSTAL included six aircraft in addition to the Terra and Aqua satellites. More than four hundred scientists

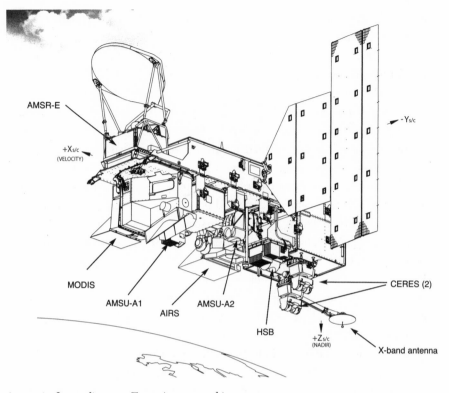

Aqua wireframe diagram. From Aqua press kit, 2004.

participated in CRYSTAL-FACE. Such wide participation by the atmospheric research community ensured that its results were also widely varied, with different investigators interested in aerosol effects on cloud composition, the impact of Saharan dust on cloud and precipitation formation over Florida, radiative impacts, and the potential for improved weather forecasts inherent in the Aqua datasets. For NASA's leadership, however, the key results were demonstration that the integrated measurement strategy represented by the A-Train of satellites would, in fact, produce useful new information. The ER-2 lidar and radar combined reproduced multilayered cloud structures in great detail, demonstrating details of structure that only ground-based instruments in the ARM system had been capable of revealing before. The lidar could reveal details of the cirrus shield that formed above convective storms while the cloud radar could detect deep structure; as the lead investigators had hoped, the two measurements overlapped and complemented each other. Commenting on the results of CRYSTAL-FACE two years later, Bill Smith noted that the integrated measurements offered by the forthcoming EOS and NPOESS instruments promised the ability to forecast clouds before they formed, impacting both weather prediction and climate studies.[43]

INTEGRATING MEASUREMENTS

One of EOS's central goals had been to demonstrate the ability of integrated Earth observations to generate new scientific knowledge. The field experiments carried out under the radiation and chemistry programs during the 1990s were one aspect of preparation for that. In all these experiments, many investigators using numerous instruments collaborated to produce a larger scientific perspective than any individual instrument team could achieve on its own. This had been necessary to resolve the question of causality of the Antarctic ozone hole, for example. For EOS's post-1991 emphasis on clouds and climate, one key instrument was the Clouds and Earth's Radiation Budget instrument, CERES, which was to fly on three vehicles: the TRMM and both EOS-AM and EOS-PM. Achievement of CERES's science goals was heavily dependent upon the use of information from other sensors, however.

Technologically, CERES was a descendant of the ERBE launched in 1984. Bruce Barkstrom and Bruce Wielicki of Langley Research Center, both members of the ERBE science team, had proposed the new instrument in response to the 1988 announcement of opportunity for EOS. From the standpoint of cloud radiative impacts, ERBE had had a number of weaknesses. First and foremost, it could not identify the radiative impact of different cloud types. ERBE had no other source of data to draw upon other than its broadband radiometers, and therefore it could not identify the contents of a scene beyond the very simple categories of "clear," "partly cloudy," "mostly cloudy," and "overcast." Identification of types of clouds was beyond its capabilities. Further, ERBE's own purpose had been to produce climatological averages. Averaging effectively removed random, daily variability in both cloud cover and instrument characteristics, enabling the production of very accurate, stable monthly data products.[44] This was a benefit, given ERBE's focus on discerning the overall impact of current cloudiness patterns on the Earth's radiation budget. But if one wished to determine the radiative impact of specific cloud types, a different approach was necessary.

To fully express cloud radiation fields, Wielicki and Barkstrom needed to sample in eight dimensions. With ERBE, the science team had only attempted to sample three, two with the scanner (the x and y values that permitted location on the Earth's surface) and time. They had eliminated the vertical dimensions by deciding to only attempt production of top-of-the-atmosphere fluxes, therefore treating the atmosphere as if it were a flat radiating surface. The angular dimensions, necessary for the problem of resolving specular anisotropy, the ERBE team had dealt with by using angular distribution models. Specular anisotropy is the reflectance property that causes sun glint, the sudden flash of intense light one sees off the ocean surface, or off a car window, as the angle between the Sun, the observer, and the surface changes. This produced biases in the data that with ERBE's simple cloud descriptions could still reach 10 percent. As Wielicki put it, "climate is a 1% game."[45] Getting the radiation field errors down to the 1 percent level meant trying to measure the angular reflectance of a complete range of cloud types in order to improve the angular models.

To do this, Wielicki and Barkstrom proposed to make one major change to the instrument itself. ERBE had consisted of a cross-track scanning instrument that had a moderate field of view, which the experiment team had needed to identify the cloudiness level of each scene as well as to provide subregional-scale albedo measurement. The other portion had been a fixed wide field of view instrument that was highly stable. This provided very accurate regional-scale albedo. But the very large scene size made it essentially impossible to identify the cloudiness of a given scene because there were no clear scenes at that scale—the atmosphere was almost never clear over an entire region. Hence, it was deleted from the CERES proposal and replaced with a second scanner. One instrument would scan in the cross-track mode, while the other would employ a new hemispheric scan. The hemispheric scan would produce the angular sampling that they wanted, although only over time. They estimated two years of data would be necessary to produce enough samples for statistically valid new models. Another change they proposed was a smaller field-of-view than the ERBE scanner had had, to permit better discrimination between clear and cloudy scenes.

Wielicki and Barkstrom also wanted to produce estimates of flux within the atmosphere with CERES, which required measurements in the vertical dimension. For this, however, they did not propose changes to the instrument. Instead, they intended to rely on data from other instruments to improve their own data products. MODIS could provide cloud altitude, cloud physical properties, and aerosol properties data, while also improving the ability to identify cloudy and clear scenes and cloud types. Cloud altitude was essential to defining the vertical radiation field; the lidar instruments proposed for and later deleted from the early large-platform version of EOS would also have provided this information as well as greater cloud particle detail. (Much later, of course, the selection of CloudSat and CALIPSO as ESSP missions promised to restore this capability.) Other vertical information they intended to get from the daily global analyses produced by numerical forecast models. These, of course, relied on vertical temperature profiles from radiosondes and increasingly during the 1990s on satellite temperature sounders.

The CERES team also needed surface flux measurements to check the accuracy of the vertical flux fields. These the two intended to get from a surface radiation network being constructed under the auspices of the World Climate Research Program (WCRP), the Baseline Surface Radiation Network, and from the Department of Energy's ARM sites. The ARM sites also provided more detailed vertical radiation fields from their variety of radiometers, lidars, and radars, albeit from a limited set of three locations. This was not all that limiting, however, as their continuous operation and daily overflight by the satellites meant that the several thousand cases necessary to achieve statistical relevance could be built up relatively quickly. The principal limitation of the ARM sites for the CERES's team's purpose was that from the satellite's point of view, the sites were surrounded by regions that provided an inconsistent background. The Salinas, Kansas, site was surrounded by agricultural fields planted in different crops, and which therefore were different colors. So Wielicki needed one more site with a consistent background. An ocean platform was the obvious choice, and the team found one in a soon-to-be abandoned Coast Guard platform in Chesapeake Bay. It was reasonably convenient to Langley Research Center, and they instrumented it for the radiation studies they needed.[46]

Because Wielicki and Barkstrom wanted to utilize data from other sources, from the outset an important consideration for CERES was the integration of different datasets. This was a complex undertaking. Because of their prior experience with ERBE, Wielicki and Barkstrom already had a clear idea of what they wanted the system to do and how data would have to move through it in order to accomplish their objectives. Barkstrom, who was the data processing system architect, believed that designing the interfaces between the various software modules first and then essentially preventing changes during the detailed programming was the most effective way to develop a large software complex, and he laid out the interface specifications for the system. After that, the programming was to be done at Langley, largely by the CERES team itself, with the first prototype system finished in 1996.

The 1992 decision to place a single CERES scanner on the TRMM gave the CERES team the opportunity to acquire data earlier and begin validating the processing system. Like the EOS platforms, TRMM had a visible/infrared imager that could be used to obtain cloud and aerosol information. It also carried a passive microwave imager, which could produce atmospheric water content and precipitation estimates, as well as its primary radar instrument for direct precipitation measurement. Vern Suomi had lobbied NASA for years to get TRMM approved, hoping it would produce the first near-global measurements of tropical precipitation for weather and climate research purposes.[47] The mission's primary focus was the study of convection, not radiation budget studies, despite Suomi's interest. But they were complementary, in that CERES would allow identification of energy flux changes associated with convection. Further, the TRMM orbit was much like that of the old ERBS, which had been designed to precess through the entire diurnal cycle to permit determination of flux changes throughout the day. This would be complementary to the EOS polar platforms, which would only view each part of the Earth at one specific local time.

A year prior to TRMM, the first version of the CERES processing system was completed. In the absence of data, the team tested it by using data that had been taken by the ERBE, AVHRR, and High Resolution Infrared Sounder (HIRS) instruments on the operational weather satellite NOAA 9 during October 1986. This provided them with estimates of the processing needs for the full CERES dataset when it appeared as well as demonstrating the performance of the algorithms.[48] Of course, it also pointed the teams to weak spots in their work in time to correct them before TRMM's launch.

TRMM's launch on 28 November 1997 gave the CERES teams their first operational data. Plate 7 illustrates the CERES heat flow data for a single day in 2003. Unfortunately, the instrument began to operate unreliably after eight months due to a failing power converter, and was shut off in hope that it could be turned on again briefly after the EOS-AM launch for cross-calibration. The instrument's operational period happened to coincide with the very strong El Niño event of 1998, however, which gave them an unexpected opportunity to respond to a prominent climate skeptic's claim that the Earth would automatically counteract any warming due to carbon dioxide.

In 2001, meteorologist Richard Lindzen of MIT and two colleagues at the Goddard Space Flight Center published a paper in the *Bulletin of the American Meteorological Society* claiming that the Earth had what they termed "an adaptive infrared iris" in the tropics. They had studied data from one of the Japanese geosynchronous meteorological satellites of cloud cover changes over the Pacific Warm Pool. Lindzen was interested in determining whether there was a relationship between sea surface temperatures and cloudiness patterns. They found that there seemed to be one, with cirrus clouds decreasing with increasing surface temperature. Because cirrus primarily absorbed outgoing heat, without reflecting much incoming sunlight, this change suggested that more heat would escape to space as the surface warmed. This would provide a large, negative feedback to the climate system. The three then constructed a simple radiative-convective model of their proposed process to examine how much their negative feedback might reduce the Earth's sensitivity to carbon dioxide–induced warming. They found that the changing cloud properties would produce a strong cooling in the tropics, effectively countering about three-quarters of the warming expected by the IPCC.[49]

CERES team member Bing Lin had found the hypothesis exciting, as it appeared to substantially mitigate the global warming problem. Lindzen's assumptions about cloud properties were also testable using the CERES data from TRMM. So Lin had examined the CERES ERBE-like data product, generated to maintain commensurability with the ERBE sensor data, using Lindzen's stated methodology. He sought to determine whether the observed radiation fields produced by TRMM matched those predicted by the iris model. But Lin found that the effect of changing cloudiness in the CERES data was opposite Lindzen's prediction. Cloudiness did decrease, but this produced a modest warming at the surface, not a strong cooling. This was due to several differences between CERES's measured cloud properties and Lindzen's assumed ones. Cloud albedo was much higher in the CERES data than in the iris model, and infrared absorption by clouds was lower. Total cloud cover was also lower in the data. When applied to Lindzen's radiative-convective model, the observed data produced a positive feedback that was equal to the greenhouse forcing of tropospheric ozone.[50]

Lindzen and his colleagues attacked these results, however, driving Bing Lin's colleague Lin Chambers to use a new CERES data product, the Single Scanner Footprint, to make a further refutation. Released in late 2001, this was the fully integrated dataset, including new angular models produced from the first two years of operation from the EOS-AM satellite. Because it also contained the imager data, it could be used to identify the specific characteristics of different cloud types. Chambers used the data to try to identify cloudy scenes with fluxes like those proposed in Lindzen's original article, but without success. In no cases were the measured cloud radiative properties similar to Lindzen's assumed ones. As a result, the strength of the feedback effect was an order of magnitude lower than predicted in his iris model, and opposite in sign. It produced a modest warming, not a strong cooling.[51]

There were other refutations of Lindzen's thesis based on various other tests. One, by Brian Soden and Richard Wetherald at the Geophysical Fluid Dynamics

Laboratory (GFDL), drew upon satellite data from the Microwave Sounding Units and Television-Infrared Observations Satellite (TIROS) Operational Vertical Sounders aboard the NOAA series of weather satellites that had been collected during and after the 1991 eruption of Mount Pinatubo. James Hansen at GISS had predicted that the eruption would cause a measurable global cooling from the radiative impact of sulfate aerosols injected into the stratosphere, and indeed it had. Over the eighteen months following the eruption, the troposphere had cooled an average 0.5 degrees C according to the weather satellite data. This had served as important verification of the climate models' ability to simulate short-term climate variations in the early 1990s.[52]

The models had also predicted that the cooling atmosphere would dry measurably, but no one had investigated whether it had or not prior to Lindzen's paper. Lindzen's iris thesis depended on the atmosphere getting drier as it warmed, and conversely wetter as it cooled; this was counter to most climatologists' and meteorologists' physical reasoning. The atmosphere was, after all, known to be wetter in summer than in winter.[53] So Soden and Wetherald decided to investigate the old satellite data to determine whether the Pinatubo eruption had caused the atmosphere to dry as expected by the mainstream models, or get wetter, as Lindzen's model indicated. They ran a series of new hindcasts of the Pinatubo eruption with the current GFDL climate model to provide a range of experiments to compare the satellite data with; as expected, both model results and the satellite data showed that the troposphere had dried as it cooled under Pinatubo's influence. The atmosphere's water vapor feedback was positive, not negative, confirming that water vapor changes would enhance global warming, not reverse it.[54]

MOVING BACK TO METEOROLOGY

In 2001, the IPCC had produced its Third Assessment Report, which had contained the body's strongest statement to date that humans were changing the Earth's climate: "there is new and stronger evidence that most of the warming observed over the last 50 years is attributable to human activities."[55] This had caused the newly elected President George W. Bush, who had become president despite the alleged scientists' conspiracy to elect Al Gore in 2000, to request a review of the document from the National Academies. This review, unsurprisingly given the two-decade-long series of such reviews and reports on climate science by the Academies, concluded that "the IPCC's conclusion that most of the observed warming of the last 50 years is likely to have been due to the increase in greenhouse gas concentrations accurately reflects the current thinking of the scientific community on this issue."[56]

This did not change the president's mind; instead, as late as 2005, he preferred to believe global warming was an environmentalist hoax, even meeting with science fiction writer Michael Crichton that year.[57] In 2003, Crichton had given a lecture at Caltech in which he had argued that belief in global warming derived from belief in space aliens; in a 2004 novel, he had explored the environmentalists' global warming hoax in detail.[58] The president found himself to be in "near-total agreement" with the novelist. This made Crichton a star with anti-environmentalists in Congress as well as the White House. He was invited to testify before the Senate Committee on Environment and Public Works the following September, leading one surprised senator to comment, "why are we having a hearing that features a fiction writer as our key witness?"[59]

Following the first Bush administration's policy of emphasizing uncertainty, the second Bush presidency's appointees spent these years revising official publications to magnify uncertainty, even deleting mention of global warming entirely from them on occasion.[60] After failing to prevent the Kyoto Protocol from going into effect, they blocked all international attempts to negotiate a successor until 2007, when the IPCC's Fourth Assessment Report claimed still greater certainty that anthropogenic warming was occurring: "Most of the observed increase in global average temperature since the mid-20th century is *very likely* due to the observed increase in anthropogenic greenhouse gas concentrations.[61] Their deliberate and widespread distortion of climate science drew repeated denunciation from Donald Kennedy, editor of *Science*. In January 2003, he declared, "The scientific evidence on global warming is now beyond doubt." For this claim, he was attacked by S. Fred Singer, who had finally given up in his war against chlorofluorocarbon (CFC) regulation to become a leader in the effort to deny the existence of global warming. This didn't stop Kennedy. In January 2006, after recounting the success of the administration's effort to block any post-Kyoto negotiations, he pointed straight to the guilty: "The climate-denial consortium, supported by a dwindling but effective industry lobbying effort, has staved off serious action. It is a disgraceful record, and the scientific community, which has been on the right side of this one, doesn't deserve to be part of what has become a national embarrassment."[62]

In January 2004, President Bush had announced a new Vision for Space Exploration. A retread of the "flags-and-footprints" space program of the 1960s, this was aimed at returning humans to the Moon sometime in the later half of the 2010s and then establishing a permanent Moon base. It required development of a new launch vehicle to replace the aging Shuttles while simultaneously repairing and returning the Shuttle to operation to complete the International Space Station, which would be abandoned around 2017. Justified by appeals to the "space frontier," to "space resources," and to the "space economy," this Vision resulted in rapid and substantial cuts to NASA's science budget, the deemphasis of atmospheric science, and the delaying of all new Earth science missions except for the ESSPs beyond 2010.[63]

The speed with which this happened caught the science community by surprise; in April 2005, Congressman Sherwood Boehlert, then chairman of the House Science Committee, intervened to force NASA leaders to explain what they were doing. His hearings went badly for NASA, with the agency's acting associate administrator for space science arguing that his intent was to transfer many of NASA's Earth science responsibilities to NOAA—plans that Congress had never heard before, or approved of, let alone actually funded.[64] In one sense, of course, this was the old policy failure still lingering from the Reagan administration. There was no still no formal mechanism for transferring new space-based monitoring capabilities from NASA to NOAA. In another sense, it reflected the desire of the new NASA leadership to get rid of the politically undesirable research program that its predecessors had built, by any means available.

NASA's Earth science establishment sought to save itself by reemphasizing meteorology. With the political unpopularity of the climate problem, and with widespread recognition that its policies of the last two decades had failed to produce much progress toward better weather prediction (still an economically desirable outcome!), this was an obvious approach. The weather was still apolitical, even if climate no longer was. The successful launch and operation of TRMM in 1997 had helped facilitate that trend, as did the EOS Aqua satellite, which served both meteorological and climate purposes. TRMM's precipitation radar had provided the first well-quantified and reliable monthly rainfall averages from the tropics (30 percent of the Earth's surface), results valuable for both meteorology and climatology. While its orbit did not permit it to provide early warning of short-lived storms in most cases, it was able to see deep inside longer-lived hurricanes and typhoons, quantifying in detail processes that had previously been difficult to measure.

In its 2002 budget, NASA initiated its last new, large Earth science mission for the decade, the Global Precipitation Mission (GPM). Improving on TRMM, this was to be a multi-satellite project with a core vehicle containing a precipitation radar like TRMM's and a set of small satellites with passive microwave sensors that could also detect precipitation. The goal of the multi-satellite approach was resolution of the old temporal coverage problem. Precipitation formed so rapidly, and often lasted such a short period, that a single satellite's orbit could never provide a very accurate daily look at precipitation patterns. TRMM had been accurate at the level of monthly averages (i.e., climatically accurate) but from the standpoint of weather had been able only to provide reliable information about multi-day phenomena. GPM was designed to achieve results applicable to the daily weather. It, however, was delayed from 2010 to 2013 by the diversion of funds to the Vision for Space Exploration; whether it survives the voracious financial appetite of the Vision, of course, remains to be seen. Nonetheless, its selection and structure reflect a renewed emphasis on meteorology, and perhaps on applications more broadly, in the first decade of the twenty-first century.

During the 1990s, numerical chemical and climate models began to gain a measure of credibility. Short-term chemical forecast models became useful in the context of field expeditions. Climate models gained respectability from both the volcanic test the Pinatubo eruption had provided as well as from the community's acceptance that the warming Hansen had forecasted in the late 1980s actually had become obvious in the late 1990s. This is not to say the models had been perfected. The cloud representation problem demonstrated at the end of the 1980s was still a problem in the early 2000s, albeit one receiving a great deal of attention. The same was true of the tropospheric aerosol problem. And while anthropogenic chlorine loading of the stratosphere peaked and began to decline as expected, no one could forecast when the stratospheric ozone layer would recover. The greenhouse gas increases that were warming the troposphere had the effect of cooling the stratosphere, more efficiently converting the remaining chlorine into active, ozone-destroying forms. While this effect had been predicted by the major climate models, the models could not foresee how long it would continue.

When Vern Suomi's U.S. Committee for the Global Atmospheric Research Program had undertaken its 1975 study of the state of knowledge of global climate, they had not been able to hang reliable numbers on even the global effects of aerosols, let alone the intricate regional variations of aerosol forcings or the indirect impact of aerosols on cloud albedo.[65] The independent MODIS, MISR, and AERONET aerosol datasets and the CERES measurements of albedo, aerosol, and cloud properties finally allowed that detailed, quantitative, and verifiable analysis to begin; it had only required a generation of effort and $10 billion or so of investment to bring about.[66] Plate 8 shows global seasonal aerosol distribution in 2001–2. While expensive, this was a bargain compared to its sibling, NASA's showcase International Space Station, which had ballooned from its $8 billion cost estimate to about $52 billion in 2005, and was expected to cost upward of $80 billion to complete—an order-of-magnitude cost overrun.

Yet NASA's atmospheric science community stood at a crossroads in the early 2000s. They had developed and deployed powerful new research capabilities in the preceding decades. They were just beginning to exploit those capabilities as the political tides turned against them. At a science meeting in early 2006, for example, Moustafa Chahine's AIRS science team finally decided that they were satisfied with their calibration and data quality and could begin doing real science. Launched in 2002, this meant their expected six- or seven-year life was already more than half over; AIRS, of course, would not be replaced by NASA, but they hoped their work would contribute to the successor instruments on the NPP and NPOESS satellites and the Infrared Atmospheric Sounding Interferometer on the new European METOPS satellites. It wasn't clear, though, whether they themselves would continue to be involved in this research area. The same challenge existed for all of the other science teams attached to EOS.

It's far too soon to fully understand the scientific legacies of NASA's Mission to Planet Earth. New research findings in a wide variety of disciplines are being published weekly in the major English-language science journals. What the most important will be cannot yet be determined. But beyond the architectural issues raised in the preceding chapter, one legacy of Mission to Planet Earth already seems clear. Remote sensing was finally becoming a mainstream scientific tool. Scientists had published 973 papers using MODIS data by late 2005, for example.[67] This was likely due to several factors: the deliberate focus on calibration and reliability NASA embarked on in the late 1970s, the personal computing revolution that made powerful computers nearly ubiquitous during the 1990s, and the open data policy that permitted anyone access to the EOS datasets.[68] Achieving that mainstream status had required an extensive investment in infrastructure, aimed largely at the credibility question. While the infrastructure elements, ground-based measuring stations, airplane and balloon instruments, and computing and data centers are often scientifically useful in their own right, the infrastructure is expensive to operate and maintain.

It remained to be seen whether the new observing technologies of the EOS era would make a successful transition to operational use. As had EOS, the NPOESS project was taking a very long road to orbit. Hearings held late in 2005 strongly suggested that it would not launch before 2011. Hearings in June 2006 indicated a 2013 launch date, meaning NPOESS was likely to take nineteen years from its approval in 1994 to its first launch. Assuming, of course, it isn't cancelled outright. In addition, to reclaim a small part of its 50 percent-plus cost overrun, four climate-related instruments were dropped from the program. Asked by House Science Committee chairman Sherwood Boehlert what his plan was to recover these measurements was, NASA administrator Michael Griffin responded that he didn't have a plan yet, but it would cost more money, whatever it turned out to be.[69]

The space science directorate's financial straits, finally, led to the quiet abandonment of Gore's Triana, which had been renamed Deep Space Climate Observatory, or DSCOVR. Designed as a Shuttle payload, after the destruction of Space Shuttle Columbia in February 2003, it had been put on hiatus while the three remaining Shuttles were returned to a flight status. By the time they were, the remaining number of authorized Shuttle flights prior to their permanent retirement was only enough to complete the International Space Station. And DSCOVR could not be inexpensively converted into a payload for an expendable rocket. In January 2006, it was formally cancelled.[70]

Conclusion

The planet that has to matter the most to us is the one we live on.

—*Sherwood Boehlert, 28 April 2005*

During its first four decades, NASA developed a sophisticated ability to study the Earth as an integrated global system. It had drawn this agenda in part from its institutional interests in planetary science as well as from larger national concerns in the 1970s about anthropogenic impacts on the global atmosphere. In the process, it generated new knowledge about the global atmosphere. By the early 1990s, the Global Tropospheric Experiment (GTE) had demonstrated that agricultural burning had large chemical impacts on the atmosphere, overturning a 1970s belief that only industrial nations threatened global chemistry. With the National Oceanic and Atmospheric Administration (NOAA), it had demonstrated human responsibility for the Antarctic ozone hole, further documenting the ability of human activities to change the global atmosphere. Finally, it had developed the technology necessary to establish a climate record for Earth as it warms.

We began with the question, why NASA? How did the space agency wind up so deeply embroiled in politically controversial science? As the foregoing narrative suggests, it did exactly what one expects a science agency to do: it investigated new scientific questions. And being an engineering agency as well, it developed new capabilities to bring to bear on those questions. Its leaders used the agency's capabilities to remain relevant during changing political conditions. And, further, politicians found it useful to have NASA play a role in examining these questions. The agency's prestige would help limit dissent, though surely not prevent it. Letting an existing agency study these questions also avoided creating an agency specifically aimed at studying environmental problems (and thus creating a constituency for such problems). NASA, which has many scientific constituencies to serve, was unlikely to be captured by any one of them for very long. Solar and astrophysicists would dearly love the funds lavished on Earth science in the 1990s; indeed, as the Earth science funding has diminished over the past half dozen years, both these fields have gained. Hence, NASA seemed an excellent choice to politicians as well as to NASA leaders.

NASA was able to accomplish what it did by drawing on the capabilities of its various research centers, and by building new capabilities into them during the 1970s as it adopted Earth science as a new mission. Goddard, Langley, Ames, and the Jet Propulsion Laboratory (JPL) all developed specializations during the 1970s; all retained them past the turn of the century. The Clouds and the Earth's Radiant Energy System (CERES) group at Langley evolved directly out of the late 1970s Earth Radiation Budget Experiment (ERBE) team, for example, representing a vital continuity of expertise. Its lidar group was even older, having started developing these instruments in 1969, albeit for the somewhat different purpose of wind tunnel instrumentation. Continuity of expertise meant that the science teams could draw on their past experiences as measurement demands tightened during the 1980s and 1990s.

NASA also drew heavily on the support and interest of university-based scientists. In addition to supplementing its in-house scientific capabilities, the university community helped it direct its research agenda toward cutting-edge science problems as well as providing support during the annual Washington budget wars. The relationship was not always a smooth one, as the conflict between NASA and the larger scientific community over the EOS architecture suggests. The conflict derived from the Big Engineering dreams of 1980s-era NASA leaders, of course, not over the scientific goals of the project. But support from the scientific community, particularly as expressed via the National Academy of Science, was crucial, and evidence of dissention regarding priorities could jeopardize the agency's plans.

SCIENCE POLITICS AT THE END OF THE CENTURY

By the end of the twentieth century, the atmospheric sciences were under assault in the United States. The problem was not that the atmospheric sciences had grown too big, as the historical literature on Big Science might imply. There are almost infinite resources available to the United States; what these are spent on is dependent on national politics. Some commentators were blaming the declining fortunes of atmospheric and climate science on the opposition of powerful corporations threatened by future science-based regulation combined with a supine public media.[1] There is some truth to this. There is no question that deliberate disinformation campaigns took place during the 1990s, that these were funded by fossil-fuel energy companies, and that politicians believed them.[2] But such campaigns can only be effective if people are already prepared to believe them. It is at the level of national politics and political ideology that we must look for the roots of the controversies faced by scientists at the turn of the century.

The scientific findings made during the 1980s and 1990s threatened core American beliefs. Environmentalism had been divisive in the United States because it threatened a host of traditional beliefs: in private property rights, in market capitalism, in the superiority of private over public enterprise, in technological progress, in American exceptionalism. To anyone who believed in the superiority of market capitalism, and particularly to those who held the central tenet of late twentieth-century American conservative dogma, that human freedom and market capitalism were inextricably linked, the finding that human industry could cause global-scale damage had been ideologically shattering.

We do not have to look further than the writings of the former chief scientist of the Weather Bureau S. Fred Singer to find evidence for these claims. In 1989, Singer accused his former colleagues of hyping the threat of ozone depletion to secure funding for their research; in 1991, he went further. In an essay denying the reality of anthropogenic warming, he wrote that the hidden agenda of these scientist-activists was "against business, the free market, and the capitalistic system."[3] Singer was hardly alone in this belief, as sociologist Myanna Lahsen documented in her 1998 dissertation.[4] Instead, he was merely one of the most outspoken critics of the new scientific view of humanity as a geologic actor, capable of changing the very conditions of life on Earth.

In the view of these political critics of late twentieth-century science, atmospheric scientists' efforts had led directly to government regulation, indeed elimination, of the chlorofluorocarbon (CFC) industry. These scientists had not only made conceivable a future in which all emissions became pollution to be subject to regulation and monitoring, they seemed well on their way to achieving this future under the Montreal and Kyoto ozone and climate treaties. To believers in what investor George Soros has called "market fundamentalism," these impingements on human (economic) freedom were intolerable.[5] NASA came under attack in the late 1990s not because its research program was too large, or because it was insufficiently scientific, but because it was ideologically threatening. Atmospheric scientists had destroyed the comforting notion that humans could have no significant impact on the Earth and its life-sustaining capabilities. This made them targets of political retaliation, as the conservative revanchment of 1994 became entrenched in the corridors of power.

In his recent history of the American economics profession, Michael Bernstein has examined economists' increasing abandonment of their field's original public purpose, the use of economic knowledge to improve the well-being of the nation's citizenry, to become "mere shills for particular corporate elites eager to seize upon public assets now increasingly 'privatized.'"[6] This did not happen in atmospheric science. In keeping with a long-standing tradition in geophysics of conducting science for public benefit—as a form of public *service*—atmospheric scientists became increasingly involved in public policy matters, although not necessarily comfortably.[7] Robert Watson, Rich Stolarski, Robert Harriss, Joe Waters, and Dixon Butler all chose their science *because* it was a form of service. It offered them a chance to contribute to the common good.

Hence, while the economics profession abandoned public purpose during the 1980s and 1990s, driven by political imperatives for privatization and the market fundamentalism of American political culture, atmospheric scientists, and climate scientists more broadly, embraced it. This led to their increasing marginalization from public policy circles with their science under attack by politicians, lobbyists, and political propagandists.[8] During the 1970s, the physics community had been attacked for desiring ever-larger portions of the public's money for research facilities that appeared to have no practical value—their research was no longer relevant to the public sphere.[9] In the 2000s, the atmospheric science community's problem was the inverse. They had become *too* relevant. A lot of Americans (and some very wealthy businesses) did not *want* to hear the knowledge atmospheric scientists had created about the world. Relevance had become a double-edged sword, making atmospheric science vulnerable to political retaliation. Atmospheric scientists thus faced the question of how to maintain the vitality of their discipline with the nation's political culture turned strongly against them.

REALISM AND ANTI-REALISM IN THE EARTH SCIENCES

Over the past two decades or so, a putative war over the epistemic status of science has raged between a relative handful of scientists who continue to believe that science produces universal truth about the universe and some science studies scholars who argue that scientific knowledge is merely a social construct. And as a social construct, it has no inherent reality—scientific knowledge is no more true than any other form of knowledge. Thus, this line of argument goes, nonscientists accept it as true knowledge only because of credentialing.[10] Much of this so-called science war is overblown and based on substantial misunderstandings on both sides.[11] Social constructivists often argue from an anti-realist position because methodologically, it allows us (and I'm one) to investigate the biases of scientists and their sciences, be they methodological, cultural, religious, political, or gender biases.

At the same time, I haven't yet met a social constructivist who actually believes that there is no reality at all, the caricature depicted of us in some of the science wars literature. Instead, social constructivists believe scientific knowledge about the world is constructed through theory building, data collection, interpretation, publication, and negotiation. One can theorize and collect data in isolation, but one can never interpret, publish, and discuss in isolation from one's peers or one's larger society. It is in these social aspects of science that knowledge is constructed. Yet Earth scientists in particular collect data about *something*. There is an Earth, it has aspects we can measure, and thus scientists develop knowledge that relates to something real. The resulting knowledge is imperfect, contingent, and conditioned by the methodological practices of individual disciplines as well as by the cultures scientists are immersed in—institutional culture, national culture, and so on. Scientific knowledge is constructed, but nonetheless it's knowledge about something that actually exists.

The constructed nature of scientific knowledge allows it to be deconstructed—challenged—by those who don't like what that knowledge implies. These attacks serve to prolong doubt about the facticity of facts, permitting vested interests to gain delays in policy action. Indeed, it's possible for these attacks to effectively destroy a fact entirely, transforming that fact into an invalid mythology. Robert Proctor and Londa Schiebinger have documented cases of what Proctor calls "agnotology," the social construction of ignorance.[12] They show how business and religious leaders recruited scientists sympathetic to them in order to destroy pieces of undesired knowledge. One can see that process at work in the attacks on ozone and climate science. Knowledge that is constructed can be deconstructed. But the fact that knowledge can be destroyed does not mean that the knowledge in question is *wrong*.

The prevailing view of scientists in the 1950s was that Earth was basically stable and unchanging; during the period of this study, scientists developed a new view of Earth as a complex, dynamic, evolutionary system. But they did not simply dream this up. A series of discoveries about the Earth (and Venus and Mars) left their older narrative about planetary evolution untenable. (Recall that nobody expected craters on Mars—it was a shocking revelation.) Older narratives of the planets could no longer be reconciled with new facts about the solar system or Earth. But the new view certainly was not more amenable to scientists professionally, or to their status within the body politic. Indeed, as I have argued, their new view of the world was, and is, extremely controversial. It was controversial inside the scientific community in the 1960s and 1970s, and since then, while the internal controversy has been laid to rest, their new synthesis of climatic evolution has put them at odds with much of the U.S. political establishment.

Precisely because the new, dynamic view was controversial, scientists did not discard their more comfortable older view of the world lightly or carelessly. Indeed, as Oreskes has shown in her study of plate tectonics, scientists are generally resistant to new ideas. In order to make new knowledge, one often has to unmake old knowledge. Scientists, like any other scholarly group, have a great deal of time and effort invested in the old knowledge that is being threatened. So they do not abandon it lightly. The plate tectonics revolution she studied and the revolution in atmospheric science this book examines occurred simultaneously. In both cases, an accumulation of facts that could not be reconciled with older, more static conceptions forced change. In Oreskes' story, the availability of new measuring techniques enabled the acceptance of plate tectonics.[13] In atmospheric science, NASA was the source of new measurement techniques as well as the planetary view that spacecraft afforded. These new techniques, in company with other technological innovations in related scientific disciplines, enabled transformation of scientists' worldview.

Philosopher of science Ronald Giere argues that scientists produce knowledge by comparing new evidence to their existing conceptual models of how things are.[14] When new evidence does not fit the conceptual model, they can reject the data as somehow flawed (as Goddard's TOMS scientists did at first), go generate more data (as Farman's Antarctic Survey did), or discard the model and devise a new one that might fit the data (both new and old data) better. The last is what

Lovelock and his collaborators did in formulating the Gaia hypothesis. And it is this slow, piece-wise process of comparing data to conceptual models has led the American scientific community far away from its old, static view of Earth.

THE PLANETARY VIEW, OR, SCIENCE PRACTICE
IN THE LATE TWENTIETH CENTURY

The planetary view immersed NASA and the American Earth sciences in a political firestorm over ozone depletion and global warming. But it also altered the practice of American Earth science. Most historians, and nearly all scientists, believe that Humboldtian science ended in the nineteenth century. His holistic view, they think, did not carry into the twentieth century, dominated as it was by laboratory science, with its allegedly objective, controlled, limited-variable experiments. The very antithesis of holism, laboratory experiments seek to reduce complexity to a single, measurable, and controllable variable. The Earth cannot be reduced to a laboratory, and scientists of the non-physicist variety all know it. To Earth scientists, the world *is* the laboratory. It must be studied in its native environment, messy as that most definitely is. Regardless of scientists' general commitment to the ideals of objectivity, controllability, and experimental repeatability, Earth scientists have not escaped the need to venture into the field. Moreover, neither have planetary scientists, a core constituency of NASA.

NASA's scientific leaders sought to bring laboratory measurements into the field. They deployed field expeditions equipped with the latest measurement technologies to study various scientific questions and used the results to build new understandings of Earth processes. They also did them to support understanding of data collected by planetary spacecraft. Rudy Hanel's Infrared Interferometer Spectrometer (IRIS) instrument went to Earth and Mars, for example. James Lovelock used his understanding of planetary conditions to propose the radical view of Earth as an integrated, living organism. Shelby Tilford and Dixon Butler explicitly sought to bring the planetary view to Earth in the 1980s with the Earth Observing System (EOS). The planetary view, then, was a fundamental part of NASA's scientific efforts, even for its local field experiments.

Making sense of the planetary view exacerbated a traditional problem of Earth science research, the problem of scale. During the agency's efforts to help the Weather Bureau achieve global forecasting, it became clear fairly quickly that while global forecasts needed global data at one remove, the individual measurements had to be reduced in spatial scale to deal with the cloud-clearance problem. One could not find cloud-free scenes in low-spatial resolution datasets. As the agency entered the fields of chemistry and climate, the problem of scale became more pronounced, and the solutions changed. To definitively link rainforest soil emissions to the global atmosphere, for example, NASA had to construct a hybrid observing system consisting of ground-based, tower-based, aircraft-based, and finally space-based measurements. Research teams then used models to integrate the data from different scales, or used them to examine the level of consistency between datasets from different scales. This use of models for data integration became a powerful new tool for observation-based research in the late twentieth century.

As the weather observing and climate simulation studies done at the Goddard Institute for Space Studies (GISS) suggest, the use of models for simulation studies was another powerful tool for late twentieth-century science. Serving as machine-assisted thought experiments, they enabled scientists to perform experiments in virtual worlds that could reveal the relative importance of various processes.[15] The late 1960s simulation studies for the Global Atmospheric Research Program (GARP) demonstrated the utility of various kinds of observations for the satellite-based global weather observation system, while Hansen, Pollack, and others used climate models in the 1980s and 1990s to gradually pick apart and examine individually the feedbacks and forcings in the climate system. These model experiments provided hypotheses testable in the real world. One could build instruments and design field experiments to look for the high-altitude tropospheric ice proposed to occur in Ramanathan's supergreenhouse effect, or the changing cloud properties proposed by Hansen's study of the shrinking diurnal temperature variation over land, or the global cooling effect of Plinian volcanism, or the cloud radiative properties necessary for Lindzen's Iris effect. Hence during the four decades examined in this study, numerical models became vital research tools, enabling scientists to make sense of large and disparate datasets and to generate new hypotheses to inform future research.

Atmospheric scientists in the late twentieth century thus returned to the early nineteenth-century methodology of Humboldt, while greatly expanding upon it using the technologies of the post–World War II era: aircraft, radiometric sensors, and computer-driven models. They adopted the laboratory sciences of spectroscopy and kinetics virtually unchanged from physics, and applied them to new questions. While physics in the twentieth century pursued ever-smaller particles with ever-larger accelerators, atmospheric scientists took on the task of studying the global atmosphere. They learned to take laboratories into the field. And inspired by the planetary view, they set out to remake the Earth sciences into a single, integrated discipline around the concept of Earth System Science.

NASA's role in the American Earth sciences in the twentieth century, then, was revolutionary. Its embrace of the planetary view, its development and deployment of new measurement techniques, its support of theoretical and model studies, and its managerial activism in discipline-building and integration, has had enormous impacts on American science. It has played a fundamental role in altering our understanding of our world and of life's place in its evolution. There are many on the human space flight side of NASA's house who decry the loss of its historic mission with the cancellation of Project Apollo, perceiving an agency adrift and rudderless. They should pay more attention. While they were mourning the lost Moon, their agency rediscovered our own Earth.

Epilogue

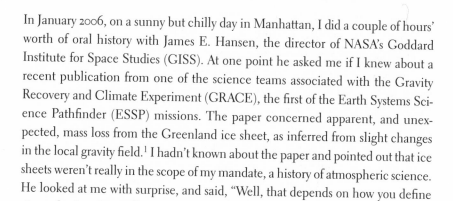

In January 2006, on a sunny but chilly day in Manhattan, I did a couple of hours' worth of oral history with James E. Hansen, the director of NASA's Goddard Institute for Space Studies (GISS). At one point he asked me if I knew about a recent publication from one of the science teams associated with the Gravity Recovery and Climate Experiment (GRACE), the first of the Earth Systems Science Pathfinder (ESSP) missions. The paper concerned apparent, and unexpected, mass loss from the Greenland ice sheet, as inferred from slight changes in the local gravity field.[1] I hadn't known about the paper and pointed out that ice sheets weren't really in the scope of my mandate, a history of atmospheric science. He looked at me with surprise, and said, "Well, that depends on how you define atmospheric science."[2]

Indeed. Despite the thickness of the tome you're holding, I've drawn my disciplinary lines rather narrowly, focusing on the gaseous part of atmospheric science, if you will. I've generally ignored the atmosphere's interaction with the Earth's land, water, and cryospheric surfaces, all of which happen to be claimed by other scientific disciplines. I may have made a mistake in being so narrow. But if I have, I'm not alone.

Hansen had a larger point that took me a few months to see. In March 2002, Earth scientists of many disciplines were shocked by the sudden collapse of the Larsen B ice shelf, a chunk of floating Antarctic ice the size of Rhode Island. Imagery from the Moderate Resolution Imaging Spectrometer (MODIS) instrument showed what had happened. Meltwater has a much lower albedo than does ice; meltwater ponds on the surface of Larsen B had acted as thermal drills, rapidly absorbing incoming sunlight, warming, and boring their way through to the base of the ice shelf. An ice shelf that climate models suggested would survive for decades thus disintegrated in weeks. Albedo has long been considered the domain of atmospheric scientists (recall that Vern Suomi was interested in the albedos of both cornfields and of Earth) but this particular effect of albedo was unexpected by that community.

This error in understanding ice and albedo matters, because the world's ice sheets hold enormous amounts of water. Glaciologist John H. Mercer had figured back in 1978 that the West Antarctic ice sheet held 5 to 6 meters' worth of sea level; the Greenland ice sheet holds slightly more.[3] If either of these ice sheets disintegrate, most of the world's coastal cities will be inundated, and hundreds of millions of people will be forced to migrate inland. The cost will be in the trillions of dollars. Climate models predict that those ice sheets will remain intact for centuries, but the Larsen B event showed that the models' treatment of ice sheet behavior is conceptually flawed. In a 2005 editorial spanning ten pages, Hansen took his own modeling community to task for this.[4] Climate models build ice sheets by depositing and compressing snow in a dry process. Modelers simply run this process in reverse to predict their breakup. What actually happens is what happened to Larsen B: meltwater drills through the ice sheet, destroying it quickly. But neither Hansen nor anyone else in his community has figured out how to model this process adequately.

This ice problem was not seen as a problem by the leaders of the climate science community until the Larsen B event. One will peruse the Intergovernmental Panel on Climate Change's (IPCC) 1995 and 2001 reports in vain for discussions of ice-enhanced rapid sea level rise. The IPCC, composed primarily of atmospheric scientists and climate modelers, laid out priority research areas that didn't touch on this ice problem. Since NASA based its downsized EOS on the IPCC's priorities, it didn't favor this research area either. Yet the issue of meltwater destruction of ice has been known to glaciologists for decades at the least; when Hansen started his campaign to get his modeler colleagues to pay more attention to ice, he used a photograph of a crevasse in the ice with a meltwater stream pouring into it to make his point clearer.[5] It was provided to him by Roger Braithwaite, University of Manchester, who took it during a field expedition. There have been field expeditions to glaciers and ice sheets since the nineteenth century; crevasses, and meltwater, have been known to science for quite a long time.[6] To put not too fine a point on it, Mercer's 1978 article was written *specifically* in the context of rapid Antarctic ice sheet melting induced by human carbon dioxide emissions.

There has clearly been a disconnect between the atmosphere and climate modeling community and glacier specialists over the past few decades. That's not particularly surprising. The Earth sciences have grown so enormously in the post-war era that the *Journal of Geophysical Research* publishes more than forty thousand pages per year, and there are many other Earth science journals. No one could possibly read it all, and scientists don't try. They read what's directly relevant to their own specialty, and, in reality, only a select few papers in their specialty. So while Francis Bretherton's 1986 committee formulated the concept of Earth System Science to try to break disciplinary boundaries down, the professional demands of science have the opposite tendency. Keeping a broad view is extraordinarily difficult.

The Larsen B event seems to have caused a number of scientists to broaden their view just a bit; the 2007 IPCC assessment released in February contained the following statement: "understanding of these effects is too limited to assess their likelihood or provide a best estimate or an *upper bound for sea level rise.*"[7] They specifically excluded sea level contributions from future "rapid dynamical changes in ice flow" in their sea level forecasts.[8] Because they cannot quantify the possibility of rapid deglaciation, they effectively assigned these possibilities zero value. In other words, after three decades of climate research by the world's most advanced nations, we cannot yet put a ceiling on the potential impact of rising seas. That's quite an assertion, given the potentially enormous cost of coastal inundation. Hansen seems to think that several meters of sea level rise may be possible this century.[9]

Much of the science discussed in the tome you're holding was *directed* in nature. In other words, it resulted from either carefully formulated, long-term research programs that were advocated by proponents for years, or, like the Upper Atmosphere Research Program (UARP), were effectively commanded into being by law. Such programs have achieved important results (else this book would be much thinner!). But as the above story suggests, there is still a strong element of random walk to modern science. Even the best scientists are not all seeing; IPCC didn't foresee the importance of ice dynamics in its first nineteen years of existence, so I was in excellent company.

From a science policy standpoint, then, it's clear that maintaining flexibility in long-term research programs is necessary. Despite vast knowledge and best intentions, American scientific leaders didn't quite see in 1988, or 1995, or 2001 what all the key scientific and policy challenges presented by global warming would turn out to be. Having resources to explore surprises seems essential. It isn't at all clear that NASA's budget outlook, at least, contains flexibility.

There's a lesson for historians here, too. In her discussion of Big Science, Catherine Westfall warned historians against allowing scientists to define our rhetoric and our research programs.[10] In a very real sense, I did exactly what she warned against in writing this book. I built it around a rather traditional, narrowly drawn definition of atmospheric science. In so doing, I've missed a great story about the belated discovery of ice-atmosphere interactions by atmospheric scientists. I hope some other historian will be able to do the topic justice in the future.

Notes

Epigraphs to Book: Project RAND, "Preliminary Design of an Experimental World-Circling Spaceship," RAND report SM-11827, 1946; Ponchitta Pierce, "Sagan: Dump Environmentally Unconscious Slobs," *Earth Summit Times*, 28 February 1992; reprinted in Tom Head, ed., *Conversations with Carl Sagan* (Jackson: University Press of Mississippi, 2006), 80.

INTRODUCTION

1. *The Day After Tomorrow*, Roland Emmerich, director, 20th Century Fox, 2004.

2. Wallace S. Broecker, Dorothy M. Peteet, and David Rind, "Does the Ocean-Atmosphere System Have More Than One Stable Mode of Operation?" *Nature* (2 May 1985): 21–26.

3. Andrew C. Revkin, "NASA Curbs Comments on Ice Age Disaster Movie," *New York Times*, 25 April 2004, electronic edition; Andrew J. Weaver and Claude Hillaire-Marcel, "Global Warming and the Next Ice Age," *Science* (16 April 2004): 400–402.

4. Erik M. Conway, *High Speed Dreams: The Technopolitics of Supersonic Transport Research* (Baltimore: Johns Hopkins University Press, 2005), 157–88.

5. The term *Big Science* was coined by Derek de Solla Price, *Little Science, Big Science* (New York: Columbia University Press, 1963). Recently, Catherine Westfall has argued that the concept is irrelevant: Westfall, "Rethinking Big Science: Modest, Mezzo, Grand Science and the Development of the Bevelac, 1971–1993," *Isis* 94, no. 1 (March 2003): 30–56. For a brief history of the demise of the Superconducting Super Collider, see Daniel J. Kevles, *The Physicists: The History of a Scientific Community in Modern America*, 2nd ed. (Cambridge, Mass.: Harvard University Press, 1995), ix–xlii.

6. Howard M. McCurdy, *The Space Station Decision: Incremental Politics and Technological Choice* (Baltimore: Johns Hopkins University Press, 1990); T. A. Heppenheimer, *The Space Shuttle Decision: NASA's Search for a Reusable Space Vehicle* (Washington, D.C.: NASA, 1999); T. A. Heppenheimer, *Development of the Space Shuttle, 1972–1981* (Washington, D.C.: Smithsonian Books, 2002); Lloyd S. Swenson, Jr., James M. Grimwood, and Charles C. Alexander, *This New Ocean: A History of Project Mercury* (Washington, D.C.: NASA, 1966); Roger Bilstein, *Stages to Saturn: A Technological History of the Apollo/Saturn* (Washington, D.C.: NASA, 1980). On

robotic space science, see, for example, Henry C. Dethloff and Ronald A. Schorn, *Voyager's Grand Tour: To the Outer Planets and Beyond* (Washington, D.C.: Smithsonian Books, 2003); Robert W. Smith, *The Space Telescope: A Study of NASA Science, Technology and Politics* (Cambridge: Cambridge University Press, 1989); Joseph Tatarewicz, *Space Technology and Planetary Astronomy* (Bloomington: Indiana University Press, 1990). For a complete list of NASA-sponsored histories, see http://history.nasa.gov/series95.html.

7. Pamela E. Mack, *Viewing the Earth: The Social Construction of the Landsat Satellite System* (Cambridge, Mass.: MIT Press, 1990).

8. Bruce Wielicki interview with author, 23 May 2005.

9. See Harvey Sapolsky, *Science and the Navy: The History of the Office of Naval Research* (Princeton, N.J.: Princeton University Press, 1990); Daniel J. Kevles, "The National Science Foundation and the Debate over Postwar Research Policy, 1942–1945," *Isis* 68 (1977): 5–26; J. Merton England, *A Patron for Pure Science: The National Science Foundation's Formative Years, 1945–1957* (Washington, D.C.: National Science Foundation, 1982); Jessica Wang, "Liberals, the Progressive Left, and the Political Economy of Postwar American Science: The National Science Foundation Debate Revisited," *HSPS* 26, no. 1 (1995): 139–66; Daniel Lee Kleinman, *Politics on the Endless Frontier: Postwar Research Policy in the United States* (Durham: Duke University Press, 1995). For the impact of the National Science Foundation, see Toby A. Appel, *Shaping Biology: The National Science Foundation and American Biological Research* (Baltimore: Johns Hopkins University Press, 2000); Dian Olson Belanger, *Enabling American Innovation: Engineering and the National Science Foundation* (West Lafayette, Ind.: Purdue University Press, 1998).

10. Walter A. McDougall, . . . *the Heavens and the Earth: A Political History of the Space Age* (New York: Basic Books, 1985), 228.

11. On the spacefaring vision, see Howard E. McCurdy, *Space and the American Imagination* (Washington, D.C.: Smithsonian Institution Press, 1997), 1–7, 29–51.

12. On supersonic transports, see Conway, *High-Speed Dreams*, 157–88; on nuclear power, see Thomas Raymond Wellock, *Critical Masses: Opposition to Nuclear Power in California, 1958–1978* (Madison: University of Wisconsin Press, 1998), 18–68; Carroll Pursell, "The Rise and Fall of the Appropriate Technology Movement in the United States, 1965–1985," *Technology and Culture* 34, no. 3 (July 1993): 629–37; Carroll Pursell, *The Machine in America: A Social History of Technology* (Baltimore: Johns Hopkins University Press, 1995), 299–319.

13. The classic work promoting alternative technologies is E. F. Schumaker, *Small is Beautiful: Economics as if People Mattered* (London: Abacus, 1974). For a recent study of counterculture technological thought, see Andrew Kirk, "Appropriating Technology: The Whole Earth Catalog and Counterculture Environmental Politics," *Environmental History* 6, no. 3 (July 2001): 374–94.

14. Kevles, *The Physicists*, 393–426.

15. J. Brooks Flippen, *Nixon and the Environment* (Albuquerque: University of New Mexico Press, 2000), details Nixon's role in environmental legislation. On anti-

environmentalism in the late 1990s, see Samuel P. Hays, *A History of Environmental Politics since 1945* (Pittsburgh: University of Pittsburgh Press, 2000), 118–21.

16. W. Henry Lambright, "NASA, Ozone, and Policy-Relevant Science," *Research Policy* (1995): 747–60.

17. Fletcher also generally supported conservation efforts. See Roger D. Launius, "A Western Mormon in Washington, DC: James C. Fletcher, NASA and the Final Frontier," *Pacific Historical Review* 64, no. 2 (1995): 217–41.

18. Conway, *High Speed Dreams*, 157–69.

19. Edward A. Parson, *Protecting the Ozone Layer: Science and Strategy* (Oxford: Oxford University Press, 2003); Richard Elliot Benedick, *Ozone Diplomacy: New Directions in Safeguarding the Planet* (Cambridge, Mass.: Harvard University Press, 1991); Maureen Christie, *The Ozone Layer: A Philosophy of Science Perspective* (New York: Cambridge University Press, 2000).

20. National Aeronautics and Space Administration Advisory Committee, *Earth Systems Science: A Program for Global Change* (Washington, D.C.: NASA, 1986). Also known as "The Bretherton Report."

21. The U.S. Global Change Research Program emphasized basic research over mitigation and adaptation strategies for climate change. See Roger A. Pielke, Jr., "Policy History of the U.S. Global Change Research Program: Part I. Administrative Development," *Global Environmental Change* 10 (2000): 9–25; Roger A. Pielke, Jr. "Policy History of the U.S. Global Change Research Program: Part II. Legislative Process," *Global Environmental Change* 10 (2000): 133–44.

22. Spencer Weart, *The Discovery of Global Warming* (Cambridge, Mass.: Harvard University Press, 2003); James Rodger Fleming, *Historical Perspectives on Climate Change* (New York: Oxford University Press, 1998).

23. James B. Pollack, "Climatic Change on the Terrestrial Planets," *Icarus* 37, no. 3 (1979): 479–553.

24. Samuel P. Hays, *Beauty, Health, and Permanence: Environmental Politics in the United States, 1955–1985* (New York: Cambridge University Press, 1986), 491–526; Samuel P. Hays, *A History of Environmental Politics since 1945* (Pittsburgh: University of Pittsburgh Press, 2000), 109–21.

25. U.S. Senate, Committee on Energy and Natural Resources, "Greenhouse Effect and Global Climate Change, part 2," 100th Cong., 1st sess., 23 June 1988, 44.

26. Chris C. Mooney, *The Republican War on Science* (New York: Basic Books, 2005), 78; Senator Inhofe's speech was given on the Senate floor on 28 July 2003, and reiterated on 3 January 2005. See http://inhofe.senate.gov/pressreleases/climateupdate .htm. The scientific consensus position as of 2001 is given in Intergovernmental Panel on Climate Change, *Climate Change 2001: Synthesis Report* (Cambridge: Cambridge University Press, 2001), 51. See also Naomi Oreskes, "The Scientific Consensus on Climate Change," *Science* (3 December 2004): 1686, doi 10.1126/science.1103618.

27. Kevles, *The Physicists*, 393–426.

28. Gregory A. Good, ed., *The Earth, the Heavens, and the Carnegie Institute of Washington*, vol. 5, *History of Geophysics* (Washington, D.C.: American Geophysical Union, 1994); in her study of continental drift, Oreskes discusses the relative decline

of field studies in the face of new methodologies. Naomi Oreskes, *The Rejection of Continental Drift: Theory and Method in American Earth Science* (New York: Oxford University Press, 1999), 290; a new interest in oceanography seems to be developing: Helen M. Rozwadowski and David K. van Keuren, eds., *The Machine in Neptune's Garden: Historical Perspectives on Technology and the Marine Environment* (New York: Science History Publications, 2004). Robert E. Kohler has done important work in exploring the relationship between laboratory and field biology: Robert E. Kohler, *Landscapes and Labscapes: Exploring the Lab-Field Border in Biology* (Chicago: University of Chicago Press, 2002).

29. Quoted in Oreskes, *The Rejection of Continental Drift*, 290.

30. Sharon Roan, *The Ozone Crisis* (New York: Wiley, 1989); Christie, *The Ozone Layer*, 52–65; and W. Henry Lambright, *NASA and the Environment: The Case of Ozone Depletion*, NASA SP-2005-4538, May 2005.

31. Susan Faye Cannon, *Science in Culture: The Early Victorian Period* (New York: Science History Publications, 1978).

32. Aaron Sachs, *The Humboldt Current: Nineteenth-Century Exploration and the Roots of American Environmentalism* (New York: Viking, 2006), 73–79.

33. Steven J. Dick, *The Biological Universe: The Twentieth-Century Extraterrestrial Life Debate and the Limits of Science* (Cambridge: Cambridge University Press, 1996), 120–26.

CHAPTER 1. ESTABLISHING THE METEOROLOGY PROGRAM

Epigraph: Harry Wexler, "The Circulation of the Atmosphere," *Scientific American* 193, no. 3 (1955): 114.

1. For NASA's genesis in the IGY, see Walter A. McDougall, . . . *the Heavens and the Earth: A Political History of the Space Age* (New York: Basic Books, 1985), 118–22, 157–76.

2. James Rodger Fleming, *Meteorology in America, 1800–1870* (Baltimore: Johns Hopkins University Press 1990), 141–62; Frederik Nebeker, *Calculating the Weather: Meteorology in the 20th Century* (New York: Academic Press, 1995), 36–39.

3. For Bjerknes's role in inspiring numerical prediction research, see Nebeker, *Calculating the Weather*, 47–56. On Bjerknes's career overall, see Robert Marc Friedman, *Appropriating the Weather: Vilhelm Bjerknes and the Construction of a Modern Meteorology* (Ithaca: Cornell University Press, 1989).

4. See Philip Duncan Thompson, "A History of Numerical Weather Prediction in the United States," *Bulletin of the American Meteorological Society* 64, no. 7 (July 1983): 7576–77; and Nebeker, *Calculating the Weather*, 58–82.

5. Nebeker, *Calculating the Weather*, 81–82.

6. Thompson, "A History of Numerical Weather Prediction in the United States," 757.

7. The most detailed study of Von Neumann's meteorological project is Kristine Harper, "Boundaries of Research: Civilian Leadership, Military Funding, and the International Network Surrounding the Development of Numerical Weather Predic-

tion in the United States" (PhD thesis, Oregon State University, 2003). See also William Aspray, *John von Neumann and the Origins of Modern Computing* (Cambridge, Mass.: MIT Press, 1990), 143–46; Nebeker, *Calculating the Weather*, 135–48.

8. Aspray, *John von Neumann and the Origins of Modern Computing*, 143–46.

9. Aspray, *John von Neumann and the Origins of Modern Computing*, 147–48. See also Nebbeker, *Calculating the Weather*, 156–58; Harper, "Boundaries of Research," 357–442.

10. Aspray, *John von Neumann and the Origins of Modern Computing*, 147–48. See also Nebbeker, *Calculating the Weather*, 156–58.

11. For a fuller discussion of the spread of general circulation model experimentation, see Paul N. Edwards, "A Brief History of Atmospheric General Circulation Modeling," in *General Circulation Model Development: Past, Present, and Future*, ed. David A. Randall (New York: Academic Press, 2000), 67–87.

12. Aspray, *John von Neumann and the Origins of Modern Computing*, 145.

13. W. W. Kellogg, R. R. Rapp, and S. M. Greenfield, "Close-In Fallout," *Journal of Meteorology* 14, no. 1 (February 1957): 1–8; William W. Kellogg, "Diffusion of Dust in the Stratosphere," *Journal of Meteorology* 13 (June 1956): 241–50.

14. David H. DeVorkin, *Science with a Vengeance: How the Military Created the US Space Sciences After World War II* (New York: Springer Verlag, 1992), 89–97.

15. William W. Kellogg, "Early Satellite Program Developments," in *The Conception, Growth Accomplishments, and Future of Meteorological Satellites*, NASA CP-2257, 1982, 1. The RAND study was "Preliminary Design of an Experimental World-Circling Spaceship," RAND report SM-11827, 1946.

16. DeVorkin, *Science with a Vengeance*, 4.

17. Berkner's role is detailed in Allan Needell, *Science, Cold War, and the American State* (Washington, D.C.: Smithsonian Institution Press, 2000).

18. Walter Sullivan, *Assault on the Unknown* (New York: McGraw Hill, 1961), 20–21.

19. The first International Polar Year had been 1882–83, and the second in 1932–33.

20. A. Hunter Dupree, *Science in the Federal Government: A History of Policies and Activities* (Baltimore: Johns Hopkins University Press, 1985), 135–41.

21. Constance McLaughlin Green and Milton Lomask, *Vanguard: A History* (Washington, D.C.: NASA, SP-4202, 1970), 97–98.

22. On the selection process, see John Naugle, *First Among Equals: The Selection of NASA Space Science Experiments* (Washington, D.C.: NASA SP-4325, 1991), 7–12; Green and Lomask, *Vanguard*, 113–32.

23. Harry Wexler to Joseph Kaplan, 31 May 1956, National Academy of Sciences Archives (hereafter NAS Archives), IGY series, "Earth Satellite Project 32, Radiation Balance of the Earth."

24. J. G. Reid to Members of the Working Group on Internal Instrumentation, 28 October 1956, and attachment, "A Proposal For Obtaining the Radiation Balance of the Earth from a Satellite," by V. E. Suomi, 17 October 1956, NAS Archives, IGY series, "Earth Satellite Project 32, Radiation Balance of the Earth."

25. J. G. Reid to Members, WGII, 15 February 1957, and attachment, NAS Archives, IGY series, "Earth Satellite Project 32, Radiation Balance of the Earth."

26. William W. Kellogg to Horace Byers et al., 7 October 1957, NAS Archives, IGY series, "Earth Satellite Project 32, Radiation Balance of the Earth."

27. Clayton Koppes, JPL and the American Space Program (New Haven: Yale University Press, 1982), 82–86.

28. James Van Allen, Origins of Magnetospheric Physics (Washington, D.C.: Smithsonian Institution Press, 1983), 54–56. See also Koppes, JPL and the American Space Program, 82–86.

29. Naugle, First Among Equals, 30.

30. See Reichelderfer to Odishaw, 6 June 1958; George Derbyshire to Hugh Odishaw, 11 June 1958, and attachment; Odishaw to Waterman, 12 June 1958; Odishaw to Waterman, 19 June 1958, NAS Archives, IGY series, file "Space Science Board: Beginning of Program 1958."

31. Naugle, First Among Equals, 31; and Minutes of the Second Meeting of the Board, 19 July 1958, NAS Archives, IGY series, file "Space Science Board, Meetings, Second."

32. Naugle, First Among Equals, 31–32.

33. McDougall, . . . the Heavens and the Earth, 228.

34. Minutes of the Third Meeting of the Space Science Board, 24–26 October 1958, NAS Archives, SSB collection, file "SSB: Meetings, Third"; see also Naugle, First Among Equals, 44–45.

35. Homer Newell, Beyond the Atmosphere: Early Years of Space Research (Washington, D.C.: NASA SP-4211, 1980), 205.

36. Silverstein to Glennan, 12 December 1958, "Re: Proposed NASA Policy on Space Flight Experiments," NASA History Office. Also Naugle, First Among Equals, 51–52.

37. Odishaw to Glennan et al., 1 February 1959, "Re: Recommendations of the Space Science Board," and attachment, National Archives and Records Administration, College Park (hereafter NARA II), RG 255, Records of the National Air and Space Administration, Homer Newell papers, box 21, file 112.

38. "National Space Sciences Program," 16 April 1959, NARA II, RG 255, Records of the National Aeronautics and Space Administration, Homer Newell papers, box 6, file 34.

39. Hearings on NASA Appropriations, House, Committee on Appropriations, Subcommittee on Independent Offices, 86th Cong., 1st sess., 29 April 1959, 110.

40. Hearings on NASA Appropriations, House, Committee on Appropriations, Subcommittee on Independent Offices, 86th Cong., 1st sess., 29 April 1959, 32.

41. Harry Wexler, "Meteorology," in Science in Space, ed. Lloyd V. Berkner and Hugh Odishaw (New York: McGraw-Hill, 1961), 147–48.

42. Sullivan, Assault on the Unknown, 411–13.

43. Edgar Cortright interview with author, 14 April 2003.

44. Douglas Aircraft Company, Inc., "Preliminary Design of an Experimental World-Circling Spaceship," Report No. SM-11827, 2 May 1946 (reprint ed. 1998).

45. Merton E. Davies and William R. Harris, "RAND's Role in the Evolution of Balloon and Satellite Observation Systems and Related U.S. Space Technology," RAND Corporation, Santa Monica, Calif., 1988, 25.

46. R. Cargill Hall, *A History of the Military Polar Orbiting Meteorological Satellite Program* (Washington, D.C.: National Reconnaissance Office, 2001), 2.

47. Richard Leroy Chapman, "A Case Study of the US Weather Satellite Program: The Interaction of Science and Politics" (PhD thesis, Syracuse University, 1967), 33.

48. Davies and Harris, "RAND's Role in the Evolution of Balloon and Satellite Observation Systems and Related U.S. Space Technology," 25.

49. Edgar Cortright to Attendees of June 20th meeting at NACA Headquarters re scientific satellite payloads, 24 June 1958, Re: Information presented concerning current and future status of meteorological payloads, Suitland Federal Records Center (hereafter Suitland FRC), RG 255-84-0637, box 1, file "Advanced Research Projects Agency."

50. Chapman, "A Case Study of the US Weather Satellite Program."

51. John Allen to T. Keith Glennan, 30 November 1960, Wexler papers, box 31, file 10; Hearings, U.S. Senate, "NASA Authorization for Fiscal Year 1961, part 2," 30 June 1960, 86th Cong., 2nd sess., 646.

52. Space Science Summer Study, Working Group on Meteorology, 29 June 1962, Berkner papers, box 68, "SSB-SS-Meteorological Rockets and Satellites." Also S. Fritz and H. Wexler, "Cloud Pictures from Satellite TIROS I," *Monthly Weather Review* 3, no. 88 (March 1960): 79–87.

53. Minutes, Committee on Meteorological Aspects of Satellites, 24–25 June 1960, NAS Archives, SSB collection, file "Committees, Meteorological Aspects of Satellites, Third Meeting."

54. William R. Bandeen and Warren P. Manger, "Angular Motion of the Spin Axis of the TIROS I Meteorological Satellite Due to Magnetic and Gravitational Torques," *Journal of Geophysical Research* 65, no. 9 (September 1960): 2992–95.

55. Precession is the change in orientation of a body's spin axis—the movement of a spinning top away from the vertical, for one example.

56. "U.S. Orbits Weather Satellite, It Televises Earth and Storms: New Era In Meteorology Seen," *New York Times*, 2 April 1960, 1.

57. "U.S. Orbits Weather Satellite, It Televises Earth and Storms," 4.

58. Wexler to Reichelderfer et al., 10 September 1958, Library of Congress (hereafter LOC), Wexler papers, box 31, file 11.

59. Berkner to Kistiakowsky, 29 June 1960, Wexler papers, box 31, file 11.

60. Reichelderfer to Glennan, 29 April 1960, Wexler papers, box 31, file 11.

61. Glennan to Frederick Mueller, 19 September 1960, Wexler papers, box 31, file 11.

62. Harry Wexler, "Draft Position on Letter and Enclosures from Dr. T. Keith Glennan dated 14 November 1960," Wexler papers, box 31, file 11.

63. Reichelderfer to Mr. Eberly, 25 November 1960, Wexler papers, box 31, file 11; David S. Johnson for the Record, 30 September 1963, Re: Withdrawal from Opera-

tional Satellite Program, NARA II, RG 51 Office of Management and Budget, series 57A, box 43, file 7.

64. Morris Tepper interview with author, 26 April 2003.

65. Edgar Cortright, 4 June 1959, Re: Some Preliminary Considerations of Meteorological Payloads (Nimbus) for the Vega, and attachment, William Stroud to Cortright, n.d., Suitland FRC, accession 255-84-0637, box 2, file "Flight Projects General #1."

66. Morris Tepper to File, 4 November 1960, Re: Meeting with Personnel of USWB this Date to Obtain Preliminary Briefing on USWB Plan for Operational Meteorological Satellite Program, Suitland FRC, accession 255-84-0637, box 2, file "Flight Projects General #1." On the Reichelderfer-Dryden friendship, see Chapman, "A Case Study of the US Weather Satellite Program," 90–91.

67. David S. Johnson for the Record, 30 September 1963, Re: Withdrawal from Operational Satellite Program, NARA II, RG 51 Office of Management and Budget, series 57A, box 43, file 7.

68. Chapman, "A Case Study of the US Weather Satellite Program," 202–3. Solar cells on earlier spacecraft had failed more quickly than expected, particularly those that had been in space during a series of high-altitude nuclear weapons tests. The relationship between the tests and a series of premature solar cell failures led satellite researchers to suspect that radiation was responsible, and Stroud's engineers had to seek a more resistant design. Chapman also assigns blame for the delays to the advanced camera system and a need to improve the horizon sensors that helped control spacecraft orientation.

69. Chapman, "A Case Study of the US Weather Satellite Program," 227.

70. Commerce and Finance Division to Files, 12 September 1963, Re: Nimbus Satellite Program, and attached routing slip dated 13 September 1963, NARA II, RG 51 Records of the Office of Management and Budget, series 57A, box 43, file 7.

71. Hall, A History of the Military Polar Orbiting Meteorological Satellite Program, 4.

72. Chapman, "A Case Study of the US Weather Satellite Program," 230; see also Hall, A History of the Military Polar Orbiting Meteorological Satellite Program, 4–5.

73. Commerce and Finance Division to Files, 12 September 1963, Re: Nimbus Satellite Program, NARA II, RG 51 Records of the Office of Management and Budget, series 57A, box 43, file 7.

74. J. Herbert Holloman to Robert C. Seamans, Jr., 27 September 1963; Secretary of Commerce (Luther Hodges) to Kermit Gordon, 2 October 1963, and attached memo by David S. Johnson, 30 September 1963; all NARA II, RG 51 Records of the Office of Management and Budget, series 57A, box 43, file 7.

75. Seamans to Holloman, 3 October 1963, NARA II, RG 51 Records of the Office of Management and Budget, series 57A, box 43, file 7.

76. Summary of Meeting on Meteorological Satellite Programs, 7 October 1963, NARA II, RG 51 Records of the Office of Management and Budget, series 57A, box 43, file 7.

77. For additional details on the operational satellite series of the late 1960s and

1970s, see James F. W. Purdom and W. Paul Menzel, "Evolution of Satellite Observations in the United States and Their Use in Meteorology," in *Historical Essays on Meteorology, 1919–1995*, ed. James Rodger Fleming (Boston: American Meteorological Society, 1996), 99–155.

78. Margaret Ellen Courain, "Technology Reconciliation in the Remote-Sensing Era of United States Civilian Weather Forecasting, 1957–1987" (PhD thesis, Rutgers, 1991), 254.

CHAPTER 2. DEVELOPING SATELLITE METEOROLOGY

Epigraph: Quoted from William W. Kellogg to the Record, 24 July 1958, NAS Archives, Space Science Board Collection, file "Committee on Meteorological Aspects of Satellites—Meetings, 1958–1960."

1. Frederik Nebekker, *Calculating the Weather: Meteorology in the 20th Century* (New York: Academic Press, 1995), 135–51.

2. William W. Kellogg to the Record, 24 July 1958, NAS Archives, Space Science Board Collection, file "Committee on Meteorological Aspects of Satellites—Meetings, 1958–1960."

3. William Nordberg, "Geophysical Observations from Nimbus I," *Science* 150 (29 October 1965): 559–72.

4. See Harry R. Glahn, "On the Usefulness of Satellite Infrared Measurements in the Determination of Cloud Top Heights and Areal Coverage," *Journal of Applied Meteorology* 5 (April 1966): 189–97.

5. Guenter Warnecke et al., "Remote Sensing of Ocean Currents and Sea Surface Temperature Changes Derived from the Nimbus II Satellite," *Journal of Physical Oceanography* 1 (January 1971): 45–60.

6. Linda Neumann Ezell, *NASA Historical Data Book*, vol. II, *Programs and Projects 1958–1968*, NASA SP-4012 (Washington, D.C.: NASA, 1988), 378–79, 392–93.

7. William W. Jones to the Record, 18 December 1963; M. I. Schneebaum to W. G. Stroud, 16 January 1964; M. I. Schneebaum to Harry J. Goett, 22 January 1964, Re: SMS Program, Suitland FRC, accession 255-75-684, box 2, file "SMS 1960–1965."

8. Richard Haley to Morris Tepper, 9 December 1964, Re: ATS Status, Suitland FRC, Accession 255-75-684, box 2, file "SMS 1960–1965."

9. V. E. Suomi and R. J. Parent, "Initial Proposal to National Aeronautics and Space Administration for an ATS Technological Experiment," 28 September 1964, SSEC Pub. 64–00–01, Schwerdtfeger Library, Space Science and Engineering Center, Madison, Wis. Robert Parent was then chairman of the electrical engineering department at the University of Wisconsin.

10. Robert White to Director, National Weather Satellite Center (Johnson), 3 February 1963, Re: Comments on TOS Systems Status Report No. 2, dated 25 January and Satellite Program Review Board Meeting, SSEC Pub. 64–00–01, Schwerdtfeger Library, Space Science and Engineering Center, Madison, Wis.

11. See Vincent J. Oliver to ATS-3 Project Office, 1 April 1968, Re: Report on "Tornado Alert" Cooperative Project; David S. Johnson to Leonard Jaffe, 10 October

1968, both Suitland FRC, accession 255-75-684, box 2, file "SMS 1967–1968"; and Tetsuya Fujita to Morris Tepper, 15 June 1968; Morris Tepper to Tetsuya Fujita, 25 July 1968, Suitland FRC, accession 255-75-684, box 2, file "Advanced Technology satellite 1963–1968."

12. K. Ninomiya, "Dynamical Analysis of Outflow from Tornado-Producing Thunderstorms as Revealed by ATS III Pictures," *Journal of Applied Meteorology* 10 (April 1971): 275–94; Kozo Ninomiya, "Mesoscale Modification of Synoptic Situations from Thunderstorm Development as Revealed by ATS III and Aerological Data," *Journal of Applied Meteorology* 10, no. 6 (December 1971): 103–21.

13. R. Cecil Gentry, Tetsuya T. Futjita, and Robert C. Sheets, "Aircraft, Spacecraft, Satellite and Radar Observations of Hurricane Gladys, 1968," *Journal of Applied Meteorology* 9, no. 6 (December 1970): 837–49.

14. See, for example, C. L. Smith, E. J. Zipser, S. M. Daggupaty, and L. Sapp, "An Experiment in Tropical Mesoscale Analysis: Part 1," *Monthly Weather Review* 103 (October 1975): 878–92; D. W. Martin and V. E. Suomi, "A Satellite Study of Cloud Clusters over the Tropical North Atlantic Ocean," *Bulletin of the American Meteorological Society* 53 (1972): 135–56.

15. R. H. Simpson to Deputy Director for Operations, NESO, 12 September 1970, Suitland FRC, accession 255-75-0684, box 2, file "General 1970–1971."

16. Vincent E. Lally, "Satellite Satellites—A Conjecture on Future Atmospheric-Sounding Systems," *Bulletin of the American Meteorological Society* 41, no. 8 (August 1960): 429–32.

17. Elisabeth Lynn Halgren, *The University Corporation for Atmospheric Research and the National Center for Atmospheric Research: An Institutional History 1960–1970* (Boulder, Colo.: UCAR, 1974).

18. Vincent E. Lally interview with William Kellogg, 13 July 1993, American Meteorological Society/University Corporation for Atmospheric Research Tape Recorded Interview Project, NCAR Archives.

19. Vincent E. Lally, "Trial Balloons in the Southern Hemisphere," *Science* 155 (27 January 1967): 458.

20. Vincent E. Lally interview with William Kellogg, 13 July 1993, American Meteorological Society/University Corporation for Atmospheric Research Tape Recorded Interview Project, NCAR Archives.

21. Described in John E. Masterson, "An In Situ Measurement System for GARP Using Balloons, Buoys, and a Satellite," preprint, paper presented at Eighth Space Congress, *Progress in Meteorology*, Cocoa Beach, Fla., 19 April 1971. Copy in Records of USC-GARP, accession 78-005, file "USC-GARP FGGE Advisory Panel: TWERLE and FGGE," NAS Archives.

22. Director of Earth Observations Programs to Chairman, Space Science and Applications Steering Committee, 5 January 1970, Re: Amendment to the Nimbus IRLS Experiment, NASA History Office, file 006172, part 2.

23. Vincent E. Lally, Aubrey P. Schumann, and Richard J. Reed, "Superpressure Balloon Flights in the Tropical Stratosphere," *Science* 166, no. 3906 (7 November 1969): 738–39.

24. James K. Angell, "Air Motions in the Tropical Stratosphere Deduced from Satellite Tracking of Horizontally Floating Balloons," *Journal of the Atmospheric Sciences* 29 (April 1972): 570–82.

25. Pierre Morel and William Bandeen, "The EOLE Experiment: Early Results and Current Objectives," *Bulletin of the American Meteorological Society* 54, no. 4 (April 1973): 298–306.

26. Lewis D. Kaplan, "Inference of Atmospheric Structure from Remote Radiation Measurements," *Journal of the Optical Society of America* 49, no. 10 (October 1959): 1004–7.

27. D. Q. Wark and H. E. Fleming, "Indirect Measurements of Atmospheric Temperature Profiles from Satellites, I: Introduction," *Monthly Weather Review* 94, no. 6 (June 1966): 351–62.

28. D. G. James, "Indirect Measurements of Atmospheric Profiles from Satellites: IV. Experiments with the Phase 1 Satellite Infrared Spectrometer," *Monthly Weather Review* 95, no. 7 (July 1967): 457.

29. M. Wolk, F. Van Cleef, and G. Yamamoto, "Indirect Measurements of Atmospheric Temperature Profiles from Satellites: V. Atmospheric Soundings from Infrared Spectrometer Measurements at the Ground," *Monthly Weather Review* 95, no. 7 (July 1967): 463–67.

30. D. Q. Wark, F. Saiedy, and D. G. James, "Indirect Measurements of Atmospheric Temperature Profiles from Satellites: VI. High-Altitude Balloon Testing," *Monthly Weather Review* 95, no. 7 (July 1967): 468.

31. D. Q. Wark and D. T. Hilleary, "Atmospheric Temperature: Successful Test of Remote Probing," *Science* 165 (19 September 1969): 1256–58.

32. William L. Smith, " An Iterative Method for Deducing Tropospheric Temperature and Moisture Profiles from Satellite Radiation Measurements," *Monthly Weather Review* 95, no. 6 (June 1967): 363–69.

33. Rudolf A. Hanel, "The Infrared Interferometer Spectrometer (IRIS) Experiment," in *The Nimbus III User's Guide*, NASA CN-127.604, 109–26; R. Hanel and B. Conrath, "Interferometer Experiment on Nimbus 3: Preliminary Results," *Science* 165 (19 September 1969): 1258–60.

34. R. A. Hanel et al., "The Nimbus 4 Infrared Spectroscopy Experiment, I: Calibrated Thermal Emission Spectra," *Journal of Geophysical Research* 77, no. 15 (20 May 1972): 2629–41; see also V. Ramanathan, "The Role of Earth Radiation Budget Studies in Climate and General Circulation Research," *Journal of Geophysical Research* 92, no. D4 (20 April 1987): 4075–95.

35. M. T. Chahine, "A General Relaxation Method for Inverse Solution of the Full Radiative Transfer Equation," *Journal of the Atmospheric Sciences* (May 1972): 741–47; J. H. Shaw et al., "Atmospheric and Surface Properties from Spectral Radiance Observations in the 4.3 Micron Region," *Journal of the Atmospheric Sciences* 27 (August 1970): 773–80; Moustafa T. Chahine interview with author, 6 November 2003.

36. M. T. Chahine, H. H. Aumann, and F. W. Taylor, "Remote Sounding of Cloudy Atmospheres. III. Experimental Verifications," *Journal of the Atmospheric Sci-*

ences (May 1977): 758–65; Moustafa T. Chahine interview with author, 6 November 2003.

37. M. L. Meeks and A. E. Lilley, "The Microwave Spectrum of Oxygen in the Earth's Atmosphere," *Journal of Geophysical Research* 68, no. 6 (15 March 1963): 1683–1703.

38. A. H. Barrett and V. H. Chung, "A Method for the Determination of High-Altitude Water-Vapor Abundance from Ground-Based Microwave Observations," *Journal of Geophysical Research* 67, no. 11 (1962): 4259; Y. H. Katz, ed., "The Application of Passive Microwave Technology to Satellite Meteorology: A Symposium," RM-3401-NASA, August 1963, 57–75.

39. Katz, ed., "The Application of Passive Microwave Technology to Satellite Meteorology: A Symposium," 142.

40. W. B. Lenoir, "Remote Sounding of the Upper Atmosphere by Microwave Measurements" (PhD thesis, MIT, 1965).

41. William B. Lenoir, John W. Barrett, and D. Cosmo Papa, "Observations of Microwave Emission by Molecular Oxygen in the Stratosphere," *Journal of Geophysical Research* 73, no. 4 (15 February 1968): 1119–26.

42. David H. Staelin, "Measurements and Interpretation of the Microwave Spectrum of the Terrestrial Atmosphere near 1 Centimeter Wavelength," *Journal of Geophysical Research* 71, no. 2 (15 June 1966): 2875–81.

43. C. Catoe, W. Nordberg, P. Thaddeus, and G. Ling, "Preliminary Results from Aircraft Flight Tests of an Electrically Scanning Microwave Radiometer," NASA TM-X-622-67-352, August 1967.

44. Karl Hufbauer tells this story from the standpoint of solar constant measurements made by the Nimbus 6 and 7 Earth Radiation Budget instruments. See Karl Hufbauer, *Exploring the Sun: Solar Science since Galileo* (Baltimore: Johns Hopkins University Press, 1991), 272–303; also Herbert Lee Kyle interview with author, 1 March 2004; William L. Smith et al., "Nimbus 6 Earth Radiation Budget Experiment," *Applied Optics* 16, no. 2 (1977): 306–18.

45. M. Patrick McCormick interview with author, 18 April 2003; Edward V. Browell interview with author, 15 April 2005.

46. William L. Smith interview with author, 3 May 2003.

CHAPTER 3. CONSTRUCTING A GLOBAL METEOROLOGY

Epigraph: Edward N. Lorenz, "Predictability: Does the Flap of a Butterfly's Wings in Brazil Set off a Tornado in Texas?" presentation at the American Association for the Advancement of Science, Washington, D.C., 29 December 1972, reprinted in Edward N. Lorenz, *The Essence of Chaos* (Seattle: University of Washington Press, 1993), 181–84.

1. Committee on Atmospheric Sciences, "The Atmospheric Sciences 1961–1971," Pub. 946 (Washington, D.C.: National Academy of Sciences, 1962), vii–viii, 10.

2. Joint Organizing Committee, "An Introduction to GARP," GARP Publication

Series Nr. 1, October 1969, 15; National Academy of Sciences, "The Atmospheric Sciences, 1961–1971," Publication Nr. 946, Washington, D.C., 1961.

3. World Meteorological Organization, *One Hundred Years of International Cooperation in Meteorology, 1873–1973* (Geneva: World Meteorological Organization, 1973), 31–38. A predecessor, the International Meteorological Organization, had been in place prior to World War II.

4. Clark A. Miller, "Scientific Internationalism in American Foreign Policy: The Case of Meteorology, 1947–1958," in *Changing the Atmosphere: Expert Knowledge and Environmental Governance*, ed. Clark A. Miller and Paul N. Edwards (Cambridge, Mass.: MIT Press, 2001), 167–218.

5. James Gleick, *Chaos: Making a New Science* (New York: Viking, 1987), 18.

6. Morris Tepper to Deputy Administrator, 27 March 1964, Suitland FRC, accession 255-84-0637, box 2, file "Flight Projects—General #4."

7. Morris Tepper to Deputy Administrator, 27 March 1964, Suitland FRC, accession 255-84-0637, box 2, file "Flight Projects—General #4."

8. Joint Organizing Committee, "An Introduction to GARP," GARP Publication Series Nr. 1, October 1969, 15; National Academy of Sciences, "The Atmospheric Sciences, 1961–1971," Publication Nr. 946, Washington, D.C., 1961.

9. National Academy of Sciences–National Research Council, *Feasibility of a Global Observation and Analysis Experiment*, Publication number 1290, 1966.

10. "Minutes of the Third ICSU/IUGG/CAS Meeting, 1–3 March 1967," Suomi papers, file "COSPAR VI Meetings, Stockholm," Schwerdtfeger Library, Space Science and Engineering Center, Madison, Wis.

11. "Minutes of the Third ICSU/IUGG/CAS Meeting, 1–3 March 1967," Suomi papers, file "COSPAR VI Meetings, Stockholm," Schwerdtfeger Library, Space Science and Engineering Center, Madison, Wis.

12. Joint Organizing Committee, "An Introduction to GARP," GARP Publication Series Nr. 1, October 1969, 15; National Academy of Sciences, "The Atmospheric Sciences, 1961–1971," Publication Nr. 946, Washington, D.C., 1961.

13. ICSU/IUGG Committee on Atmospheric Sciences, "Study Conference on a Global Atmospheric Research Programme," 16 March 1967, Suomi papers, file "COSPAR VI Meetings, Stockholm," Schwerdtfeger Library, Space Science and Engineering Center, Madison, Wis.

14. Author interview with Milton Halem, 13 May 2003; author interview with Morris Tepper, 26 April 2003.

15. Author interview with Milton Halem, 13 May 2003; author interview with Morris Tepper, 26 April 2003; Memo dated 22 September 1970, NASA History Office, file "GARP/GATE Documentation, 009761"; John Naugle to Robert M. White, 31 August 1970, NASA History Office, file "GARP/GATE Documentation, 009761."

16. Author interview with Milton Halem, 13 May 2003.

17. Halem interview, 13 May 2003; Jule Charney, Milton Halem, and Robert Jastrow, "Use of Incomplete Historical Data to Infer the Present State of the Atmosphere," *Journal of the Atmospheric Sciences* 26, no. 5, part 2 (September 1969): 1160–63.

18. Charney, Halem, and Jastrow, "Use of Incomplete Historical Data to Infer the Present State of the Atmosphere," 1160–63.

19. Milton Halem and Robert Jastrow, "Analysis of GARP Data Requirements," *Journal of the Atmospheric Sciences* 27, no. 1 (January 1970): 177.

20. Halem interview, 13 May 2003.

21. Robert Jastrow and Milton Halem, "Simulation Studies Related to GARP," *Bulletin of the American Meteorological Society* 51, no. 6 (June 1970): 490, 498–501.

22. Jastrow and Halem, "Simulation Studies Related to GARP," 496.

23. Jastrow and Halem, "Simulation Studies Related to GARP," 495.

24. Summaries of the simulation studies are Akira Kasahara, "Simulation Experiments for Meteorological Observing Systems for GARP," *Bulletin of the American Meteorological Society* 53, no. 3 (March 1971): 252–64; Andre Robert and Robert Jastrow, "Observing System Simulation Studies: Report to the Joint Organizing Committee," 28 June–4 July 1972, manuscript, Jule Charney Papers, MC 184, box 10, folder 323, MIT Archives.

25. Edward N. Lorenz, "Deterministic Nonperiodic Flow," *Journal of the Atmospheric Sciences* (March 1963): 130–48. See also James Gleick's discussion of Lorenz's discovery: Gleick, *Chaos*, 18–21.

26. Andre Robert and Robert Jastrow, "Observing System Simulation Studies: Report to the Joint Organizing Committee," 28 June–4 July 1972, manuscript, Jule Charney Papers, MC 184, box 10, folder 323, MIT Archives.

27. Richard C. J. Somerville oral history, 22 February 2008; R. C. J. Somerville et al., "The GISS Model of the Global Atmosphere," *Journal of the Atmospheric Sciences* 31 (1974): 84–117; Somerville et al., "A Search for Short-Term Meteorological Effects of Solar Variability in an Atmospheric Circulation Model," *Journal of Geophysical Research* 81 (1976): 1572–76.

28. "Report on the First Session of the Study Group on Tropical Disturbances," Madison, Wis., 21 October–8 November 1968, Jule Charney papers, MC 184, box 4, folder 194, MIT Archives.

29. National Research Council, "Plan for U.S. Participation in the Global Atmospheric Research Program" (Washington, D.C.: National Academy of Sciences, 1969), 54–55.

30. The approval letter is Peter M. Flanigan to Secretary of Commerce, 16 March 1970, GARP collection, box 76-004, file "USC-GARP: Proposed US Participation Document Review 1970," National Academies Archives.

31. Numbers drawn from "GATE News U.S. Press Kit," 9 May 1974, file 009760 [GARP/GATE Documentation], NASA History Office.

32. Norman Durocher to Record, 22 December 1971, Re: Second Session of GARP Tropical Experiment Board, 14–16 December 1971, file 009761 [GARP/GATE Documentation], NASA History Office collection.

33. These were called Airborne Integrated Data Systems and were standard equipment on post-1970 transoceanic aircraft.

34. Paul R. Julian and Robert Steinberg, "Commercial Aircraft as a Source of

Automated Meteorological Data for GATE and DST," *Bulletin of the American Meteorological Society* 56, no. 2 (February 1975): 243–51.

35. Joachim P. Kuettner, "GATE Final International Scientific Plans: General Description and Central Program of GATE," *Bulletin of the American Meteorological Society* 55, no. 7 (July 1974): 711–19.

36. D. E. Parker, "The Synoptic-Scale Subprogram," *Bulletin of the American Meteorological Society* 55, no. 7 (July 1974): 720–23; D. R. Rodenhuis, "The Convection Subprogram," *Bulletin of the American Meteorological Society* 55, no. 7 (July 1974): 724–31; H. Hoeber, "The Boundary-Layer Subprogram for GATE," *Bulletin of the American Meteorological Society* 55, no. 7 (July 1974): 731–34; H. Kraus, "The Radiation Subprogram of GATE," *Bulletin of the American Meteorological Society* 55, no. 7 (July 1974): 734–37; S. G. H. Philander, "The Oceanographic Subprogram of GATE," *Bulletin of the American Meteorological Society* 55, no. 7 (July 1974): 738–44.

37. Joachim P. Kuettner and David E. Parker, "GATE: Report on the Field Phase," *Bulletin of the American Meteorological Society* 57, no. 1 (January 1976): 11–28.

38. James R. Greaves, "The Data Systems Tests: The Final Phase," *Bulletin of the American Meteorological Society* 60, no. 7 (July 1979): 791–94 gives the schedule as finally carried out.

39. On TWERLE, see The TWERLE Team, "The TWERL Experiment," *Bulletin of the American Meteorological Society* 58, no. 9 (September 1977): 936–48. The article is, unusually for this journal, attributed to the team instead of to the individual authors because the authors wished to dedicate it to two University of Wisconsin graduate students, Charles Blair and John Kruse, who perished in an aircraft accident during preparation for the experiment. The TWERLE science team was Paul Julian, Vincent Lally, William Kellogg, Vern Suomi, and Charles Cote.

40. Space Science and Engineering Center, *Space Capsule* 2, no. 3 (April 1972).

41. Thomas H. R. O'Neill to the Record, 28 November 1977, file "USC-GARP FGGE Advisory Panel, FGGE Quality Control," box no. 84-009-2, National Academies Archives.

42. A. Desmarais et al., "The NMC Report on the Data Systems Test," Camp Springs, Md., March 1978; Thomas H. R. O'Neill to Verner E. Suomi, 12 December 1977, box 84-009-2, file "USC-GARP FGGE Advisory Panel, FGGE Quality Control," National Academies Archives. See also M. Steven Tracton and Ronald D. McPherson, "On the Impact of Radiometric Sounding Data upon Operational Numerical Weather Prediction at NMC," *Bulletin of the American Meteorological Society* 58, no. 11 (November 1977): 1201–9; M. S. Tracton, A. J. Desmarais, R. J. Van Haaren, and R. D. McPherson, "The Impact of Satellite Soundings on the National Meteorological Center's Analysis and Forecast System—the Data Systems Test Results," *Bulletin of the American Meteorological Society* 108, no. 5 (May 1980): 543–86.

43. "Report on Informal FGGE Quality Control Meeting," typescript, 10–12 January 1978, box 84-009-2, file "USC-GARP FGGE Advisory Panel, FGGE Quality Control," National Academies Archives.

44. James R. Greaves et al., "A 'Special Effort' to Provide Improved Sounding and

Cloud-Motion Wind Data for FGGE," *Bulletin of the American Meteorological Society* 60, no. 2 (February 1979): 124–27.

45. U.S. Committee for GARP, "Record of Actions: Meeting of the U.S. Committee for the Global Atmospheric Research Program," 13–14 September 1973, 8 August 1974, file "USC-GARP Meeting 13–14 September 1973," box 84-009-2, National Academies Archives.

46. Verner E. Suomi to Russel C. Drew, 23 October 1973, file "USC-GARP Meeting 13–14 September 1973," box 84-009-2, National Academies Archives.

47. Seasat-A Press Kit, NASA History Office file 006352; also Linda Ezell, *NASA Historical Data Book*, vol. III, *Programs and Projects 1969–1978* (Washington, D.C.: NASA, 1985), 342–44.

48. L. Bengtsson, "Advances and Prospects in Numerical Weather Prediction," *Quarterly Journal of the Royal Meteorological Society* (July 1991, Part B): 855–902.

49. Bengtsson, "Advances and Prospects in Numerical Weather Prediction," 855–902; see also George Ohring, "Impact of Satellite Temperature Sounding Data on Weather Forecasts," *Bulletin of the American Meteorological Society* 60, no. 10 (October 1979): 1142–47.

50. For this argument, see J. R. Eyre and A. C. Lorenc, "Direct Use of Satellite Sounding Radiances in Numerical Weather Prediction," *Meteorology Magazine* 118 (1989): 13–16.

51. William L. Smith, "Atmospheric Soundings from Satellites—False Expectation or the Key to Improved Weather Prediction," *Quarterly Journal of the Royal Meteorological Society* 117 (1991): 267–97.

52. W. L. Smith, H. B. Howell, and H. M. Woolf, "The Use of Interferometric Radiance Measurements for Sounding the Atmosphere," *Journal of the Atmospheric Sciences* (April 1979): 566–75; W. L. Smith et al., "GHIS—The GOES High-Resolution Interferometer Sounder," *Journal of Applied Meteorology* (December 1990): 1189–1203.

53. Purdom and Menzel, "Evolution of Satellite Observations in the United States and their Use in Meteorology"; also National Research Council, *From Research to Operations in Weather Satellites and Numerical Weather Prediction: Crossing the Valley of Death* (Washington, D.C.: National Academy Press, 2000), 43–53.

54. A brief summary of NOAA's fortunes under Reagan is Robert Fleagle, *Global Environmental Change: Interactions of Science, Policy, and Politics in the United States* (Westport, Conn.: Praeger, 1994), 115–16. On the instrument upgrades to HIRS, see Purdom and Menzel, "Evolution of Satellite Observations in the United States and their Use in Meteorology," 99–155.

55. Spencer Weart, *The Discovery of Global Warming* (Cambridge, Mass.: Harvard University Press, 2003), 118.

CHAPTER 4. PLANETARY ATMOSPHERES

Epigraph: Conway Leovy, interview with author, 22 February 2006.

1. Steven J. Dick, *The Biological Universe: The Twentieth-Century Extraterrestrial*

Life Debate and the Limits of Science (Cambridge: Cambridge University Press, 1996), 120–26.

2. James R. Fleming, *Historical Perspectives on Climate Change* (New York: Oxford University Press, 1998), 113–28; see also Spencer Weart, *The Discovery of Global Warming* (Cambridge, Mass.: Harvard University Press, 2003), 24. Weart provides an expanded review of this subject on his Web site: www.aip.org/history/climate/index .html, accessed October 2005.

3. Weart, *The Discovery of Global Warming*, 20–38.

4. Ronald E. Doel, *Solar System Astronomy in America: Communities, Patronage, and Interdisciplinary Research, 1920–1960* (New York: Cambridge University Press, 1996), 57–62.

5. Ronald A. Schorn, *Planetary Astronomy: From Ancient Times to the Third Millennium* (College Station: Texas A&M University Press, 1998), 120.

6. Keay Davidson, *Carl Sagan: A Life* (New York: John Wiley & Sons, 1999), 102.

7. Davidson, *Carl Sagan*, 102.

8. Carl Sagan, "The Radiation Balance of Venus," JPL Technical Report No. 32–34, 15 September 1960.

9. William W. Kellogg and Carl Sagan, "The Atmospheres of Mars and Venus: A Report by the Ad Hoc Panel on Planetary Atmospheres of the Space Science Board," Publication no. 944 (Washington, D.C., 1961), 43–46.

10. Kellogg and Sagan, "The Atmospheres of Venus and Mars," 46.

11. F. T. Barath et al., "Mariner 2 Microwave Radiometer Experiment and Results," *The Astronomical Journal* 69, no. 1 (February 1964): 49–58.

12. Mikhail Ya. Marov and David H. Grinspoon, *The Planet Venus* (New Haven: Yale University Press, 1998), 40.

13. Andrew P. Ingersoll, "The Runaway Greenhouse: A History of Water on Venus," *Journal of the Atmospheric Sciences* 26 (November 1969): 1191–98.

14. S. I. Rasool and C. De Bergh, "The Runaway Greenhouse and the Accumulation of CO_2 in the Venus Atmosphere," *Nature* 226 (13 June 1970): 1037–39.

15. Marov and Grinspoon, *The Planet Venus*, 43–45.

16. Marov and Grinspoon, *The Planet Venus*, 63.

17. Marov and Grinspoon, *The Planet Venus*, 68–76.

18. Sill published first; Louise Young made the suggestion to her husband, astronomer Andrew T. Young, in whose name the resulting *Icarus* article was published although he clearly identified her as the source of the idea. See Schorn, *Planetary Astronomy*, 259.

19. James Pollack et al., "Aircraft Observations of Venus's Near-Infrared Reflection Spectrum: Implications for Cloud Composition," *Icarus* 23 (1974): 8–26; James Pollack et al., "A Determination of the Composition of the Venus Clouds from Aircraft Observations in the Near Infrared," *Journal of the Atmospheric Sciences* 32 (1975): 376–90.

20. James B. Pollack et al., "Calculations of the Radiative and Dynamical State of the Venus Atmosphere," *Journal of the Atmospheric Sciences* 32 (June 1975): 1025–37.

21. Richard O. Fimmel, Lawrence Colin, and Eric Burgess, *Pioneer Venus: A Planet Unveiled*, NASA SP-518, Ames Research Center, 1994, 18.

22. Fimmel, Colin, and Burgess, *Pioneer Venus*, 19–25, 39–40.

23. See Fimmel, Colin, and Burgess, *Pioneer Venus*, 192–93; Marov and Grinspoon, *The Planet Venus*, 314–27.

24. There are many histories of life on Mars. I used: William Sheehan, *The Planet Mars: A History of Observation and Discovery* (Tucson: University of Arizona Press, 1997), 98–113, 130–45; Schorn, *Planetary Astronomy*, 206–9.

25. Schorn, *Planetary Astronomy*, 208–9.

26. Schorn, *Planetary Astronomy*, 209.

27. Arvydas Kliore et al., "Occulation Experiment: Results of the First Direct Measurement of Mars's Atmosphere and Ionosphere," *Science* 149 (10 September 1965): 1243–48.

28. Crofton B. Farmer and Daniel D. LaPorte, "The Detection and Mapping of Water Vapor in the Martian Atmosphere," *Icarus* 16 (1972): 34–46; also Schorn, *Planetary Astronomy*, 210–11.

29. Edward Clinton Ezell and Linda Neuman Ezell, *On Mars: Exploration of the Red Planet 1958–1978*, NASA SP-4212, 1984, 346–47.

30. B. Conrath et al., "Atmospheric and Surface Properties of Mars Obtained by Infrared Spectroscopy on Mariner 9," *Journal of Geophysical Research* 78, no. 20 (10 July 1973): 4267–78.

31. William K. Hartmann and Odell Raper, *The New Mars: The Discoveries of Mariner 9*, NASA SP-337, 1974, 36–37.

32. R. P. Turco, O. B. Toon, T. P. Ackerman, J. B. Pollack, and Carl Sagan, "Nuclear Winter: Global Consequences of Multiple Nuclear Explosions," *Science* (23 December 1983): 1283–92; also Paul R. Ehrlich, Carl Sagan, Donald Kennedy, and Walter Orr Roberts, *The Cold and the Dark: The World after Nuclear War* (New York: W. W. Norton, 1984), 83–85. Also see Lawrence Badash, "Nuclear Winter: Scientists in the Political Arena," *Physics in Perspective* (2001): 76–105.

33. For example, see Bruce Murray, Michael Malin, and Ronald Greeley, *Earthlike Planets: Surfaces of Mercury, Venus, Earth, Moon, Mars* (San Francisco: S. H. Freeman, 1981), 113–22; Schorn, *Planetary Astronomy*, 242.

34. William K. Hartmann and Odell Raper, *The New Mars: The Discoveries of Mariner 9*, NASA SP-337, 1974, 94–103.

35. Carl Sagan, O. B. Toon, and P. J. Gierasch, "Climatic Change on Mars," *Science* 181 (14 September 1973): 1045–49; Carl Sagan and George Mullen, "Earth and Mars: Evolution of Atmospheres and Surface Temperatures," *Science* 177 (7 July 1972): 52–56.

36. By the mid-1980s, these missing glaciations had largely been found, and Milankovitch's analysis had been largely accepted. A detailed review of this from a scientific perspective is A. Berger, "Milankovitch Theory and Climate," *Reviews of Geophysics* 26, no. 4 (November 1988): 624–57. See also Weart, *The Discovery of Global Warming*, 46–50, 76–77.

37. Obliquity is the angle the rotational axis makes with the plane of a planet's

orbit. For both Earth and Mars, this changes slowly—the planets wobble like tops. For Earth, the period of the wobble is 41,000 years, for Mars, it is about 50,000.

38. Sagan, Toon, and Gierasch, "Climatic Change on Mars," 1045–49.

39. Clayton Koppes, *JPL and the American Space Program* (New Haven: Yale University Press, 1982), 228–30; Bruce Murray interview with Blaine Baggett, 7 February 2002.

40. C. B. Farmer and P. E. Doms, "Global Seasonal Variation of Water Vapor on Mars and the Implications for Permafrost," *Journal of Geophysical Research* 84, no. B6 (10 June 1979): 2881–88; see also Bruce M. Jakosky and Crofton B. Farmer, "The Seasonal and Global Behavior of Water Vapor in the Mars Atmosphere: Complete Global Results of the Viking Atmospheric Water Detector Experiment," *Journal of Geophysical Research* 87, no. B4 (10 April 1972): 2999–3019, and Crofton B. Farmer, "Liquid Water on Mars," *Icarus* 28 (1976): 279–89.

41. See, for example, Jaffrey S. Kargel, "Proof for Water, Hints of Life?" *Science* 306 (3 December 2004): 1689–91; S. W. Squyres et al., "In Situ Evidence for an Ancient Aqueous Environment at Meridiani Planum, Mars," *Science* 306 (3 December 2004): 1709–14.

42. Sagan and Mullen, "Earth and Mars," 52–56.

43. James E. Lovelock, *Gaia: A New Look at Life on Earth* (New York: Oxford University Press, 1974), 1–9.

44. Lovelock, *Gaia*, 6–7.

45. Lovelock, *Gaia*, 35–36.

46. Lovelock, *Gaia*, 35.

47. Lovelock, *Gaia*, 72.

48. J. E. Lovelock, "A Physical Basis for Life Detection Experiments," *Nature* 207 (1965): 568+; see also D. R. Hitchcock and J. E. Lovelock, "Life Detection by Atmospheric Analysis," *Icarus* 7, no. 2 (1967): 149+.

49. Lynn Margulis and J. E. Lovelock, "Biological Regulation of the Earth's Atmosphere," *Icarus* 21 (1974): 471–89; see also Lovelock, *Gaia*, 13–32.

50. Margulis and Lovelock, "Biological Regulation of the Earth's Atmosphere," 478–79. The stromatolite data they drew from P. E. Cloud, Jr., "Atmospheric and Hydrospheric Evolution on the Primitive Earth," *Science* 160 (1968): 729–36, and Cloud, "Premetazoa Evolution and the Origin of Metazoa," in *Evolution and Environment*, ed. E. T. Drake (New Haven: Yale University Press, 1968), 3–72.

51. Recent evidence suggests that the climate crisis that Lovelock thought should have happened actually did, and the Earth "went iceball" for between 35 and 100 million years in the late pre-Cambrian. Its recovery seems to have been a product of volcanic outgassing over eons raising the greenhouse effect sufficiently to break the iceball climate. See Robert E. Kopp et al., "The Paleoproterozoic Snowball Earth: A Climate Disaster Triggered by the Evolution of Oxygenic Photosynthesis," *Proceedings of the National Academy of Sciences* 102, no. 32 (9 August 2005): 11131–36.

52. Lovelock, *Gaia*, 10; Margulis and Lovelock, "Biological Regulation of the Earth's Atmosphere," 471.

53. Margulis and Lovelock, "Biological Regulation of the Earth's Atmosphere," 478–79.

54. Margulis and Lovelock, "Biological Regulation of the Earth's Atmosphere," 478–79.

55. Lovelock, *Gaia*, 48–63; Margulis and Lovelock, "Biological Regulation of the Earth's Atmosphere," 487.

56. Homer Newell, *Beyond the Atmosphere*, NASA SP-4211 (1980), 327–28.

CHAPTER 5. NASA ATMOSPHERIC RESEARCH IN TRANSITION

Epigraph: W. C. Wang et al., "Greenhouse Effects due to Man-Made Perturbations of Trace Gases," *Science* 194, no. 4266 (12 November 1976): 689.

1. Scott Hamilton Dewey, *Don't Breathe the Air: Air Pollution and U.S. Environmental Politics, 1945–1970* (College Station: Texas A&M University Press), 1–14.

2. See J. Brooks Flippen, *Nixon and the Environment* (Albuquerque: University of New Mexico Press, 2000) for a study of the politics behind these laws.

3. William Bryant, "The Re-Vision of Planet Earth: Space Flight and Environmentalism in Post-Modern America," *American Studies* (Fall 1995): 43–63; Sheila Jasanoff, "Image and Imagination: The Formation of Global Environmental Consciousness," in *Changing the Atmosphere: Expert Knowledge and Environmental Governance*, ed. Clark A. Miller and Paul N. Edwards (Cambridge, Mass.: MIT Press, 2001), 309–37.

4. MIT, *Man's Impact on the Global Environment: Report of the Study of Critical Environmental Problems* (Cambridge, Mass.: MIT Press, 1970).

5. Kellogg tells his own story in William W. Kellogg, "Mankind's Impact on Climate: The Evolution of an Awareness," *Climatic Change* 10, no. 2 (April 1987): 113–36.

6. James Rodger Fleming, *Historical Perspectives on Climate Change* (New York: Oxford University Press, 1998), 106–28. A somewhat different perspective is given by Spencer Weart, "The Discovery Of the Risk of Global Warming," *Physics Today* (January 1997): 34–40.

7. Kellogg, "Mankind's Impact on Climate," 121; MIT, *Man's Impact on the Global Environment*, 40–107.

8. See Syukuro Manabe and Richard T. Wetherald, "Thermal Equilibrium of the Atmosphere with a Given Distribution of Relative Humidity," *Journal of the Atmospheric Sciences* (May 1967): 241–59; MIT, *Man's Impact on the Global Environment*, 72–73.

9. MIT, *Man's Impact on the Global Environment*, 14.

10. Thomas Vonder Haar interview with author, 11 December 2002; Thomas H. Vonder Haar and V. E. Suomi, "Satellite Observations of the Earth's Radiation Budget," *Science* 163, no. 3868 (1969): 667+; Thomas H. Vonder Haar and Verner E. Suomi, "Measurement of the Earth's Radiation Budget from Satellites During a Five-Year Period. Part I: Extended Time and Space Means," *Journal of the Atmospheric Sciences* 28, no. 3 (April 1971). Vonder Haar's dissertation had first discussed the issue

in 1968: Vonder Haar, "Variations of the Earth's Radiation Budget" (PhD thesis, University of Wisconsin, 1968).

11. Julius London, "A Study of the Atmospheric Heat Balance: Final Report, Contract AF 19(122)-165," New York University, 1957.

12. MIT, *Inadvertent Climate Modification: Report of the Study of Man's Impact on Climate*, MIT 201 (MIT Press, 1971), 15–16.

13. National Aeronautics and Space Administration, *Remote Measurement of Pollution*, SP-285, Washington, D.C., 1971.

14. See, for example, Paul Crutzen and Veerabhadran Ramanathan, "The Ascent of Atmospheric Sciences," *Science* (13 October 2000): 299–304.

15. Summarized from David E. Newton, *The Ozone Dilemma* (Denver: ABC Clio, 1995), 25–26.

16. Halstead Harrison, "Stratospheric Ozone with Added Water Vapor: Influence of High-Altitude Aircraft," *Science* (13 November 1970): 734–36. Harrison told his own story in an online document: Halstead Harrison, "Boeing Adventures, with Digressions," 5 March 2003, copy in author's reference collection.

17. McDonald cited F. Urbach, ed., *The Biologic Effects of Ultraviolet Radiation* (New York: Pergamon Press, 1969), A. Hollaender, ed., *Radiation Biology*, vol. 2 (New York: McGraw Hill, 1965), and H. F. Blum, *Carcinogenesis by Ultraviolet Light* (Princeton, N.J.: Princeton University Press, 1959), as his sources for the UV-skin cancer link.

18. The *New York Times* reported the story, sans byline but complete with space aliens, and buried the piece far in the back. See "Scientist Calls SST Skin Cancer Hazard," *New York Times*, 3 March 1971, 87. On McDonald and the extraterrestrial hypothesis, see Steven J. Dick, *The Biological Universe* (New York: Cambridge University Press, 1996), 293–95.

19. Lydia Dotto and Harold Schiff, *The Ozone War* (Garden City, N.Y.: Doubleday, 1978). The authors credit the conference to Joe Hirschfelder from the University of Wisconsin, who believed that the Department of Transportation was trying to pull a snow job on the panel. See Dotto and Schiff, *The Ozone War*, 45.

20. Dotto and Schiff, *The Ozone War*, 45.

21. In atmospheric chemistry, NO_x represents the total of nitric oxide (NO) and nitrogen dioxide (NO_2) in a sample. There are many other oxides of nitrogen, but they are less active in ozone formation.

22. Dotto and Schiff discuss the conference in a great deal of detail. See Dotto and Schiff, *The Ozone War*, 39–68.

23. The paper is: Paul Crutzen, "The Influence of Nitrogen Oxides on the Atmospheric Ozone Content," *Quarterly Journal of the Royal Meteorological Society* 96 (1970): 320.

24. Two separate accounts of this sequence exist. The first is Dotto and Schiff, *The Ozone War*, 59–68. In a retrospective article published in 1992, Johnston gives a somewhat different account from his voluminous notes: Harold S. Johnston, "Atmospheric Ozone," *Annual Review of Physical Chemistry* (1992): 1–32. I have synthesized this account from both sources, relying on Johnston where they differ.

25. Senator Clinton P. Anderson to James Fletcher, 10 June 1971, Washington National Records Center Suitland, MD, accession 255-77-0677, box 52, file "NASA's Advanced SST Program, Catalytic Destruction of Ozone."

26. Herbert Friedman to Homer Newell, 1 September 1971, Suitland, accession 255-77-0677, box 52, file "NASA's Advanced SST Program, Catalytic Destruction of Ozone." The author of this letter was a researcher at the Naval Research Laboratory and chair of the Ad Hoc Panel on the NO_x-Ozone Problem, formed by the National Research Council on an administration request.

27. Harold Johnston, "Reduction of Stratospheric Ozone by Nitrogen Oxide Catalysts from Supersonic Transport Exhaust," *Science* (6 August 1971): 517–22.

28. Johnston, "Atmospheric Ozone," *Annual Review of Physical Chemistry* (1992): 23.

29. A. J. Grobecker, S. C. Coroniti, and R. H. Cannon, Jr., "The Effects of Stratospheric Pollution by Aircraft" (Washington, D.C.: Department of Transportation, December 1974), vi.

30. J. Mormino, D. F. Sola, and C. W. Patten, "Historical Perspective of the Climatic Impact Assessment Program," DOT-TST-76–41, December 1975.

31. Climate Impact Assessment Program, "The Natural Stratosphere of 1974: CIAP Monograph Number 1," DOT-TST-75–51, September 1975, 1-37–1-38. On the impulsive quality of the injection, see J. S. Chang and W. Duewer, "On the Possible Effect of NO_x injection in the Stratosphere due to Past Atmospheric Nuclear Weapons Tests," AIAA Paper 73–538, June 1973, 8; on the inability to find any weapons-testing related depletion, see P. Goldsmith, A. F. Tuck, J. S. Foot, E. L. Simmons, and R. L. Newson, "Nitrogen Oxides, Nuclear Weapon Testing, Concorde and Stratospheric Ozone," *Nature* (31 August 1973): 545–51.

32. A. J. Grobecker, S. C. Coroniti, and R. H. Cannon, Jr., "Report of Findings: The Effects of Stratospheric Pollution by Aircraft," DOT-TST-75–50, December 1974, 95.

33. On the CIAP report's reception, see Lydia Dotto and Harold Schiff, *The Ozone War* (Garden City, N.Y.: Doubleday, 1978), 69–89. On the Lewis Research Center's work, see Conway, *High Speed Dreams: The Technopolitics of Supersonic Transport Research* (Baltimore: Johns Hopkins University Press, 2005), 174–75; Richard Niedzwicki interview with Conway, 22 May 2001.

34. "SST Cleared on Ozone," *The Washington Post*, 25 January 1975.

35. T. M. Donahue, "SST and Ozone Depletion," *Science* 187 (1975): 1144–45; see also "Reply," *Science* 187 (1975): 1145; Dotto and Schiff, *The Ozone War*, 69–89.

36. Dotto and Schiff, *The Ozone War*, 76. Niedzwiecki achieved his expected reduction in nitrogen oxides emission almost three decades later. See Conway, *High Speed Dreams*, 238–42.

37. Morris Tepper to Associate Administrator, 16 August 1971, Re: Proposed Plan for a NASA Study of the Effects of Supersonic Flight on Ozone in the Lower Stratosphere; William C. Spreen, "Minutes of Meeting of the Ad Hoc Group on the Catalytic Destruction of Ozone," 1 October 1971; both NASA History Office, file "Atmosphere Ozone (through 1987), #009797."

38. Bill Grose interview, 3 April 2001, Langley Research Center historical archives.

39. See testimony of Charles W. Matthews, U.S. Senate, Committee on Aeronautical and Space Sciences, NASA *Authorization for Fiscal Year 1973*, part 2, 92nd Cong., 2nd sess., 22 March 1972, 798–801.

40. This was almost certainly prompted by the advocacy of three scientists within the agency who believed the Shuttle might have a problem: Robert Hudson of the Johnston Space Flight Center, James King at JPL, and I. G. Poppoff at Ames. See Dotto and Schiff, *The Ozone War*, 127. Hudson managed the contract with Stolarski and Cicerone, according to Stolarski. The Shuttle assessment is R. J. Cicerone, D. H. Stedman, R. S. Stolarski, A. N. Dingle, and R. A. Cellarius, "Assessment of Possible Environmental Effects of Space Shuttle Operations," NASA CR-129003, 3 June 1973.

41. Author interview with Richard Stolarski, 26 April 2001, LaRC Historical Archives; Johnston, "Atmospheric Ozone," 26–27.

42. Harvey Herring, "Research on Stratospheric Pollution," 13 February 1974, NASA HQ History Office, Low papers, file 13638.

43. Harvey Herring, "Research on Stratospheric Pollution," 13 February 1974, NASA HQ History Office, Low papers, file 13638.

44. M. J. Molina and F. S. Rowland, "Stratospheric Sink for Chlorofluoromethanes: Chlorine Atom Catalyzed Destruction of Ozone," *Nature* 249 (1974): 810–12; see also F. S. Rowland and Mario J. Molina, "Chlorofluoromethanes in the Environment," *Reviews of Geophysics and Space Physics* (February 1975): 1–35.

45. See also Committee on Aeronautical and Space Sciences, "Stratospheric Ozone Depletion: Hearings before the Subcommittee on the Upper Atmosphere," Senate, 94th Cong., 1st sess., 8 September 1975.

46. John E. Naugle to the Record, 10 December 1974, Re: Freon Problem, NASA History Office, file "Atmosphere Ozone (through 1977)," #009747; James Fletcher to H. Guyford Stever, 20 January 1975, Suitland FRC, RG 255, accession 80-0608, box 9, "Committees: NAS/NAE Climatic Impact Committee"; and Hearings before the Subcommittee on the Upper Atmosphere, Committee on Aeronautical Space Sciences, "Stratospheric Ozone Research and Effects," Senate, 94th Cong., 2nd sess., 25 February 1976.

47. Hearings before the Subcommittee on the Upper Atmosphere, Committee on Aeronautical Space Sciences, "Stratospheric Ozone Research and Effects," Senate, 94th Cong., 2nd sess., 25 February 1976, 25.

48. Stolarski interview, 26 April 2001.

49. Robert T. Watson interview with author, 14 August 2004. For an example of kinetics work, see R. T. Watson, "Rate Constants for Reactions of ClOx of Atmospheric Interest," *Journal of Physical and Chemical Reference Data* 6, no. 3 (1977): 871–917; Moustafa Chahine interview with author, 6 November 2003.

50. James G. Anderson interview with author, 25 November 2003.

51. M. Patrick McCormick interview with author, 18 April 2003.

52. M. Patrick McCormick interview with author, 18 April 2003.

53. "UARS Mission Instruments," NASA History Office, Washington, D.C.

54. "UARS Mission Instruments," NASA History Office, Washington, D.C.

55. Daniel Albritton interview with author, 31 October 2003; I. G. Poppoff, R. C. Whitten, R. P. Turco, and L. A. Capone, "An Assessment of the Effect of Supersonic Aircraft Operations on the Stratospheric Ozone Content," NASA RP-1026 (August 1978), 44–45.

56. Richard Elliot Benedick, *Ozone Diplomacy: New Directions in Safeguarding the Planet* (Cambridge, Mass.: Harvard University Press, 1991), 13.

57. Robert T. Watson interview with author, 14 April 2004.

58. WMO et al., "The Stratosphere 1981: Theory and Measurements," Geneva, May 1981, 3-53.

59. WMO et al., "The Stratosphere 1981: Theory and Measurements," Geneva, May 1981, 3-3–3-4; Robert T. Watson interview with author, 14 April 2004.

60. Tilford interview with author, 18 February 2004; N. W. Spencer to TIROS Project Manager, 25 July 1980, Suitland FRC, RG 255-87-298, box 4, file 819.6.

61. Crofton B. Farmer interview with author, 20 May 2004; Larry Simmons interview with author, 31 January 2005, JPL Archives.

62. R. T. Watson, H. K. Roscoe, and P. T. Woods, "The Balloon Intercomparison Campaigns: An Introduction and Overview," *Journal of Atmospheric Chemistry* 10 (1990): 99–110.

63. Farmer interview, 20 May 2004.

CHAPTER 6. ATMOSPHERIC CHEMISTRY

Epigraph: Quoted from Ponchitta Pierce, "Sagan: Dump Environmentally Unconscious Slobs," *Earth Summit Times*, 28 February 1992; reprinted in *Conversations with Carl Sagan*, ed. Tom Head (Jackson: University Press of Mississippi, 2006), 81.

1. Mars Observer Mission Failure Investigation Board, *Report of the Mars Observer Mission Failure Investigation Board*, vol. 1, 31 December 1993.

2. For an overview of this period, see Amy Paige Snyder, "NASA and Planetary Exploration," in *Exploring the Unknown: Selected Documents in the History of the U.S. Civil Space Program*, vol. V, *Exploring the Cosmos*, ed. John M. Logsdon et al., NASA SP-2001-4407 (Washington, D.C.: NASA, 2001), 280–93.

3. A brief summary of the shuttle program is Ray A. Williamson, "Developing the Space Shuttle," in *Exploring the Unknown: Selected Documents in the History of the U.S. Civil Space Program*, vol. IV, *Accessing Space*, ed. John M. Logsdon et al., NASA SP-4407 (Washington, D.C.: NASA, 1999), 161–93. On the Reagan administration's search for a Shuttle replacement, see Ivan Bekey, "Exploring Future Space Transportation Possibilities," in *Exploring the Unknown: Selected Documents in the History of the U.S. Civil Space Program*, vol. IV, *Accessing Space*, ed. John M. Logsdon et al., NASA SP-4407 (Washington, D.C.: NASA, 1999), 503–512. On Shuttle costs, see Roger Pielke, Jr. and Radford Byerly, Jr., "The Space Shuttle Program: Performance versus Promise," in *Space Policy Alternatives*, ed. Radford Byerly, Jr. (Boulder, Colo.: Westview Press, 1992), 223–45.

4. For a detailed discussion of the impact of shuttle delays on Galileo, see Bruce C. Murray, *Journey into Space: The First Thirty Years of Space Exploration* (New York: Norton, 1989), 203–20.

5. Jack Fishman and Paul J. Crutzen, "The Origin of Ozone in the Troposphere," *Nature* 274 (31 August 1978): 855–58.

6. Fishman and Crutzen, "The Origin of Ozone in the Troposphere," 855–58.

7. Robert J. McNeal interview with author, 25 February 2005.

8. Robert J. McNeal interview with author, 25 February 2005.

9. Robert J. McNeal interview with author, 25 February 2005; Robert C. Harriss interview with author, 31 January 2005.

10. National Research Council, *Global Tropospheric Chemistry: A Plan for Action* (Washington, D.C.: National Academy Press, 1984), ix.

11. National Research Council, *Global Tropospheric Chemistry*, 3, 7–8.

12. National Research Council, *Global Tropospheric Chemistry*, 22–25.

13. National Research Council, *Global Tropospheric Chemistry*, 26.

14. Robert J. McNeal, "NASA Global Tropospheric Experiment," *EOS* 64, no. 38 (20 September 1983): 561–62.

15. Gerald L. Gregory, James M. Hoell, Sherwin M. Beck, David S. McDougal, Jerome A. Meyers, and Dempsey B. Bruton, "Operational Overview of Wallops Island Instrument Intercomparison: Carbon Monoxide, Nitric Oxide, and Hydroxyl Instrumentation," *Journal of Geophysical Research* 90, no. D7 (20 December 1985): 12808–18.

16. Gregory, Hoell, Beck, McDougal, Meyers, and Bruton, "Operational Overview of Wallops Island Instrument Intercomparison," 12808–18; Hoell interview with author, 13 April 2005.

17. Harriss used the term ABLE to represent each geographic area: Atlantic Boundary Layer Experiment, Amazon Boundary Layer Experiment, Arctic Boundary Layer Experiment. McNeal used it to represent the Atmospheric Boundary Layer Experiment. See Harriss interview with author, 31 January 2005.

18. Edward V. Browell interview with author, 15 April 2005.

19. R. C. Harriss et al., "Atmospheric Transport of Pollutants from North America to the North Atlantic Ocean," *Nature* 308 (19 April 1984): 722–24.

20. H. G. Reichle et al., "Middle and Upper Tropospheric Carbon Monoxide Mixing Ratios as Measured by a Satellite-Borne Remote Sensor during November 1981," *Journal of Geophysical Research—Atmospheres* 91, no. D10 (20 September 1986): 865–67.

21. McNeal interview with author; see also J. Fishman, P. Minnis, and H. G. Reichle, "Use of Satellite Data to Study Tropospheric Ozone in the Tropics," *Journal of Geophysical Research—Atmospheres* 91, no. D13 (20 December 1986): 14451–65.

22. G. W. Sachse et al., "Carbon Monoxide over the Amazon Basin During the 1985 Dry Season," *Journal of Geophysical Research* 93, no. D2 (20 February 1988): 1422–30, and R. C. Harriss et al., "The Amazon Boundary Layer Experiment (ABLE 2A): Dry Season 1985," *Journal of Geophysical Research* 93, no. D2 (20 February 1988): 1351–60.

23. R. C. Harris et al., "The Arctic Boundary Layer Expedition (ABLE 3A): July–August 1988," *Journal of Geophysical Research* 97, no. D15 (30 October 1992): 16383–94.

24. Harris et al., "The Arctic Boundary Layer Expedition (ABLE 3A)," 16383–94.

25. Harris et al., "The Arctic Boundary Layer Expedition (ABLE 3A)," 16383–94.

26. Harris et al., "The Arctic Boundary Layer Expedition (ABLE 3A)," 16383–94.

27. On SAFARI, see Meinrat O. Andreae, Jack Fishman, and Janette Lindesay, "The Southern Tropical Atlantic Region Experiment (STARE): Transport and Atmospheric Chemistry near the Equator—Atlantic (TRACE A) and Southern African Fire-Atmosphere Research Initiative (SAFARI): An Introduction," *Journal of Geophysical Research* 101, no. D10 (30 October 1996): 23519–20.

28. Jack Fishman et al., "NASA GTE TRACE A Experiment (September–October 1992)," *Journal of Geophysical Research* 101, no. D19 (30 October 1996): 23865–79.

29. Henry Stommel and Elizabeth Stommel, "The Year without a Summer," *Scientific American* (June 1979): 176–202.

30. James B. Pollack, Owen B. Toon, Carl Sagan, Audrey Summers, Betty Baldwin, and Warren Van Camp, "Volcanic Explosions and Climatic Change: A Theoretical Assessment," *Journal of Geophysical Research* 81, no. 6 (20 February 1976): 1071–83; also O. Brian Toon interview with author, 13 February 2004. The group detailed their sulfate aerosols work in R. C. Whitten, O. B. Toon, and R. P. Turco, "The Stratospheric Sulfate Aerosol Layer: Processes, Models, Observations, and Simulations," *Pure and Applied Geophysics* 118 (1980): 86–127.

31. M. P. McCormick, Patrick Hamill, T. J. Pepin, W. P Chu, T. J. Swissler, and L. R. McMaster, "Satellite Studies of the Stratospheric Aerosol," *Bulletin of the American Meteorological Society* 60, no. 9 (September 1979): 1038–46.

32. McCormick, Hamill, Pepin, Chu, Swissler, and McMaster, "Satellite Studies of the Stratospheric Aerosol," 1038–46.

33. M. Patrick McCormick, G. S. Kent, G. K. Yue, and D. M. Cunnold, "Stratospheric Aerosol Effects from Soufriere Volcano as Measured by the SAGE Satellite System," *Science* (4 June 1982): 1115–18.

34. "Researchers Track Volcanic Plume," *Langley Researcher*, 30 May 1980, 4–5; M. P. McCormick interview with author, 7 April 2004.

35. See W. I. Rose, Jr. and M. F. Hoffman, "The 18 May 1980 Eruption of Mount St. Helens: The Nature of the Eruption, with an Atmospheric Perspective," in *Atmospheric and Potential Climatic Impact of the 1980 Eruptions of Mount St. Helens*, ed. Adarsh Deepak, NASA CP-2240, 1–14; Owen B. Toon, "Volcanoes and Climate," in *Atmospheric and Potential Climatic Impact of the 1980 Eruptions of Mount St. Helens*, ed. Adarsh Deepak, NASA CP-2240, 15–36; E. C. Y. Inn, J. F. Vedder, E. P. Condon, and D. O'Hara, "Precursor Gases of Aerosols in the Mount St. Helens Eruption Plumes at Stratospheric Altitudes," in *Atmospheric and Potential Climatic Impact of the 1980 Eruptions of Mount St. Helens*, ed. Adarsh Deepak, NASA CP-2240, 47–54; M. P. McCormick, "Ground-Based and Airborne Measurements of Mount St. Helens Stratospheric Effluents," in *Atmospheric and Potential Climatic Impact of the 1980 Eruptions of Mount St. Helens*, ed. Adarsh Deepak, NASA CP-2240, 125–30;

R. P. Turco, O. B. Toon, R. C. Whitten, R. G. Keese, and P. Hamill, *Atmospheric and Potential Climatic Impact of the 1980 Eruptions of Mount St. Helens*, ed. Adarsh Deepak, NASA CP-2240, 161–90.

36. Michael R. Rampino and Stephen Self, "The Atmospheric Effects of El Chichon," *Scientific American* (January 1984): 48–57.

37. M. P. McCormick et al., "Airborne and Ground-Based Lidar Measurements of the El Chichon Stratospheric Aerosol from 90N to 65S," *Geofisica Internacional* 23–22 (1984): 187–221.

38. William G. Mankin and M. T. Coffey, "Increased Stratospheric Hydrogen Chloride in the El Chichon Cloud," *Science* (12 October 1984): 170–72.

39. David J. Hofmann, "Perturbations to the Global Atmosphere Associated with the El Chichon Volcanic Eruption of 1982," *Reviews of Geophysics* 25, no. 4 (May 1987): 743–59.

40. Matthias Dörries, "In the Public Eye: Volcanology and Climate Change Studies in the 20th Century," *Historical Studies in the Biological and Physical Sciences* 37, no. 1 (2006): 87–124.

41. The ozone hole was discovered in 1982 but the Survey did not publish their findings until 1985. See J. C. Farman, B. G. Gardiner, and J. D. Shanklin, "Large Losses of Total Ozone in Antarctica reveal seasonal ClO_x/NO_x interaction," *Nature* 315 (16 May 1985): 207–10.

42. Sharon L. Roan, *Ozone Crisis: The 15 Year Evolution of a Sudden Global Emergency* (New York: Wiley & Sons, 1989), 131.

43. Tuck interview with author, 3 November 2003.

44. Edward A. Parson, *Protecting the Ozone Layer: Science and Strategy* (New York: Oxford University Press, 2003), 84–85.

45. Roan, *Ozone Crisis*, 132.

46. Joe Waters interview with author, 8 February 2005.

47. Roan, *Ozone Crisis*, 137–39.

48. Roan, *Ozone Crisis*, 138–39.

49. Adapted from Parson, *Protecting the Ozone Layer*, 149.

50. Parson, *Protecting the Ozone Layer*, 147–50; Roan, *Ozone Crisis*, 138–39.

51. Michael R. Gunson and Rodolphe Zander, "An Overview of the Relevant Results from the ATMOS Missions of 1985 and 1992," in *The Role of the Stratosphere in Global Change*, ed. Marie-Lise Chanin (Berlin: Springer-Verlag, 1993), 387–401; G. C. Toon, C. B. Farmer, and R. H. Norton, "Detection of Stratospheric N_2O_5 by Infrared Remote Sounding," *Nature* (13 February 1986): 570–71. Also Jack Kaye interview with author, 12 April 2004.

52. Adrian Tuck interview with author, 3 November 2003.

53. NO_y represents reactive odd nitrogen compounds, including the NO_x group and its oxidation products.

54. Sharon Roan describes this controversy in her popular book, *Ozone Crisis*, 173–79.

55. See, for example, the introduction to the AAOE mission definition document: NASA, *Airborne Antarctic Ozone Experiment*, Ames Research Center MS 245-5 (July

1987), 1–2. Copy from NASA HQ History Office, file "Airborne Antarctic Ozone Experiment."

56. Tuck interview with author, 3 November 2003.

57. Roan, *Ozone Crisis*, 212–13; see also Steven Hipskind interview with author, 18 December 2001; Adrian Tuck interview with author, 3 November 2003.

58. A. F. Tuck, R. T. Watson, E. P. Condon, J. J. Margitan, and O. B. Toon, "The Planning and Execution of ER-2 and DC-8 Aircraft Flights Over Antarctica, August and September 1987," *Journal of Geophysical Research* 94, no. D9 (30 August 1989): 11181–222.

59. J. G. Anderson interview with author, 26 November 2003.

60. J. G. Anderson, W. H. Brune, and M. H. Proffitt, "Ozone Destruction by Chlorine Radicals within the Antarctic Vortex: The Spatial and Temporal Evolution of ClO-O$_3$ Anti-correlation Based on in situ ER-2 Data," *Journal of Geophysical Research* 94, no. D9 (30 August 1989): 11465–479.

61. D. W. Fahey et al., "Measurements of Nitric Oxide and Total Reactive Nitrogen in the Antarctic Stratosphere: Observations and Chemical Implications," *Journal of Geophysical Research* 94, no. D14 (30 November 1989): 16665–82; G. C. Toon et al., "Infrared Aircraft Measurements of Stratospheric Composition Over Antarctica During September 1987," *Journal of Geophysical Research* 94, no. D14 (30 November 1989): 16571–98.

62. Parson, *Protecting the Ozone Layer*, 151.

63. Parson, *Protecting the Ozone Layer*, 150; G. D. Hayman et al., "Kinetics of the Reaction ClO+ClO→Products and its Potential Relevance to Antarctic Ozone," *Geophysical Research Letters* 13, no. 12 (November Supp. 1986): 1347–50; R. A. Cox and G. D. Hayman, "The Stability and Photochemistry of Dimers of the ClO Radical and Implications for Antarctic Ozone Depletion," *Nature* (28 April 1988): 796–806.

64. Parson, *Protecting the Ozone Layer*, 152. Also see G. P. Brasseur et al., "Group Report: Changes in Antarctic Ozone," in *The Changing Atmosphere*, ed. F. S. Rowland and I. S. A. Isaksen (New York: John Wiley & Sons, 1988), 235–58.

65. On this, see Parson, *Protecting the Ozone Layer*, 142; Watson interview with author, 14 April 2004.

66. See Benedick, *Ozone Diplomacy*, 88.

67. World Meteorological Organization, *Atmospheric Ozone 1985: Assessment of Our Understanding of the Processes Controlling Its Present Distribution and Change*, Global Ozone Research and Monitoring Project Report No. 16, 1986, 1–26, 605–48. See also Parson, *Protecting the Ozone Layer*, 68–69.

68. Parson, *Protecting the Ozone Layer*, 156–57 makes this point very clearly.

69. Parson, *Protecting the Ozone Layer*, 153.

70. Parson, *Protecting the Ozone Layer*, 154–55.

71. Robert T. Watson, F. Sherwood Rowland, and John Gille, "Ozone Trends Panel Press Conference," NASA Headquarters, 15 March 1988, Langley Research Center document CN-157273, 1988; see also "Executive Summary of the Ozone Trends Panel," 15 March 1988, Langley Research Center doc. CN-157277, 1988.

72. W. H. Brune, D. W. Toohey, J. G. Anderson, W. L. Starr, J. F. Vedder, and

E. F. Danielsen, "In Situ Northern Mid-Latitude Observations of ClO, O₃, and BrO in the Wintertime Lower Stratosphere," *Science* (28 October 1988): 558–62.

73. S. Solomon et al., "Observations of the Nighttime Abundance of OClO in the Winter Stratosphere above Thule, Greenland," *Science* (28 October 1988): 550–55.

74. Tuck interview with author, 3 November 2003.

75. Richard Turco, Alan Plumb, and Estelle Condon, "The Airborne Arctic Stratospheric Expedition: Prologue," *Geophysical Research Letters* 17, no. 4 (March 1990, Supplement): 313–16.

76. Turco, Plumb, and Condon, "The Airborne Arctic Stratospheric Expedition: Prologue," 313–16; W. H. Brune et al., "In Situ Observations of ClO in the Arctic Stratosphere: ER-2 Aircraft Results from 59N to 80N Latitude," *Geophysical Research Letters* 17, no. 4 (March 1990 Supplement): 505–8; D. S. McKenna et al., "Calculations of Ozone Destruction during the 1988/1989 Arctic Winter," *Geophysical Research Letters* 17, no. 4 (March 1990, Supplement): 553–56; M. H. Proffitt et al., "Ozone Loss in the Arctic Polar Vortex Inferred from High-Altitude Aircraft Measurements," *Nature* 347 (6 September 1990): 31–36.

77. Parson, *Protecting the Ozone Layer*, 206.

78. J. R. Herman et al., "A New Self-Calibration Method Applied to TOMS and SBUV Backscattered Ultraviolet Data to Determine Long-Term Global Ozone Change," *Journal of Geophysical Research* 96, no. D4 (20 April 1991): 7531–45; Richard Stolarski et al., "Total Ozone Trends Deduced from Nimbus 7 TOMS Data," *Geophysical Research Letters* 18, no. 6 (June 1991): 1015–18.

79. Richard Stolarski et al., "Measured Trends in Stratospheric Ozone," *Science* (17 April 1992): 342–49.

80. James G. Anderson and Owen B. Toon, "Airborne Arctic Stratospheric Expedition II: An Overview," *Geophysical Research Letters* 20, no. 22 (19 November 1993): 2499–2502; see also Adrian Tuck interview with author, 3 November 2003.

81. Anderson and Toon, "Airborne Arctic Stratospheric Expedition II," 2499–2502; see also Adrian Tuck interview with author, 3 November 2003.

82. For a review of the state of aerosol science, see Susan Solomon, "Stratospheric Ozone Depletion: A Review of Concepts and History," *Reviews of Geophysics* (August 1999): 275–316. For the Browell work, see Edward V. Browell et al., "Ozone and Aerosol Changes during the 1991–1992 Airborne Arctic Stratospheric Expedition," *Science* (27 August 1993): 1155–58.

83. Solomon, "Stratospheric Ozone Depletion," 293–95.

84. Rush Limbaugh, *The Way Things Ought to Be* (New York: Pocket Books, 1993), 154–57; R. A. Maduro and R. Schauerhammer, *The Holes in the Ozone Scare* (Washington, D.C.: 21st Century Science Associates, 1992); and S. Fred Singer, "My Adventures in the Ozone Layer," *National Review* (30 June 1989): 37.

85. See Parson, *Protecting the Ozone Layer*, 215.

86. On the non-measurement of CFCs in the stratosphere, see R. Maduro, "That the Sky Isn't Falling," *Executive Intelligence Review* 27, no. 18 (1989): 19; R. S. Bennett, *The Wall Street Journal*, 24 March 1993, A-15; and D. L. Ray and L. Guzzo, *Environmental Overkill* (Washington, D.C.: Regnery Gateway, 1993), 35. On the insistence of

volcanic (and sea spray) origin for most stratospheric chlorine, see Maduro and Schauerhammer, *The Holes in the Ozone Scare*; and Singer, "My Adventures in the Ozone Layer," 37.

87. F. Sherwood Rowland, "President's Lecture: The Need for Scientific Communication with the Public," *Science* (11 June 1993): 1571–76.

88. J. W. Waters et al., "Stratospheric ClO and Ozone from the Microwave Limb Sounder on the Upper-Atmosphere Research Satellite," *Nature* 362 (15 April 1993): 597–602.

89. R. Zander et al., "Increase of Carbonyl Fluoride in the Stratosphere and Its Contribution to the 1992 Budget of Inorganic Fluorine in the Upper Stratosphere," *Journal of Geophysical Research* 99, no. D8 (20 August 1994): 16737–43.

90. M. R. Gunson et al., "Increase in Levels of Stratospheric Chlorine and Fluorine Loading between 1985 and 1992," *Geophysical Research Letters* 21, no. 20 (1 October 1994): 2223–26.

91. See also N. Oreskes, K. Shraderfrechette, and K. Belitz, "Verification, Validation, and Confirmation of Numerical Models in the Earth Sciences," *Science* 263 (4 February 1994): 641–46.

CHAPTER 7. THE QUEST FOR A CLIMATE OBSERVING SYSTEM

Epigraph: Roger Revelle and Hans E. Suess, "Carbon Dioxide Exchange between Atmosphere and Ocean and the Question of an Increase of Atmospheric CO_2 during the Past Decades," *Tellus* 9 (1957): 19–20.

1. Robert G. Fleagle, *Global Environmental Change: Interactions of Science, Policy, and Politics in the United States* (Westport, Conn.: Praeger, 1994), 95; Mark Bowen, *Thin Ice: Unlocking the Secrets of Climate in the World's Highest Mountain* (New York: Henry Holt, 2005), 132–33.

2. John D. Cox, *Climate Crash: Abrupt Climate Change and What it Means for Our Future* (Washington, D.C.: Joseph Henry Press, 2005), 55; Reid A. Bryson, "A Perspective on Climatic Change," *Science* (17 May 1974): 753–60; Spencer Weart, *The Discovery of Global Warming* (Cambridge, Mass.: Harvard University Press, 2000), 71–75; S. J. Johnsen et al., "Oxygen Isotope Profiles through the Antarctic and Greenland Ice Sheets," *Nature* 235 (25 February 1972): 429–34; W. Dansgaard et al., "One Thousand Centuries of Climatic Record from Camp Century on the Greenland Ice Sheet," *Science* (17 October 1969): 377–80.

3. J. Hansen et al., "Climate Modeling in the Global Warming Debate," in *General Circulation Model Development*, ed. David A. Randall (New York: Academic Press, 2000), 127–64.

4. Hansen credits Kiyoshi Kawabata with this demonstration. See Hansen et al., "Climate Modeling in the Global Warming Debate," 130. The 1976 study was W. C. Wang et al., "Greenhouse Effects due to Man-Made Perturbations of Trace Gases," *Science* (12 November 1976): 685–90. Also see Hansen interview with author, 19 January 2006.

5. Hansen et al., "Climate Modeling in the Global Warming Debate," 131–32.

6. James E. Hansen interview with author, 20 January 2006. The resulting publication is National Research Council, *Carbon Dioxide and Climate: A Second Assessment* (Washington, D.C.: National Academy Press, 1982).

7. National Research Council, *Carbon Dioxide and Climate*, Washington, D.C., 1979, vii.

8. J. Hansen et al., "Climate Impact of Increasing Carbon Dioxide," *Science* 213, no. 4511 (28 August 1981): 957–66.

9. Hansen et al., "Climate Impact of Increasing Carbon Dioxide," 966.

10. Hansen interview with author, 19 January 2006; also see Bowen, *Thin Ice*, 133–36; Weart, *Discovery of Global Warming*, 143–44.

11. J. Hansen et al., "Climate Sensitivity: Analysis of Feedback Mechanisms," *Climate Processes and Climate Sensitivity*, Geophysical Monograph 29 (Washington, D.C.: American Geophysical Union, 1984), 130–63.

12. Hansen et al., "Climate Sensitivity," 130–63; David Rind and Dorothy Peteet, "Terrestrial Conditions at the Last Glacial Maximum and CLIMAP Sea Surface Temperature Estimates: Are They Consistent?" *Quaternary Research* 24 (1985): 1–22. This has been discussed recently in a very good popular work as well: Bowen, *Thin Ice*, 253, 277–78; on CLIMAP, also see John Imbrie and Katherine Palmer Imbrie, *Ice Ages: Solving the Mystery* (Cambridge, Mass.: Harvard University Press, 1979), 162–67.

13. Hansen et al., "Climate Sensitivity," 131 (emphasis in original).

14. The study had chosen a fixed fifteen-year response time for the oceans, which had the effect of biasing the warming low. See Hansen et al., "Climate Sensitivity," 131; and Carbon Dioxide Assessment Committee, *Carbon Dioxide and Climate* (Washington, D.C.: National Academy of Science Press, 1983).

15. Luis W. Alvarez, *Alvarez: Adventures of a Physicist* (New York: Basic, 1987), 252–58.

16. Trevor Palmer, *Perilous Planet Earth: Catastrophes and Catastrophism through the Ages* (New York: Cambridge University Press, 2003), 55–59; Stephen Jay Gould, "Uniformity and Catastrophe," in *Ever Since Darwin: Reflections in Natural History* (New York: W. W. Norton, 1977), 147–52; A. Hallam, *Great Geological Controversies*, 2nd ed. (New York: Oxford University Press, 1989), 30–64.

17. Alvarez, *Alvarez*, 262.

18. Frances Fitzgerald, *Way Out There In the Blue: Reagan, Star Wars and the End of the Cold War* (New York: Simon & Schuster, 2000), 179–82; also see Lawrence Badash, "Nuclear Winter: Scientists in the Political Arena," *Physics in Perspective* 3 (2001): 76–105.

19. David S. Meyer, *A Winter of Discontent: The Nuclear Freeze Movement and American Politics* (Westview, Conn.: Praeger, 1990), 69–74, discusses the impact of nuclear war rhetoric on the formation of the nuclear freeze movement.

20. Fitzgerald, *Way Out There In the Blue*, 179–82.

21. Paul J. Crutzen and John W. Birks, "The Atmosphere after a Nuclear War: Twilight At Noon," *Ambio* 118, nos. 2–3 (1982): 114–25; Paul R. Ehrlich, Carl Sagan, Donald Kennedy, and Walter Orr Roberts, *The Cold and the Dark: The World after Nuclear War* (New York: W. W. Norton, 1984), 83–85.

22. R. P. Turco, O. B. Toon, T. P. Ackerman, J. B. Pollack, and Carl Sagan, "Nuclear Winter: Global Consequences of Multiple Nuclear Explosions," *Science* (23 December 1983): 1283–92; also Ehrlich, Sagan, Kennedy, and Roberts, *The Cold and the Dark*, 83–85.

23. Turco, Toon, Ackerman, Pollack, and Sagan, "Nuclear Winter," 1283–92.

24. Turco, Toon, Ackerman, Pollack, and Sagan, "Nuclear Winter," 1292.

25. Ehrlich, Sagan, Kennedy, and Roberts, *The Cold and the Dark*, xiii–xvii.

26. Ehrlich, Sagan, Kennedy, and Roberts, *The Cold and the Dark*, xiii–xvii.

27. Ehrlich, Sagan, Kennedy, and Roberts, *The Cold and the Dark*, 89–94.

28. Ehrlich, Sagan, Kennedy, and Roberts, *The Cold and the Dark*, 100–101.

29. Turco, Toon, Ackerman, Pollack, and Sagan, "Nuclear Winter," 1283–92; Paul R. Ehrlich et al., "Long-Term Biological Consequences of Nuclear War," *Science* (23 December 1983): 1293–1300; Ehrlich, Sagan, Kennedy, and Roberts, *The Cold and the Dark*.

30. William D. Carey, "A Run Worth Making," *Science* (23 December 1983): 1281.

31. See, for example, Lawrence Badash, *Scientists and the Development of Nuclear Weapons* (Atlantic Highlands, N.J.: Humanities Press, 1995).

32. On Sagan's activism, see Lawrence Badash, "Nuclear Winter: Scientists in the Political Arena," *Physics in Perspective* (2001): 76–105.

33. Draft Proposal for the George C. Marshall Institute, sent to Bill Nierenberg, December 12, 1984, box 75, file 6, MC 13, William Aaron Nierenberg Papers, Scripps Institution of Oceanography Archives (hereafter SIO Archives), La Jolla, Calif.

34. National Aeronautics and Space Administration, *Climate Observing System Studies: An Element of the NASA Climate Research Program*, NASA TM-84040, September 1980.

35. National Aeronautics and Space Administration, *Climate Observing System Studies*, 1–24.

36. National Aeronautics and Space Administration, *Climate Observing System Studies*; Reid A. Bryson, "A Perspective on Climatic Change," *Science* (17 May 1974): 753–60; Karl Hufbauer, *Exploring the Sun: Solar Science since Galileo* (Baltimore: Johns Hopkins University Press, 1991), 284–92; T. H. Vonder Haar and W. H. Wallschlaeger, *Design Definition Study of the Earth Radiation Budget Satellite System* (Fort Collins: Colorado State University, 1978).

37. National Aeronautics and Space Administration, *Climate Observing System Studies*, 13–15.

38. National Aeronautics and Space Administration, *Climate Observing System Studies*, xv.

39. National Aeronautics and Space Administration, *Climate Observing System Studies*, 101–4.

40. National Research Council, *Earth Observations From Space: History, Promise, Reality* (Washington, D.C.: National Academies Press, 1995), 104–6; Colin Norman, "Reagan Budget Would Reshape Science Policies," *Science* 211, no. 4489 (27 March 1981): 1399–1402; Colin Norman, "Science Budget: Coping with Austerity," *Science*

216, no. 4535 (19 February 1982): 944–48; Colin Norman, "Turbulent Times for NOAA," *Science* 226, no. 4679 (7 December 1984): 1172–74; M. Mitchell Waldrop, "What Price Privatizing Landsat," *Science* 219, no. 4585 (11 February 1983): 752–54.

41. Howard McCurdy, *The Space Station Decision: Incremental Politics and Technological Choice* (Baltimore: Johns Hopkins University Press, 1990), 40–41, 66–67.

42. Deputy Administrator (Hans Mark) to Administrator (James Beggs), 25 March 1982, box 6, Earth Observing System collection, NASA HQ History Office; Shelby Tilford to Deputy Associate Administrator, 12 March 1982, box 6, Earth Observing System collection, NASA HQ History Office.

43. David Atlas to Shelby Tilford, 6 April 1982 box 6, Earth Observing System collection, NASA HQ History Office; David Atlas to Cosmic Radiations Branch/Dr. Steve Holt, 6 April 1982, Re: Proposal for an Applications Oriented space station, box 6, Earth Observing System collection, NASA HQ History Office.

44. JPL, "Global Change: Impacts on Habitability," JPL doc. D-95, 7 July 1982, 1 (hereafter Goody Report).

45. Goody Report, 3.

46. Goody Report, 10.

47. Goody Report, 9.

48. Goody Report, 11.

49. Kenneth Pederson to Associate Administrator for Space Science and Applications (Burt Edelson), 1 July 1982, Re: Global Habitability as a U.S. Initiative at UNI SPACE 82, box 6, Earth Observing System Collection, NASA HQ History Office.

50. Office of Technology Assessment, "UNISPACE 1982: A Context for International Cooperation and Competition," Washington, D.C., 1983.

51. Shelby Tilford interview with author, 11 February 2004.

52. Associate Administrator for Space Science and Applications Special Assistant to Associate Administrator for Space Science and Applications, 31 August 1982, Re: Earth Remote Sensing Satellite System, box 9, Earth Observing System collection, NASA HQ History Office; P. G. Thome, 1 September 1982, Re: System Z (partially handwritten notes), box 9, Earth Observing System collection, NASA HQ History Office.

53. NASA, *From Pattern to Process: The Strategy of the Earth Observing System, EOS Science Steering Committee Report v. II*, Washington, D.C., 1984, 103–10.

54. NASA, *Earth Systems Science: Overview, A Program for Global Change*, Report of the Earth System Sciences Committee, 1986, 4.

55. NASA, *Earth Systems Science*, 15–17.

56. Task Group on Earth Sciences, "Space Science in the Twenty-First Century: Mission to Planet Earth," National Academy Press, 1988, 7.

57. McCurdy, *The Space Station Decision*, 169–76.

58. Roger A. Pielke, Jr. has recounted this story from the slightly different perspective of the foundation of the U.S. Global Change Research Program. See Roger A. Pielke, Jr., "Policy History of the US Global Change Research Program: Part II. Legislative Process," *Global Environmental Change* 10 (2000): 133–44; Roger A. Pielke, Jr.,

"Policy History of the US Global Change Research Program: Part I. Administrative Development," *Global Environmental Change* 10 (2000): 9–25.

59. Burt Edelson to Distribution, 24 June 1983, box 6, Earth Observing System Collection, NASA HQ History Office; Committee on Global Change, *Toward an Understanding of Global Change: Initial Priorities for U.S. Contributions to the International Geosphere-Biosphere Program* (Washington, D.C.: National Academies Press, 1988).

60. Committee on Global Change, *Toward an Understanding of Global Change*; this was also the subject of the famous work by Roger Revelle and Hans Suess in the 1950s. See James Rodger Fleming, *Historical Perspectives on Climate Change* (New York: Oxford University Press, 1998), 124–28.

61. F. Webster and M. Fieux, "TOGA Overview," in *NATO Advanced Research Workshop on Large Scale Oceanographic Experiments and Satellites*, ed. C. Gautier and M. Flieux (Boston: D. Reidel Publishing Company, 1984), 17–24.

62. Senate, Hearings, Committee on Energy and Natural Resources, "Greenhouse Effect and Global Climate Change, part 2," 100th Cong., 1st sess., 23 June 1988, 1.

63. Senate, Hearings, Committee on Energy and Natural Resources, "Greenhouse Effect and Global Climate Change, part 2," 100th Cong., 1st sess., 23 June 1988, 39–41.

64. Senate, Hearings, Committee on Energy and Natural Resources, "Greenhouse Effect and Global Climate Change," 100th Cong., 1st sess., 9 November 1987, 52.

65. J. Hansen et al., "Global Climate Changes as Forecast by Goddard Institute for Space Studies Three-Dimensional Model," *Journal of Geophysical Research* 93, no. D8 (20 August 1988): 9341–64.

66. Senate, Hearings, Committee on Energy and Natural Resources, "Greenhouse Effect and Global Climate Change, part 2," 100th Cong., 1st sess., 9 November 1987, 54.

67. Senate, Hearings, Committee on Energy and Natural Resources, "Greenhouse Effect and Global Climate Change, part 2," 100th Cong., 1st sess., 9 November 1987, 51, 64.

68. The transcript contains the pre-print of a paper submitted to the *Journal of Geophysical Research*; the published paper is J. Hansen et al., "Global Climate Changes as Forecast by Goddard Institute for Space Studies Three-Dimensional Model," *Journal of Geophysical Research* 93, no. D8 (20 August 1988): 9341–64.

69. Senate, Hearings, Committee on Energy and Natural Resources, "Greenhouse Effect and Global Climate Change, part 2," 100th Cong., 1st sess., 23 June 1988, 21–30.

70. Anthony Ramirez, "A Warming World: What it will Mean," *Fortune* (4 July 1988): 102–6, reprinted in Senate, Hearings, Committee on Energy and Natural Resources, "Greenhouse Effect and Global Climate Change, part 2," 100th Cong., 1st sess., 23 June 1988, 24–29.

71. Richard A. Kerr, "Hansen vs. the World on the Greenhouse Threat," *Science* 244 (2 June 1989): 1041–43.

72. J. T. Houghton, G. J. Jenkins, and J. J. Ephraums, eds., *Climate Change: The IPCC Scientific Assessment* (New York: Cambridge University Press, 1990), iii, v.

73. Committee on Earth Sciences, "Our Changing Planet: A U.S. Strategy for Global Change Research," n.d. (but 1989).

74. U. S. Senate, Committee on Commerce, Science, and Transportation, Hearing, "National Global Change Research Act of 1989," 22 February 1989, 1–4.

75. Leslie Roberts, "Global Warming: Blaming the Sun," *Science* 246 (24 November 1989): 992–93; William A. Nierenberg to the Editor, *Science* (January 1990): 14. On the Marshall Institute's foundation, see Robert Jastrow to Robert Walker, 1 December 1986, box 21, file George C. Marshall Institute, accession 2001-01, William A. Nierenberg Papers, SIO Archives; Draft Proposal for the George C Marshall Institute, Sent to Bill Nierenberg, December 12, 1984, MC 13, William A. Nierenberg Papers box 75, folder 6, SIO Archives.

76. Marjorie Sun, "Global Warming becomes Hot Issue for Bromley," *Science* 246 (3 November 1989): 569; the Mitchell work is J. F. B. Mitchell, C. A. Senior, and W. J. Ingram, "CO_2 and Climate: A Missing Feedback," *Nature* 341, no. 6238 (1989): 132–34.

77. D. Allan Bromley, *The President's Scientists: Reminiscences of a White House Science Advisor* (New Haven: Yale University Press, 1994), 142–63; Sun, "Global Warming becomes Hot Issue for Bromley," 569.

78. See Dick Schmalensee to Michael Boskin, 13 February 1990, Re: Toward an Administration Strategy on Global Warming, file Global Climate Change — Background Material, Juanita Duggan papers, George H. W. Bush Presidential Library [OA/ID 04722].

79. Earth Observing System Reference Handbook, NASA TM-104948, Goddard Space Flight Center, 1989.

80. Memorandum of Understanding among the United States National Aeronautics and Space Administration, the European Space Agency, the Science and Technology agency of Japan, and the Canadian Space Agency on Cooperation in Earth Observations Using Polar Platforms," 13 September 1988, box 2, EOS collection, NASA HQ History Office.

81. James P. Odom to Director, Space Station Program, 7 May 1988, Re: Polar Orbiting Platform Requirements, box 2, EOS collection, NASA HQ History Office.

CHAPTER 8. MISSIONS TO PLANET EARTH

Epigraph: Norman Augustine (Chair), *Report of the Advisory Committee on the Future of the U.S. Space Program*, NASA, Washington, D.C., December 1990.

1. Space Studies Board, *Earth Observations from Space: History, Promise, Reality* (Washington, D.C.: National Academy of Science, 1995), 52. The panel drew the term *management turbulence* from the Augustine Committee, which was formed to examine the U.S. civil space program in 1990. See Advisory Committee on the Future of the U.S. Space Program, *Report of the Advisory Committee on the Future of the U.S.*

Space Program, Washington, D.C., National Aeronautics and Space Administration, 1990.

2. The space program's heavy political burden is reflected in Walter A. McDougall, . . . *the Heavens and the Earth* (New York: Basic Books, 1985), and Howard McCurdy, *The Space Station Decision: Incremental Politics and Technological Choice* (Baltimore: Johns Hopkins University Press, 1990).

3. Daniel J. Kevles, *The Physicists: The History of a Scientific Community in Modern America*, rev. ed. (Cambridge, Mass.: Harvard University Press, 1995), ix–xlii.

4. In adjusted 1999 dollars. "Real Year" figures from National Aeronautics and Space Administration, "Aeronautics and Space Report of the President: Fiscal Year 2003 Activities," Washington, D.C., 1999, 139; adjusted using the Office of Management and Budget Implicit Price Deflator for fiscal year 2000.

5. Samuel P. Hays, *A History of Environmental Politics since 1945* (Pittsburgh: University of Pittsburgh Press, 2000), 118–19.

6. Bryan Burrough, *Dragonfly: An Epic Adventure of Survival in Outer Space* (New York: Harper Perennial, 1998), 244. *Battlestar Galactica* was a short-lived 1978 television series on ABC featuring a several-miles-long interplanetary battleship/aircraft carrier. The companion book was Glen A. Larson and Robert Thurston, *Battlestar Galactica* (New York: Berkley Books, 1978). It was revived in 2003.

7. Howard E. McCurdy, *Faster Better Cheaper: Low-Cost Innovation in the U.S. Space Program* (Baltimore: Johns Hopkins University Press, 2001), 44.

8. Committee on Earth Sciences, *Strategy for Earth Explorers in Global Earth Sciences* (Washington, D.C.: National Academy Press, 1988), 11–13; Committee on Earth Studies, *Earth Observations from Space: History, Promise, Reality* (Washington, D.C.: National Academy Press, 1995), 44.

9. Committee on Earth Sciences, *Strategy for Earth Explorers in Global Earth Sciences*, 26.

10. A. Raval and V. Ramanathan, "Observational Determination of the Greenhouse Effect," *Nature* 342 (14 December 1989): 758–61.

11. James Hansen, William Rossow, and Inez Fung, "The Missing Data on Global Climate Change," *Issues in Science and Technology* 7, no. 1 (1990): 62–69.

12. Hansen, Rossow, and Fung, "The Missing Data on Global Climate Change," 69; Butler interview, 24 August 2004.

13. Committee on Earth Studies, *Earth Observations from Space*, 133–42; C. Max and J. Vesecky, "Small Satellites: How Might DoD and DoE-originated Instrument Concepts Be Used in the Global Change Research Program?" The MITRE Corporation, 3 August 1991. On advocacy for smaller satellites and space probes in general, see McCurdy, *Faster Better Cheaper*.

14. U.S. Senate, Subcommittee on Science, Technology, and Space, "NASA's Space Science Programs and the Mission to Planet Earth," 102nd Cong., 1st sess., 24 April 1991, 1, 35; U.S. Senate, Subcommittee on Science, Technology, and Space, "NASA's Earth Observing System," 102nd Cong., 2nd sess., 26 February 1992, 2–5.

15. Frieman to EOS Engineering Review Advisory Committee Members, 3 May 1991, Re: Impressions of the Committee and its Work, file NASA-EOS Engineering

Review Panel Correspondence May 1991–June 1991 (folder 4), box 124, Frieman Papers (MC-77), SIO Archives.

16. George E. Brown to Edward Frieman, 15 May 1991, file NASA-EOS Engineering Review Panel Correspondence May 1991–June 1991 (folder 4), box 124, Frieman Papers (MC-77), SIO Archives; Edward Frieman interview with author, August 2005.

17. Committee on Earth Studies, *Earth Observations from Space*, 48.

18. Committee on Earth Studies, *Earth Observations from Space*, 48; J. T. Houghton, G. J. Jenkins, and J. J. Ephraums, eds., *Climate Change: The IPCC Scientific Assessment* (Cambridge: Cambridge University Press, 1990), xii.

19. V. Ramanathan et al., "Cloud-Radiative Forcing and Climate: Results from the Earth Radiation Budget," *Science* 243 (6 January 1989): 57–63.

20. Ramanathan et al., "Cloud-Radiative Forcing and Climate," 57–63.

21. V. Ramanathan et al., "Climate and the Earth's Radiation Budget," *Physics Today* (May 1989): 22–32.

22. R. D. Cess et al., "Interpretation of Cloud-Climate Feedback as Produced by 14 Atmospheric General Circulation Models," *Science* 245 (4 August 1989): 513–16.

23. NASA-EOS Engineering Review Panel Correspondence, March 1991–June 1991 (folder 4), box 124, Edward Freiman Papers (MC 77), SIO Archives.

24. Congressional concern over EOS's pacing is reflected in George E. Brown to Edward Frieman, 15 May 1991, NASA-EOS Engineering Review Panel Correspondence, March 1991–June 1991 (folder 4), box 124, Edward Freiman Papers (MC 77), SIO Archives; and attachment.

25. U.S. Senate, Committee on Commerce, Science, and Transportation, Hearing, "NASA's Earth Observing System," 102nd Cong., 2nd sess., 26 February 1992, 11–12.

26. For a discussion of this issue, see J. T. Houghton et al., *Climate Change 1995: The Science of Climate Change, Contribution of Working Group I to the Second Assessment Report of the Intergovernmental Panel on Climate Change* (Cambridge: Cambridge University Press, 1996), 146–48; the 2001 IPCC assessment presents considerably more detail on the difficulty of achieving reliable corrections for both radiosonde and satellite temperature records. See J. T. Houghton et al., *Climate Change 2001: The Scientific Basis, Contribution of Working Group I to the Third Assessment Report of the Intergovernmental Panel on Climate Change* (Cambridge: Cambridge University Press, 2001), 119–21.

27. This launch order controversy is reflected in the 1992 congressional hearings: U.S. Senate, Committee on Commerce, Science, and Transportation, Hearing, "NASA's Earth Observing System," 102nd Cong., 2nd sess., 26 February 1992, 48–49.

28. See Gary Taubes, "Earth Scientists Look NASA's Gift Horse in the Mouth," *Science* 259 (12 February 1993): 912–14. On the AIRS challenge, see Berrien Moore III testimony in U.S. Senate, Committee on Commerce, Science, and Transportation, Hearing, "NASA's Earth Observing System," 102nd Cong., 2nd sess., 26 February 1992, 50, 53.

29. Pierre Morel to J. Casani, 5 June 1992, folder 6, box 123, Edward Frieman

Papers, MC 77, SIO Archives; Taubes, "Earth Scientists Look NASA's Gift Horse in the Mouth," 912–14.

30. See U.S. Senate, Hearing, Committee on Commerce, Science and Transportation, "Nomination of Daniel S. Goldin to be Administrator of the National Aeronautics and Space Administration," 102nd Cong., 2nd sess., 27 March 1992, 12.

31. On "faster-better-cheaper," see McCurdy, *Faster Better Cheaper.*

32. Committee on Earth Studies, *Earth Observations from Space,* 49–51.

33. For a brief history of SAGE, see Paul N. Edwards, *The Closed World: Computers and the Politics of Discourse in Cold War America* (Cambridge, Mass.: MIT Press, 1997), 75–112.

34. On the original conception for EODIS, see Dixon Butler interview with author, 24 August 2004.

35. Bruce Barkstrom interview with author, 2003.

36. NRC, "Panel to Review EOSDIS Plans: Final Report," National Academy Press, 1994, 2.

37. On the gradual emergence of the Internet, see Janet Abbate, *Inventing the Internet* (Cambridge, Mass.: MIT Press, 1999), 181–220.

38. NRC, "Panel to Review EOSDIS Plans: Final Report," National Academy Press, 1994, 32.

39. "EOS Re-Baselining/System Drivers Review Presentation to NASA Administrator," file "EOS Rebaselining/System Drivers Review," box 44, Daniel S. Goldin papers, Center for Legislative History, National Archives and Records Administration, Washington, D.C.; Butler interview with author.

40. Charles Kennel interview with author, 12 November 2004.

41. NASA Headquarters Press Release 94–3, 6 January 1994.

42. Al Gore, Jr., *Earth in the Balance* (New York: Penguin Group, 1992).

43. Gore's attitude is clearly expressed in the 1992 Senate hearings on EOS. See U.S. Senate, Committee on Commerce, Science, and Transportation, "NASA's Earth Observing System," 102nd Cong., 2nd sess., 26 February 1992, 2–3.

44. The EOS twenty-four measurements are presented in E. J. Barron et al., "Overview," *EOS Science Plan* (Greenbelt, Md., 1999), 10.

45. Michael King, "Editor's Corner," *The Earth Observer,* March–April 1995; Presidential Decision Directive NSTC-2, 5 May 1994, Re: Convergence of U.S. Polar Orbiting Operation Environmental Satellite Systems [*sic*].

46. Samuel P. Hays, *A History of Environmental Politics Since 1945* (Pittsburgh: University of Pittsburgh Press, 2000), 109–21.

47. The hearing transcripts are: *Scientific integrity and public trust: The science behind federal policies and mandates: Case study 1, stratospheric ozone, myths and realities:* Hearing before the subcommittee on energy and environment of the committee on science, U.S. House of Representatives, 104th Cong., 1st sess., September 20, 1995, Washington, D.C.: U.S. Government Printing Office, 1996; *Scientific integrity and public trust: The science behind federal policies and mandates: Case study 2, climate models and projections of potential impacts of global climate change:* Hearing before the subcommittee on energy and environment of the committee on science,

U.S. House of Representatives, 104th Cong., 1st sess., November 16, 1995, Washington, D.C.: U.S. Government Printing Office, 1996.

48. Robert C. Harriss interview with author, 31 January 2005; Charles Kennel interview with author, 12 November 2004.

49. Robert S. Walker to Bruce Alberts, 6 April 1995, folder 3, box 17, Edward Frieman papers (MC 77), SIO Archives, San Diego, Calif.; "Leading Scientists Look again at Beleaguered EOS," *Space Business News*, 28 June 1995; U.S. House of Representatives, Committee on Science, "NASA Aeronautics and Space Administration Authorization Act, fiscal year 1996," Report 104–233, 104th Cong., 1st sess., 4 August 1995, 37–41; and "The Tenth EOS Investigators Working Group Meeting," *The Earth Observer*, July–August 1995; Chris Mooney, *The Republican War on Science* (New York: Basic Books, 2005), 62.

50. NASA, "NRC Board on Sustainable Development Workshop to Review the USGCRP and MTPE/EOS," July 1995, file "Sustainable Development Workshop," box 48, Daniel S. Goldin papers, Center for Legislative History, NARA, Washington, D.C.

51. The partnership idea was intended to foster more applications-oriented research. NASA's Harriss explained later that Goldin hoped involving private or non-profit businesses in data product generation would result in economically useful products that could then be used to justify sustaining EOS's capabilities. See Harriss interview with author, 31 January 2005.

52. National Research Council, "A Review of the U.S. Global Change Research Program and NASA's Mission to Planet Earth/Earth Observing System" (Washington, D.C.: National Academies Press, 1995), 3, 20.

53. National Research Council, "A Review of the U.S. Global Change Research Program and NASA's Mission to Planet Earth/Earth Observing System," 62–65.

54. U.S. House of Representatives, "National Aeronautics and Space Administration Authorization Act, Fiscal Year 1996," House Report 104–233, 104th Cong., 1st sess., 4 August 1995, 41.

55. Ross Gelbspan, *The Heat is On: The Climate Crisis, the Cover-Up, the Prescription*, updated ed. (New York: Perseus Books, 1998), 63. For example, Representative Dana Rohrabacher of California referred to global warming as "trendy science that is propped up by liberal/left politics rather than good science" and "at best unproven and at worst liberal clap trap" at an 8 June 1995 press conference on the fiscal year 1996 Energy and Environment Authorization bill. Faxed copy of statement in folder 7, box 123, Edward Frieman papers, MC 77, SIO Archives. See also Mooney, *Republican War on Science*, 62–64. *Scientific integrity and public trust: The science behind federal policies and mandates: Case study 1, stratospheric ozone, myths and realities*: Hearing before the subcommittee on energy and environment of the committee on science, U.S. House of Representatives, 104th Cong., 1st sess., September 20, 1995, Washington, D.C.: U.S. Government Printing Office, 1996; *Scientific integrity and public trust: The science behind federal policies and mandates: Case study 2, climate models and projections of potential impacts of global climate change*: Hearing before the subcom-

mittee on energy and environment of the committee on science, U.S. House of Representatives, 104th Cong., 1st sess., November 16, 1995, Washington, D.C.: U.S. Government Printing Office, 1996.

56. J. T. Houghton, L. G. Meira Filho, B. A. Callander, N. Harris, A. Kattneberg, and K. Maskell, eds., *Climate Change 1995: The Science of Climate Change* (Cambridge: Cambridge University Press, 1996), 147–48. This controversy was intense in the popular press due to the efforts of contrarian scientists to publicize this "disproof" of global warming. It was resolved in a series of papers published in 2005 that showed that the original analysis had a sign error—the data showed expected warming trend when this was corrected. See Carl A. Mears and Frank J. Wentz, "The Effect of Diurnal Correction on Satellite-Derived Lower Tropospheric Temperature," *Sciencexpress* (11 August 2005), doi 10.1126/science.1114772; Thomas R. Karl et al., eds., "Temperature Trends in the Lower Atmosphere: Steps for Understanding and Reconciling Differences," *Climate Change Science Program*, 2006 and Richard A. Kerr, "No Doubt About It, the World Is Warming," *Science* (12 May 2006): 825.

57. On the Global Climate Coalition, see Gelbspan, *The Heat is On*, 78–79; William K. Stevens, *The Change in the Weather: People, Weather, and the Science of Climate* (New York: Random House, 1999), 232–34.

58. See "Open Letter to Ben Santer," *Bulletin of the American Meteorological Society* 9, no. 77 (September 1996): 1961–66.

59. On the evolution of the denial effort, see Naomi Oreskes, Erik M. Conway, and Matthew Shindell, "From Chicken Little to Dr. Pangloss: William Nierenberg, Global Warming, and the Social Deconstruction of Scientific Knowledge," *Historical Studies in the Natural Sciences* 38, no. 1 (Winter 2008): 113–56; Oreskes and Conway, "Challenging Knowledge: How Climate Science Became a Victim of the Cold War," in *Agnotology: The Making and Unmaking of Ignorance*, ed. Robert N. Proctor and Londa Schiebinger (Palo Alto: Stanford University Press, 2008), 55–89.

60. Kennel interview with author; U.S. House of Representatives, Committee on Science, "NASA Aeronautics and Space Administration Authorization Act, fiscal year 1996," Report 104–233, 104th Cong., 1st sess., 4 August 1995, 37–41.

61. NASA Press Release 97–46, "New Missions Selected to Study Earth's Forests and Gravity Field Variability," 18 March 1997; Byron Tapley et al., "The Gravity Recovery and Climate Experiment: Mission Overview and Early Results," *Geophysical Research Letters* v. 31, doi:10.1029/2004GLO19920, 2004.

62. Satellites at L-1 actually orbit the point in space slowly.

63. Joel Achenbach, "For Gore Spacecraft, All Systems Aren't Go," *The Washington Post*, 8 August 2001, A-01; NASA Press release 98–46, "Earth-Viewing Satellite Would Focus on Educational, Scientific Benefits," 13 March 1998.

64. On the merits of Triana, see James J. Duderstadt, Mark Abbot, Eric J. Barron, and Raymond Jeanloz, "Letter Report: Review of Scientific Aspects of the NASA Triana Mission," 3 March 2000, http://www.nap.edu, copy in author's collection.

65. Asrar to Alberts, 14 October 1999, attached as Appendix 1 to James J. Duderstadt, Mark Abbot, Eric J. Barron, and Raymond Jeanloz, "Letter Report: Review of

Scientific Aspects of the NASA Triana Mission," 3 March 2000, http://www.nap.edu, copy in author's collection.

66. Roger Pielke, Jr. and Robert Harriss, "Science Policy and the NASA Triana Mission," *Science* (14 April 2000); Space Science Board, *Space Science in the Twenty-First Century: Imperatives for the Decades 1995 to 2015: Mission to Planet Earth* (Washington, D.C.: National Academy Press, 1988).

67. A few examples will suffice to make this point: Candace Crandall to William Nierenberg (email from the Science and Environmental Policy Project), 6 March 1998, Re: Time to Complain to Nature, file "UCSD/EPRI Workshop," box 18, William Aaron Nierenberg papers, accession 2001-01, SIO Archives; William A. Nierenberg interview with Myanna Lahsen, 31 May 1996; William A. Nierenberg to Richard S. Lindzen, 16 November 1994, file "Richard Lindzen," box 20, William Aaron Nierenberg papers, accession 2001-01, SIO Archives; Jeffrey Salmon to Board of Directors and attached minutes, 4 December 1995, Re: Minutes of Board Meeting, file "Marshall Institute, Board Meeting 3 November 1995," box 21, William Aaron Nierenberg papers, accession 2001-01, SIO Archives. On the larger conspiratorial nature of the climate change skeptics, see Myanna Lahsen, "Climate Rhetoric: Constructions of Climate Science in the Age of Environmentalism" (PhD thesis, Rice University, 1998), esp. chapter 8. Myanna H. Lahsen, "The Detection and Attribution of Conspiracies: The Controversy Over Chapter 8," in *Paranoia Within Reason: A Casebook on Conspiracy as Explanation*, ed. George H. Marcus (Chicago: University of Chicago Press, 1999).

68. Ghassem Asrar interview with author, 13 April 2005.

69. Committee on Global Change Research, *Global Environmental Change: Research Pathways for the Next Decade* (National Academy Press, 1998), 17.

70. Board on Atmospheric Sciences and Climate, *From Research to Operations in Weather Satellites and Numerical Weather Prediction: Crossing the Valley of Death* (Washington, D.C.: National Academy Press, 2000), 8–9, 49; Committee on Earth Studies, *Ensuring the Climate Record from the NPP and NPOESS Meteorological Satellites*, electronic publication http://www7.nationalacademies.org/ssb/bib_2000.html, copy in author's collection.

71. Committee on Global Change Research, *Global Environmental Change: Research Pathways for the Next Decade* (National Academy Press, 1998), 29.

CHAPTER 9. ATMOSPHERIC SCIENCE IN THE MISSION TO PLANET EARTH

Epigraph: James E. Hansen, "Dangerous Anthropogenic Interference: A Discussion of Humanity's Faustian Climate Bargain and the Payments Coming Due," presented 26 October 2006, University of Iowa, available at www.columbia.edu/~jeh1/.

1. Erik M. Conway, *High Speed Dreams: The Technopolitics of Supersonic Aviation* (Baltimore: Johns Hopkins University Press, 2005), 229–38.

2. National Research Council, *Abrupt Climate Change: Inevitable Surprises* (Washington, D.C.: National Academies Press, 2002), 19–72; John D. Cox, *Climate*

Crash: Abrupt Climate Change and What It Means for our Future (Washington, D.C.: Joseph Henry Press, 2005), 113–28.

3. James E. Hansen, "Dangerous Anthropogenic Interference: A Discussion of Humanity's Faustian Climate Bargain and the Payments Coming Due," lecture delivered to the Department of Physics and Astronomy, University of Iowa, 26 October 2004.

4. Elizabeth Kolbert, *Field Notes from a Catastrophe: Man, Nature, and Climate Change* (New York: Bloomsbury Books, 2006), 156–70. After leaving office, Bush's first treasury secretary, Paul O'Neill, gave the details of the administration's rejection to journalist Ron Suskind. See Suskind, *The Price of Loyalty: George W. Bush, the White House, and the Education of Paul O'Neill* (New York: Simon and Schuster, 2004), 98–102, 109–14, 118–28.

5. Timothy Donaghy et al., *Atmosphere of Pressure: Political Interference in Federal Climate Science* (Washington, D.C.: Union of Concerned Scientists and the Government Accountability Project, 2007) documents government-wide suppression; details of this first emerged from attempted suppression of NASA's James Hansen; see Andrew C. Revkin, "Climate Expert Says NASA Tried to Silence Him," *New York Times*, 29 January 2006, 1; Andrew C. Revkin, "Lawmaker Condemns NASA over Scientist's Accusations of Censorship," *New York Times*, 31 January 2006. Mark Bowen, *Censoring Science: Inside the Political Attack on Dr. James Hansen and the Truth of Global Warming* (New York: Dutton, 2007).

6. James E. Hansen and Andrew A. Lacis, "Sun and Dust versus Greenhouse Gases: An Assessment of their Relative Roles in Climate Change," *Nature* 346 (23 August 1990): 713–19.

7. Hansen and Lacis, "Sun and Dust versus Greenhouse Gases," 713–19. Charlson made this point clear in a more detailed article two years later. R. J. Charlson et al., "Climate Forcing by Anthropogenic Aerosols," *Science* 255 (24 January 1992): 423–30.

8. Michael D. King interview with author, 15 December 2004; and M. D. King et al., "Optical Properties of Marine Stratocumulus Clouds Modified by Ships," *Journal of Geophysical Research* 98, no. D2 (20 February 1993): 2729–39.

9. James Hansen et al., "How Sensitive is the World's Climate?" *National Geographic's Research and Exploration* 9, no. 2 (1993): 142–58; J. Hansen, M. Sato, and R. Ruedy, "Long-Term Changes of the Diurnal Temperature Cycle: Implications about Mechanisms of Global Climate Change," *Atmospheric Research* 37 (1995): 175–209.

10. Hansen, Sato, and Ruedy, "Long-Term Changes of the Diurnal Temperature Cycle: Implications about Mechanisms of Global Climate Change," 175–209.

11. M. Patrick McCormick, Larry W. Thomason, and Charles R. Trepte, "Atmospheric Effects of the Mt. Pinatubo Eruption," *Nature* 373 (2 February 1995): 399–404.

12. James Hansen, Andrew Lacis, Reto Reudy, and Makiko Sato, "Potential Climate Impact of Mount Pinatubo Aerosol," *Geophysical Research Letters* 19, no. 2 (24 January 1992): 215–18.

13. D. J. Carson, "Climate Modeling: Achievements and Prospects," *Quarterly*

Journal of the Royal Meteorological Society 125 (January 1999 Part A): 1–27; quote p. 10; McCormick, Thomason, and Trepte, "Atmospheric Effects of the Mt. Pinatubo Eruption," 399–404; Hansen interview with author, 16 January 2006.

14. James M. Hoell, Jr., D. Davis, S. Liu, R. Newell, M. A. Shipham, H. Akimoto, R. J. McNeal, R. J. Bendura, and J. W. Drewry, "Pacific Exploratory Mission—West A (PEM—West A): September–October 1991," *Journal of Geophysical Research* 101, no. D1 (1996): 1641–53; James M. Hoell, Jr., D. D. Davis, S. C. Liu, R. E. Newell, H. Akimoto, R. J. McNeal, and R. J. Bendura. "The Pacific Exploratory Mission—West Phase B: February–March 1994," *Journal of Geophysical Research* 102, no. D23 (1997): 28223–239.

15. James L. Raper, Mary M. Kleb, Daniel J. Jacob, Douglas D. Davis, Reginald E. Newell, Henry E. Fuelberg, Richard J. Bendura, James M. Hoell, and Robert J. McNeal, "Pacific Exploratory Mission in the Tropical Pacific: PEM—Tropics B, March–April 1999," *Journal of Geophysical Research* 106, no. D23 (2001): 32401–425.

16. J. M. Hoell, Jr., D. D. Davis, D. J. Jacob, M. O. Rodgers, R. E. Newell, H. E. Fuelberg, R. J. McNeal, J. L. Raper, and R. J. Bendura, "Pacific Exploratory Mission in the Tropical Pacific: PEM—Tropics A, August–September 1996," *Journal of Geophysical Research-Atmospheres* 104, no. D5 (1999): 5567–83.

17. See Conway, *High Speed Dreams*, 229–38, on the Atmospheric Effects of Stratospheric Aircraft program; and National Research Council, *The Atmospheric Effects of Stratospheric Aircraft: An Evaluation of NASA's Interim Assessment* (Washington, D.C.: National Academy of Sciences, 1994).

18. Paul A. Newman et al., "An overview of the SOLVE/THESEO 2000 campaign," *Journal of Geophysical Research* 107, no. D20, doi:10.1029/2001JD001303, 2002, p. SOL 1–22.

19. Y. J. Kaufman et al., "Smoke, Clouds, and Radiation—Brazil (SCAR-B) Experiment," *Journal of Geophysical Research* 103, no. D24 (27 December 1998): 31783–808.

20. B. Holben, T. Eck, I. Slutsker, D. Tanre, J. Buis, E. Vermote, J. Reagan, Y. Kaufman, T. Nakajima, F. Lavevau, I. Jankowiak, and A. Smirnov, "Aeronet, a Federated Instrument Network and Data-Archive for Aerosol Characterization," *Remote Sensing of The Environment* 66 (1998): 1–16; Michael King interview with author, 15 December 2004.

21. Michael Prather interview with author, September 2003; Owen B. Toon and Richard C. Miake-Lye, "Subsonic Aircraft: Contrail and Cloud Effects Special Study (SUCCESS)," *Geophysical Research Letters* 25, no. 8 (1998): 1109–12.

22. C. Prabhakara, D. P. Kratz, J. M. Yoo, G. Dalu, and A. Vernekar. "Optically Thin Cirrus Clouds: Radiative Impact on the Warm Pool," *Journal of Quantitative Spectroscopy and Radiative Transfer* 49, no. 5 (1993): 467–83.

23. Spencer Weart, *The Discovery of Global Warming* (Cambridge, Mass.: Harvard University Press, 2003), 66. Also see the expanded online edition at http://www.aip.org/history/climate/aerosol.htm, accessed 3 July 2006.

24. Some early literature on contrails and cirrus is discussed in Intergovernmental Panel on Climate Change, *Aviation and the Global Atmosphere* (New York: Cam-

bridge University Press, 1999), 95–96; see also Patrick Minnis interview with author, 21 April 2005.

25. Gerald M. Stokes and Stephen E. Schwartz, "The Atmospheric Radiation Measurement Program: Programmatic Background and Design of the Cloud and Radiation Test Bed," *Bulletin of the American Meteorological Society* 75, no. 7 (July 1994): 1201–21. Also see Ann Finkbeiner, *The Jasons: The Secret History of Science's Postwar Elite* (New York: Penguin Books, 2006), 139–40.

26. Owen B. Toon and Richard C. Miake-Lye, "Subsonic Aircraft: Contrail and Cloud Effects Special Study (SUCCESS)," *Geophysical Research Letters* 25, no. 8 (1998): 1109–12; W. L. Smith, S. Ackerman, H. Revercomb, H. Huang, D. H. DeSlover, W. Feltz, L. Gumley, and A. Collard, "Infrared Spectral Absorption of Nearly Invisible Cirrus Clouds," *Geophysical Research Letters* 25, no. 8 (1998): 1137–40; Patrick Minnis, David F. Young, and Donald P. Garber. "Transformation of Contrails into Cirrus During SUCCESS," *Geophysical Research Letters* 25, no. 8 (1998): 1157–60; Patrick Minnis interview with author, 21 April 2005. See also Intergovernmental Panel on Climate Change, *Aviation and the Global Atmosphere* (New York: Cambridge University Press, 1999), 87–99.

27. Philip B. Russell, Peter V. Hobbs, and Larry L. Stowe, "Aerosol Properties and Radiative Effects in the United States East Coast Haze Plume: An Overview of the Tropospheric Aerosol Radiative Forcing Observational Experiment," *Journal of Geophysical Research* 104, no. D2 (1999): 2213–22; B. Holben, T. Eck, I. Slutsker, D. Tanre, J. Buis, E. Vermote, J. Reagan, Y. Kaufman, T. Nakajima, F. Lavevau, I. Jankowiak, and A. Smirnov, "AERONET, a Federated Instrument Network and Data-Archive for Aerosol Characterization," *Remote Sensing of The Environment* 66 (1998): 1–16.

28. Philip B. Russell, Peter V. Hobbs, and Larry L. Stowe, "Aerosol Properties and Radiative Effects in the United States East Coast Haze Plume: An Overview of the Tropospheric Aerosol Radiative Forcing Observational Experiment (TARFOX)," *Journal of Geophysical Research* 104, no. D2 (27 January 1999): 2213–22.

29. J. A. Curry, P. V. Hobbs, M. D. King, D. A. Randall, and P. Minnis. "Fire Arctic Clouds Experiment," *Bulletin of the American Meteorological Society* 81, no. 1 (2000): 5–29; Ghassem Asrar interview with author, 13 April 2005; Dixon Butler interview with author, 23 August 2004.

30. Stephen K. Cox et al., "FIRE—The First ISCCP Regional Experiment," *Bulletin of the American Meteorological Society* 68, no. 2 (February 1987): 114–18.

31. Cox et al., "FIRE—The First ISCCP Regional Experiment," 114–18.

32. Cox et al., "FIRE—The First ISCCP Regional Experiment," 114–18.

33. Cox et al., "FIRE—The First ISCCP Regional Experiment," 114–18.

34. Patrick Minnis interview with author, April 2005.

35. D. M. Winker, R. H. Couch, and M. P. McCormick, "An Overview of LITE: NASA's Lidar In-space Technology Experiment," *Proceedings of the IEEE* 84, no.2 (February 1996): 164–80.

36. NASA Press Release 98–226, "Future Missions to Study Clouds, Aerosols, Volcanic Plumes," released 22 December 1998; NASA Press Release 99–57, "Cloudsat to Revolutionize Study of Clouds and Climate," released 30 April 1999; see also Graeme

L. Stephens et al., "The Cloudsat Mission and the A-Train: A New Dimension of Space-Based Observations of Clouds and Precipitation," *Bulletin of the American Meteorological Society* (December 2002): 1771–90; Graeme Stephens interview with author, 14 October 2005.

37. NASA Press Release 99–16, 8 February 1999, archived at ftp://ftp.hq.nasa.gov/pub/pao/pressrel/1999/99–016.txt.

38. Michael D. King interview with author, 15 December 2004.

39. Michael D. King et al., "Remote Sensing of Smoke, Land, and Clouds from the NASA ER-2 During SAFARI 2000," *Journal of Geophysical Research* 108, no. D13, 8502 doi:10.1029/2002JD003207, 2003.

40. King et al., "Remote Sensing of Smoke, Land, and Clouds from the NASA ER-2 During SAFARI 2000."

41. Robert J. Swap, Harold J. Annegarn, J. Timothy Suttles, Michael D. King, Steven Platnick, Jeffrey L. Privette, and Robert J. Scholes, "Africa Burning: A Thematic Analysis of the Southern African Regional Science Initiative (Safari 2000)," *Journal of Geophysical Research* 108, no. D13 (2003).

42. Swap, Annegarn, J. Suttles, King, Platnick, Privette, and Scholes, "Africa Burning."

43. M. J. McGill et al., "Combined Lidar-Radar Remote Sensing: Initial Results from CRYSTAL-FACE," *Journal of Geophysical Research* 109, D07203, doi:10.1029/2003F004030, 2004; W. Smith et al., "Atmospheric Properties Observed with NAST-I During CRYSTAL-FACE," presentation at the CRYSTAL-FACE Science Team Meeting, 24–28 February 2003, courtesy W. L. Smith Sr., copy in author's collection; Eric Jensen, David Starr, and Owen B. Toon, "Mission Investigates Tropical Cirrus Clouds," *EOS Transactions* 85, no. 5 (3 February 2004): 45, 50; William L. Smith, Sr. interview with author, 18 April 2005.

44. Bruce A. Wielicki, Robert D. Cess, Michael D. King, David A. Randall, and Edwin F. Harrison, "Mission to Planet Earth: Role of Clouds and Radiation in Climate," *Bulletin of the American Meteorological Society* 76, no. 11 (November 1995): 2125–59; Bruce A. Wielicki, Bruce R. Barkstrom, Edwin F. Harrison, Robert B. Lee III, G. Louis Smith, and John E. Cooper, "Clouds and the Earth's Radiant Energy System (CERES): An Earth Observing System Experiment," *Bulletin of the American Meteorological Society* 77, no. 5 (May 1996): 853–68; Bruce A. Wielicki et al., "Clouds and the Earth's Radiant Energy System: Algorithm Overview," *IEEE Transactions on Geoscience and Remote Sensing* 36, no. 4 (July 1998): 1127–41. Also see Wielicki interview with author, 19 February 2004.

45. Bruce A. Wielicki, "CERES Overview: Clouds and the Earth's Radiant Energy System," presentation to CERES Data Products Workshop, Norfolk, Va., 29–30 January 2003.

46. Wielicki, "CERES Overview."

47. Joanne Simpson and Christian Kummerow, "The Tropical Rainfall Measuring Mission and Vern Suomi's Vital Role," *10th Conference on Atmospheric Radiation: A Symposium with Tributes to the Works of Verner E. Suomi*, 28 June–2 July 1999, 44–51.

48. Wielicki et al., "Clouds and the Earth's Radiant Energy System," 1127–41.

49. Richard S. Lindzen, Ming-Dah Chaou, and Arthur Y. Hou, "Does the Earth Have an Adaptive Infrared Iris?" *Bulletin of the American Meteorological Society* 82, no. 3 (March 2001): 417–32; David Herring, "Does the Earth Have an Iris Analog?" NASA Earth Observatory, 12 June 2002, http://earthobservatory.nasa.gov/Study/Iris/, copy in author's collection.

50. Bing Lin et al., "The Iris Hypothesis: A Negative or Positive Cloud Feedback?" *Journal of Climate* (1 January 2002): 3–7. The radiative forcing of tropospheric ozone was estimated at approximately 0.3 watts/meter2 in the 2001 IPCC assessment. See Robert T. Watson et al., *Climate Change 2001: Synthesis Report* (New York: Cambridge University Press, 2001), 48.

51. Lin H. Chambers, Bing Lin, and David F. Young, "Examination of New CERES Data for Evidence of Tropical Iris Feedback," *Journal of Climate* (15 December 2002): 3719–26; Lin Chambers et al., "Reply," *Journal of Climate* (15 September 2002): 2716–17; M. D. Chou et al., "Comments on 'The iris hypothesis: A negative or positive cloud feedback,'" *Journal of Climate* (15 September 2002): 2713–15.

52. Anthony D. Del Genio, "The Dust Settles on Water Vapor Feedback," *Science* 296 (26 April 2002): 665–66; Brian J. Soden et al., "Global Cooling after the Eruption of Mount Pinatubo: A Test of Climate Feedback by Water Vapor," *Science* 296 (26 April 2002): 727–30.

53. Hansen had pointed out this widely-held fact in his 1990 review, "Sun and Dust in the Greenhouse," 717.

54. Del Genio, "The Dust Settles on Water Vapor Feedback," 665–66; Brian J. Soden et al., "Global Cooling after the Eruption of Mount Pinatubo: A Test of Climate Feedback by Water Vapor," *Science* 296 (26 April 2002): 727–30.

55. Intergovernmental Panel on Climate Change, *Climate Change 2001: Synthesis Report* (Cambridge: Cambridge University Press, 2001), 51.

56. Committee on the Science of Climate Change, "Climate Change Science: An Analysis of Some Key Questions" (Washington, D.C.: National Academies Press, 2001), 3.

57. For an indication of President Bush's personal view of global warming, as opposed to his public statements, see Fred Barnes, *Rebel in Chief: Inside the Bold and Controversial Presidency of George W. Bush* (New York: Crown Forum, 2006), 22–23.

58. Michael Crichton, "Aliens Cause Global Warming," lecture at California Institute of Technology, 17 January 2003; *State of Fear* (New York: HarperCollins, 2004).

59. Michael K. Janofsky, "Michael Crichton, Novelist, Becomes Senate Witness," *New York Times*, 29 September 2005, online edition; Senator Jeffords Statement at EPW Hearing on Science and Environment, at http://epw.senate.gov/pressitem.cfm?id=246511&party=dem, accessed 3 July 2006.

60. Paul Rauber, "See No Evil," *Sierra* (January 2006): 36–37; CBS News, 60 Minutes broadcast 19 March 2006, "Rewriting the Science"; Union of Concerned Scientists, "Scientific Integrity in Policymaking," June 2004, 5–8; on Singer's efforts to deny the reality of global warming, see, for example, S. Fred Singer, "Global Warming: Do

We Know Enough to Act?" in *Environmental Protection: Regulating for Results*, ed. Kenneth Chilton and Melinda Warren (Boulder, Colo.: Westview Press, 1991), esp. 45–46; S. Fred Singer, "Hot Words on Global Warming (Letter to the Editor)," *Wall Street Journal*, 15 January 1996 (eastern edition), 13; S. Fred Singer, "Climatic Change: Hasty Action Unwarranted," *Consumer Research Magazine* 80 (1997): 16–20; S. Fred Singer, "Global Warming: What We're Not Told (Letter to the Editor)," *The Washington Post*, 26 January 1998, 22.

61. IPPC, "2007: Summary for Policymakers," in *Climate Change 2007: The Physical Science Basis. Contribution of Working Group 1 to the Fourth Assessment Report of the Intergovernmental Panel on Climate Change*, ed. S. Solomon, D. Qin, M. Manning, Z. Chen, M. Marquis, K. B. Averyt, M. Tignor, and H. I. Miller (Cambridge: Cambridge University Press, 2007), 10.

62. Donald Kennedy, "New Year, New Look, Old Problem," *Science* (6 January 2006): 15; Donald Kennedy, "The New Gag Rules," *Science* (17 February 2006): 917; Donald Kennedy, "More Silliness on the Hill," *Science* (12 May 2006); Donald Kennedy, "Ice and History," *Science* (24 March 2006): 1673; Donald Kennedy, "Acts of God?" *Science* (20 January 2006): 303; Donald Kennedy, "The Hydrogen Solution," *Science* (13 August 2004): 917; Donald Kennedy, "The Fight of the Decade," *Science* (17 June 2005): 1713; Donald Kennedy, "Climate Change and Climate Science," *Science* (11 June 2004): 1565; Kennedy was attacked by S. Fred Singer in a letter in 2003, Singer, "Science Editor Bias on Climate Change?" *Science* (1 August 2003): 595, and Kennedy's "Response," *Science* (1 August 2003); Donald Kennedy, "The Climate Divide," *Science* (21 March 2003): 1813; Donald Kennedy, "The Policy Drought on Climate Change," *Science* (17 January 2003): 309. On Singer's role in opposing global warming, see Naomi Oreskes and Erik Conway, "Challenging Knowledge: How Climate Science Became a Victim of the Cold War," in *Agnotology: The Making and Unmaking of Ignorance*, ed. Robert N. Proctor and Londa Schiebinger (Stanford: Stanford University Press, 2008), 55–89.

63. Andrew Lawler, "NASA Starts Squeezing to Fit Missions Into Tight Budget," *Science* (9 December 2005): 1594–95; Andrew Lawler, "Earth Observation Program 'At Risk,' Academy Warns," *Science* (29 April 2005): 614–15; Andrew Lawler, "Balancing the Right Stuff," *Science* (22 April 2005): 484–87; Andrew Lawler, "Crisis Deepens as Scientists Fail to Rejigger Space Research," *Science* (12 May 2006): 824; Andrew Lawler, "A Space Race To the Bottom Line," *Science* (17 March 2006): 1540–43; Andrew Lawler, "NASA Agrees to Review What's On the Chopping Block," *Science* (10 March 2006): 1359–60. George W. Bush, "A Renewed Spirit of Discovery," 14 January 2004; John Marburger III, speech to the 44th Robert H. Goddard Memorial Symposium, 15 March 2006.

64. Lawler, "Earth Observation Program 'At Risk,' Academy Warns," 614–15; Congressman Sherwood Boehlert, "Opening Statement on NASA Earth Sciences," House Committee on Science, 28 April 2005 at http://www.house.gov/science/press/109/109 -67.htm, accessed 3 July 2006.

65. Yoram J. Kaufman, Didier Tanre, and Olivier Boucher, "A Satellite View of Aerosols in the Climate System," *Nature* 419 (12 September 2002): 215–23; U.S. Com-

mittee for the Global Atmospheric Research Program, *Understanding Climatic Change: A Program for Action* (Washington, D.C.: National Academy Press, 1975), 44.

66. Some recent examples include S. Kinne et al., "Monthly Averages of Aerosol Properties: A Global Comparison among Models, Satellite Data, and AERONET Ground Data," *Journal of Geophysical Research* 108, no. D20, doi: 10.1029/2001JD001253, 2003; M. H. Zang et al., "Comparing Clouds and Their Seasonal Variations in 10 Atmospheric General Circulation Models with Satellite Measurements," *Journal of Geophysical Research* 110, no. D15S02, doi: 10.1029/2004JF005021, 2005; Minghua Ahang et al., "Introduction to Special Section on Toward Reducing Cloud-Climate Uncertainties in Atmospheric General Circulation Models," *Journal of Geophysical Research* 110, no. D15S01, doi 10.1029/2005JD005923, 2005.

67. Vince Salomonson and Holli Riebeek, "January 2006 MODIS Science Team Meeting Overview," *The Earth Observer* 18, no. 2 (March–April 2006): 12.

68. On the adoption of the open data policy, see Erik M. Conway, "Drowning in Data: Satellite Oceanography and Information Overload in the Earth Sciences," *Historical Studies in the Physical and Biological Sciences* 37, no. 1 (2006): 127–51.

69. Kenneth Chang, "Officials Report Progress in Weather Satellite Effort," *New York Times*, 9 June 2006; Jacqueline Ruttimann, "US Satellite System Loses Climate Sensors," *Nature* 4415 (June 2006): 798; Eli Kintisch, "Stormy Skies for Polar Satellite Program," *Science* 312 (2 June 2006): 1296–97; testimony of Conrad C. Lautenbacher, Jr., 8 June 2006; testimony of Ronald M. Sega, 6 June 2006; and webcast archived at: http://www.house.gov/science/hearings/full06/June%208/index.htm; accessed 10 June 2006.

70. Andrew Lawler, "NASA Terminates Gore's Eye on Earth," *Science* 311 (6 January 2006): 26.

CONCLUSION

Epigraph: Opening statement of Sherwood Boehlert, House of Representatives, Committee on Science, 109th Cong., 1st sess., 28 April 2005, 13.

1. Chris Mooney has cast this as a largely partisan issue in Mooney, *The Republican War on Science* (New York: Basic Books, 2005), 9–11, 78–101; Ross Gelbspan, "Snowed," *Mother Jones* (May–June 2005), online edition.

2. Union of Concerned Scientists, *Smoke, Mirrors and Hot Air: How ExxonMobil uses Big Tobacco's Tactics to Manufacture Uncertainty on Climate Science*, January 2007.

3. S. Fred Singer, "Global Warming: Do We Know Enough to Act?" in *Environmental Protection: Regulating for Results*, ed. Kenneth Chilton and Melinda Warren (Boulder, Colo.: Westview Press, 1991), 29–50.

4. Myanna H. Lahsen, "Climate Rhetoric: Constructions of Climate Science in the Age of Environmentalism" (PhD thesis, Rice University, 1998), 150–168, 371–412.

5. George Soros, "The Capitalist Threat," *The Atlantic Monthly* 279 (February 1997): 45–58.

6. Michael A. Bernstein, *A Perilous Progress: Economists and Public Purpose in Twentieth-Century America* (Princeton, N.J.: Princeton University Press, 2001), 191.

7. Naomi Oreskes, "Weighing the Earth from a Submarine: The Gravity Monitoring Cruise of the U.S.S. S-21," in *The Earth, the Heavens, and the Carnegie Institute of Washington*, ed. Gregory A. Good (Washington, D.C.: American Geophysical Union, 1994), 65; Nathan Reingold, *Science in the Federal Government* (Washington, D.C.: Smithsonian Institution Press, 1979), 218–19.

8. In January 2006, Hansen again hit the front page of the *New York Times*, with the revelation that NASA Public Affairs was trying to prevent him from speaking to the press. See Andrew C. Revkin, "Climate Expert Says NASA Tried to Silence Him," *New York Times*, 29 January 2006, 1; Andrew C. Revkin, "Lawmaker Condemns NASA over Scientist's Accusations of Censorship," *New York Times*, 31 January 2006.

9. Daniel J. Kevles, *The Physicists: The History of a Scientific Community in Modern America*, rev. ed. (Cambridge, Mass.: Harvard University Press, 1997), 419–26.

10. Sheila Jasanoff, *The Fifth Branch: Science Advisors as Policymakers* (Cambridge, Mass.: Harvard University Press, 1990), 12–14.

11. The classic attack on social constructivism is Paul R. Gross and Norman Levitt, *Higher Superstition: The Academic Left and its Quarrels with Science*, reprint ed. (Baltimore: Johns Hopkins University Press, 1997); for some of the misunderstandings at the root of the attack against social construction, see Donald MacKenzie, "The Science Wars and the Past's Quiet Voices," *Social Studies of Science* 29, no. 2 (April 1999): 199–213. See also Ronald N. Giere, *Science Without Laws* (Chicago: University of Chicago Press, 1999), 1–7.

12. Robert N. Proctor, *Cancer Wars: How Politics Shapes What We Know and Don't Know about Cancer* (New York: Basic Books, 1995), 8; Londa Schiebinger, *Plants and Empire: Colonial Bioprospecting in the Atlantic World* (Cambridge, Mass.: Harvard University Press, 2004), 1–4. Robert N. Proctor, "Agnotology: A Missing Term to Describe the Cultural Production of Ignorance (and its Study)," in *Agnotology: The Making and Unmaking of Ignorance*, ed. Robert N. Proctor and Londa Schiebinger (Stanford: Stanford University Press, 2008), 1–33.

13. Oreskes, "Weighing the Earth from a Submarine," 308.

14. Mark Bowen, *Censoring Science: Inside the Political Attack on Dr. James Hansen and the Truth of Global Warming* (New York: Dutton, 2007).

15. N. Oreskes, K. Shrader-Frechette, and K. Belitz, "Verification, Validation, and Confirmation of Numerical Models in the Earth Sciences," *Science* 263 (4 February 1994): 641–46.

EPILOGUE

1. I. Velicogna and J. Wahr, "Greenland Mass Balance from GRACE," *Geophysical Research Letters* 32, no. 18 (30 September 2005).

2. James Hansen interview with author, January 2006.

3. J. H. Mercer, "West Antarctic Ice Sheet and CO_2 Greenhouse Effect: A Threat of Disaster," *Nature* 271 (26 January 1978): 321–25.

4. James E. Hansen, "A Slippery Slope: How Much Global Warming Constitutes 'Dangerous Anthropogenic Interference'?: An Editorial Essay," *Climatic Change* (2005): 269–79.

5. Hansen, "A Slippery Slope," 269–79; James E. Hansen, "Is There Still Time to Avoid 'Dangerous Anthropogenic Interference' with Global Climate? A Tribute to Charles David Keeling," paper presented at American Geophysical Union, San Francisco, Calif., 6 December 2005, courtesy the author; James E. Hansen, "Dangerous Anthropogenic Interference: A Discussion of Humanity's Faustian Climate Bargain and the Payments Coming Due," paper presented at the University of Iowa, 26 October 2004.

6. On the history of glacial physics, see Bruce Hevly, "The Heroic Science of Glacier Motion," *Osiris: A Research Journal devoted to the History of Science and its Cultural Influences* 11 (1996): 66–86.

7. Intergovernmental Panel on Climate Change, *Climate Change 2007: The Physical Science Basis*, Summary for Policymakers, 11, emphasis added; available at http://ipcc-wg1.ucar.edu/wg1/docs/WG1AR4_SPM_Approved_05Feb.pdf.

8. Intergovernmental Panel on Climate Change, *Climate Change 2007*, 11; the Technical Summary contains more detail: S. Solomon et al., "Technical Summary," *Climate Change 2007: The Physical Science Basis. Contribution of Working Group I to the Fourth Assessment Report of the Intergovernmental Panel on Climate Change,* ed. S. Solomon, D. Qin, M. Manning, Z. Chen, M. Marquis, K. B. Averyt, M. Tignor, and H. L. Miller (Cambridge: Cambridge University Press, 2007), 44, 70–71.

9. Jim Hansen, "The Threat to the Planet," *New York Review of Books* 53, no. 12 (13 July 2006); Jim Hansen, "Global Warming: Is There Still time to Avoid Disastrous Human-Made Climate Change?" discussion on 23 April 2006, National Academy of Sciences, Washington, D.C.

10. Catherine Westfall, "Rethinking Big Science: Modest, Mezzo, Grand Science and the Development of the Bevalac, 1971–1993," *Isis* 94, no. 1 (March 2003): 56; also see Paul Forman, "Independence, not Transcendence, for the Historian of Science," *Isis* 82 (1991): 71–86.

Index

Page numbers in *italics* refer to images and figures.